Mammals of the National Parks

Richard G. Van Gelder is Curator of Mammals at the American Museum of Natural History. He is the author of numerous books and articles, both popular and professional, including *Biology of Mammals*; *Monkeys and Apes*; *Bats*; and *Animals and Man, Past, Present, and Future*.

Mammals of the
National Parks

RICHARD G. VAN GELDER

The Johns Hopkins University Press/Baltimore and London

for Russell, Gordon, and Leslie

The Johns Hopkins University Press, Baltimore, Maryland 21218
The Johns Hopkins Press Ltd., London

Library of Congress Cataloging in Publication Data

Van Gelder, Richard George, 1928–
  Mammals of the national parks.

  Bibliography: pp. 307–8.
  Includes index.
  1. Mammals—United States. 2. National parks and reserves—
United States. I. Title.
QL717.V36   599.0973   81–17162
ISBN 0–8018–2688–8   AACR2
ISBN 0–8018–2689–6 (pbk.)

Contents

PART 2: Mammals

Acknowledgments

I am greatly indebted to the National Park Service and its employees for their kind and patient cooperation in the preparation of this book. In many of the parks I was given access to the "observation cards" of sightings of animals, and in some the most experienced personnel took the time to talk with me and to answer my questions about specific species. When the manuscript was completed, each park account was sent to the park for critical comments and updating. What errors remain are, of course, my own. I do not have the names of all the National Park Service employees who assisted me in the preparation of this book, but I am grateful to all, named and anonymous. Among those employees and others who provided information and help are: C. L. Andress, J. M. Anselmo, C. C. Axtell, D. Banks, R. Black, G. E. Blinn, R. G. Bruce, C. Bryant, S. Canter, D. Carter, J. F. Cassel, K. Chambers, T. Coffield, S. Coleman, S. Croll, L. Cutliff, J. Dalle-Molle, G. Davis, E. Dillon, S. M. Engfer, P. Flanagan, W. Gardiner, D. Graber, M. Hellickson, T. E. Henry, R. Huggins, L. Johnson, G. Kaye, R. Kerbo, J. A. Kushlan, L. L. Loupe, L. F. McClanahan, R. H. Maeder, R. W. Marks, M. A. Martin, D. May, M. Mayer, B. Moorehead, R. Morey, V. Naylor, M. M. Neiss, C. W. Neiss, M. S. Nicholson, L. L. Olson, V. J. Olson, N. J. Pachta, B. B. Page, J. J. Page, J. J. Palmer, P. L. Parry, J. Patton, S. Paulson, J. Pearson, D. Peters, J. T. Peters, R. Peters, R. W. Peters, C. Peterson, V. Phelps, R. Rasp, A. Red Owl, K. A. Reeves, M. Revere, J. Riddle, R. Riegelhuth, E. Roberts, W. B. Robertson, R. Rothe, C. J. Roy, J. A. Sanders, S. Snow, W. Stephenson, D. Stevens, H. Tanski, W. G. Truesdell, J. Wagoner, H. C. Warren, R. Wasem, H. Werner, N. Whelan, H. D. Wickware, H. Wills, C. Wilson, L. Winter, J. W. Wright, K. C. Wylie, T. Wylie.

All of the maps and drawings were drawn, with her usual attention to detail, by Frances W. Zweifel, to whom I am grateful, as always. Muriel V. Williams was most helpful with the final typing. The staff of The Johns Hopkins University Press has been extremely kind and understanding, especially Anders Richter and Jane Warth, whose editorial assistance has been most valuable, and James Johnston, who selected the photographs. I am grateful to the U.S. Fish and Wildlife Service for some of the photographs and to the American Museum of Natural History for others. And last, but certainly not least, my thanks must go to my family, who, in one form or another, have "lived" with this project for several years. My wife, Rosalind, was extremely tolerant of a husband who, at one point, returned from a week of scientific meetings for one night, left for Africa for two weeks, returned for one night, and left for six weeks in the national parks, and who disappeared to his office for most weekends to write. Our children, Russell, Gordon, and Leslie, participated with me in visits to thirty-six of the parks and monuments. During a period when I was recuperating from severe illness, they served as my feet, eyes, and ears in the parks, and they hiked hundreds of miles to obtain information for me. In all, I traveled more than thirty thousand miles to gather information for this book, and I hope that, in part, the pleasure of the visits to the parks that we shared in some way repays Russell, Gordon, Leslie, and Rosalind for the help that they gave me.

Photographs and color plates are reproduced by courtesy of the following sources:

U.S. Fish and Wildlife Service: plates 3b and 4b, and p. 278; Geno A. Amundson, p. 283; C. J. Bayer, p. 231; V. Berns, plate 5b; Frank L. Blake, p. 166; Dave Erickson, plate 3a; Luther C. Goldman, pp. 163, 172, 259, 279, and 287; E. P. Haddon, pp. 242, 263, 285, and 292; Harry Harmon, p. 261; George Harrison, plate 5a; E. R. Kalmbach, p. 229; N. H. Kent, p. 189; Joe Mazzoni, p. 298; Jon R. Nickles, plates 1a and 2a; D. Ohlen, p. 200; Victor B. Scheffer, pp. 169, 218, 247, 266, and 268; Rex Gary Schmidt, pp. 213 and 253; R. V. Shiver, plate 4a; LeRoy W. Sowl, p. 249; Bob Stevens, plate 1b; Walter P. Taylor, p. 251; Ralph Town, p. 289; and E. R. Warren, pp. 175 and 216

the American Museum of Natural History: pp. 183, 221, and 297

the author: plates 2b, 6a, 6b, 7a, 7b, 8a, and 8b

Introduction

There are more than three hundred areas under the jurisdiction of the National Park Service, but the brightest stars of these earth-bound constellations are clearly the forty-eight places called "national parks." Each year millions of people travel to enjoy the national parks for a few hours to several weeks at a time. Although most people think of national parks as great natural preserves—and indeed many of them are—the official designation of an area as a national park means primarily that it has been established by a law passed by the U.S. Congress and signed by the president. Other parts of the national park system may receive their mandate in other fashions; national monuments, for example, require only a presidential order. Although the National Park Service does not discriminate in the administration of the various areas, there is little doubt that the national parks receive a major share of attention and visitation by tourists.

Because the idea of the national park is so fixed in most people's minds, this book includes mention of all of the national parks, even though some are of little interest to the observer of mammals. A great many national monuments are of mammalogical interest, and a number of them have been included, largely because they are either monuments that are extensively visited or ones that represent some particular facet of North American ecology. There are others that could have been included by reason of size, especially, but their ecosystems and their mammalian fauna are generally also represented in some national park.

In preparing this book, I personally visited all of the national parks on the mainland of North America that had been designated as parks by the summer of 1980. Some I had visited many times, starting in 1936, but I visited all of them at least once between 1978 and 1980. Late in 1980, President Jimmy Carter signed into law bills designating many new national parks and other areas, mainly in Alaska. To have tried to visit these parks for first-hand knowledge in order to include them in this book would have meant a delay of at least a year. After analyzing the probable fauna of these parks, it seemed that the parks would add very few species of mammal to the ones already included in the extensive accounts of Alaska parks; thus the new ones are not dealt with in detail. Indeed, early in 1981, little specific information was available about the new national parks from the National Park Service. The locations of the new Alaskan parks can be seen on the map of the national parks, but there are no separate text accounts for each park. In the absence of good information about the distribution and abundance of mammals in these new parks, I have not included them in the checklists of parks that follow the mammal accounts. The only conspicuous land mammal that might be seen in some of these parks, but that does not occur in the ones of which I have written, is the Arctic Fox. The Arctic Fox can be distinguished from the only other Alaskan fox, the Red Fox, by its tail, which is not white-tipped. Persons planning to visit any of these parks should write to the Alaska Regional Office, National Park Service, 540 West 5th Avenue, Anchorage, Alaska 99501, for information. These parks are:

Gates of the Arctic National Park
Kenai Fjords National Park
Kobuk Valley National Park
Lake Clark National Park
Wrangell—St. Elias National Park

There are more than four hundred species of mammal in North America, and the majority of them occur in one or more national parks. Mammals are warm-blooded animals with backbones. They suckle their young with milk and, at some stage of their development, are covered with hair. Most mammals are small, nocturnal, and secretive. Few of us ever get a glimpse of, much less a good look at, the great variety of shrews, mice, rats, and bats that live around us, yet these make up most of the kinds of mammals that occur in the parks. Park visitors usually see the larger mammals or the mammals that are active during the day—mainly members of the squirrel family and the hoofed animals. This book is designed not for the mammal specialist, but for the general park visitor who sees an animal and wants to know more about it. Falling in between the most visible species and the least visible ones are a large number of less frequently seen species that are also of great interest to the general public. Many of these are predatory—the Coyotes, wolves, Bobcats, and weasels—or they are omnivorous in their food selection—the Opossum, Raccoon, and skunks. These too have been included.

Finding mammals is usually a matter of chance. Some species have become habituated to human presence and are readily found. Chipmunks and Golden-mantled Ground Squirrels are usually around lookouts and other visitor sites in many of the western parks. For other species, the chance of seeing an individual can be increased by visiting the habitat that it frequents, but although other species may be present, seeing an individual is a rarity. As there is no assurance of seeing a mammal, and the opportunities of seeing an individual may bear no relation to its actual abundance, I have used general terms in the text. When a species is said to be "readily observed" or "commonly" or "usually seen," I mean that in a two- or three-day visit to the particular park, and to the area or habitat specified, there is better than a fifty-fifty chance of seeing an individual. When an animal is said not to be "commonly" or "usually seen," or is "scarce" or "seldom seen," it means that even in a week-long visit you are not likely to see the animal. "Rare" and "seldom seen" mean just that, and for some of the animals so designated park rangers who have spent years in the area may never have seen one.

Reference to abundance, however, may not relate to observability. Abundance is expressed in terms relative to that particular species. In some parks Bobcats may be relatively abundant—compared with Bobcats elsewhere in the range of this species—even though they are seldom seen because they are nocturnal and tend to avoid human beings. Wolves on Isle Royale are relatively abundant, yet not one visitor in a thousand sees a wolf. On the other hand, prairie dogs may exist in relatively small numbers in some parks, but, because they tend to remain in one place and are out during the daytime, they are readily seen.

Human activity has also affected the presence or absence of certain species. Wolves and Grizzly Bears, for example, were exterminated from some areas even before they were established as parks. Attempts have been made in some parks to restore species that once occurred there by transplanting (reintroducing) animals of the same species (but not always the same subspecies) from other places. In a few parks, species not native to the area have been intentionally or unintentionally introduced, and they may now be an established part of the fauna. Whether or not a species actually occurs in a park may, in some cases, be questionable or unknown. In the checklists of mammals, the following abbreviations or symbols indicate these situations:

(E)   Species extirpated within the park

(I)   Introduced species (not native to the park)

(RI)   Reintroduced species (formerly native to the park, extirpated, and then restocked from some other population of the same species)

(?)   Status or presence of the species is uncertain.

The chances of seeing mammals, other than by chance encounters, are increased by knowing something about their habitats, habits, and periods of activity. Midday, especially when it is hot in midsummer, is not a good time to look for mammals. Dawn is an especially good time to see mammals; you may often encounter nocturnal species on their way to their daytime resting places. At daybreak in Shenandoah National Park I once counted more than forty White-tailed Deer in less than five miles; at 9 A.M. over the same route not one was to be seen. Raccoons, jack rabbits, Coyotes, Bobcats, and foxes are among the mammals that are most often seen at dawn. An ideal mammal-watching schedule would be to rise about a half hour before sunrise and to drive slowly on the roads for an hour or hour and a half to look for rabbits and carnivores, as well as some of the hoofed animals. Then return back to camp for breakfast, spend the morning and the late afternoon hiking through areas suitable for seeing tree and ground squirrels, marmots, and chipmunks. At dusk it might be worth driving along the roads again, although it could be even more rewarding to spend the half hour before bedtime on this drive. As an alternative, stroll around campgrounds with a flashlight just before retiring.

Remember that the mammals in the national parks are wild. Despite how charming and cute they may appear to you, or ideas of their tameness or friendliness that you may have received from films or television shows, all mammals are potentially dangerous. Even the "friendly" chipmunk is capable of biting your finger to the bone. Overly friendly mammals should be regarded with suspicion, and this is particularly true of foxes, skunks, Coyotes, and Raccoons, which are sometimes carriers of rabies. Ground squirrels and prairie dogs usually have a large population of fleas, among which may be a carrier of bubonic plague and other diseases. The larger mammals are even more dangerous, and any of the hoofed animals can kick hard enough to break bones. Cow Moose with calves are extremely protective of their young and may chase people and bite or kick them. Bears—Black, Brown, and Grizzly—represent an especial hazard because of their large size and formidable teeth and claws. They are best viewed from afar, and if encountered on a trail, should be given the right of way. Mother bears with cubs are particularly dangerous. Parks with bear populations usually have information bulletins on how to deal with them, and visitors are discouraged from using trails that pass through areas where Grizzly Bears are known to be active.

It is against National Park Service regulations to feed any animals in the national parks—either intentionally or unintentionally. Food should be stored in tight containers out of the reach of animals, and all crumbs and garbage should be disposed of in animal-proof containers as well. Back-country visitors should suspend their food out of reach of bears. There is usually a great temptation to feed chipmunks, ground squirrels, tree squirrels, Raccoons, and marmots, especially if they are near a campsite. Feeding the animals is not only illegal, but it also generally does not help them. Many individuals that become dependent upon human largesse suffer from improper nutrition. If they are species that normally store food for the winter, they may fail to do so because they are so preoccupied with the popcorn and peanuts—neither of which are storable—offered them by people. And, of course, when the main visitor season is over, these animals are hard put to survive.

Mammals especially respond to movement and sound. You will see more mammals and have them under observation longer if you learn to move slowly and quietly. If you run toward an animal, it will either flee or attack. If you stand still as soon as you see an animal, and then move very slowly in its direction, you can probably approach close enough to get a good view of it. Shouting and whistling usually scare it away. Carry a pair of binoculars (7 power is a useful magnification) for safe, distant observation of mammals.

Maps have been prepared for the national parks and monuments where there is good mammal-viewing, and these appear with the appropriate park account. They were drawn from the maps published by the National Park Service for use in 1979 and 1980. Since then, a few of the parks have had, mostly minor, boundary changes, and these maps have not been updated. The information about the mammals in these parks is based on the parks as they appear on the maps. The maps show most of the main roads in each park, as well as the location of the visitor centers, some of the campgrounds, buildings, and major features, such as rivers and mountains. More complete maps may be obtained by writing to the superintendent of each park at the address shown at the beginning of each account. Maps are also available free at the entrances to the parks. Symbols used on the maps of the parks are:

★   Indicator of site or area for viewing mammals
▲   Campground
◯   Mountain
∘   Locality or special feature
▢   Building
—   Road
---   Trail

The national parks provide excellent opportunities for photographing animals, but the visitor should not expect to come away with photographs of the quality that might be found in popular magazines or on exhibit at the parks, unless he or she is willing to go to some expense to obtain special equipment. Probably the most useful adjunct to a camera for animal photography is a telephoto lens. Telephoto lenses are available in a great variety of sizes and prices, and the visitor interested in specializing in animal photography would do well to start with a good quality zoom lens with a maximum range of around 200 mm. (for a 35 mm. camera). This will give the photographer a magnification of about four times that of the standard (50 mm.) camera lens. The greater the magnification of the lens, however, the steadier he or she must stand, and, usually, more light is necessary for proper film exposure. The former problem can be overcome by using a tripod or other steadying device, such as a gunstock; the latter by using "fast" film, which requires less light. It is often difficult to focus on an animal, especially with "split-image" focusing devices, because there are no straight lines for orientation on mammals. Focusing should be done on the animal's eye. If the eye is in focus, even if the rest of the animal is not, the picture will be useful; if the eye is out of focus, the picture will not be satisfactory. Another tip is to try to lower yourself to eye level with the animal to avoid the usual appearance of looking down on the creature. A low camera angle will increase the drama of your picture. Do not change film in direct sunlight, and if no other shade is available, use your body to protect your film. Try to keep your film as cool as possible and never store it in a car that is left in direct sunlight; the temperature inside may reach well over 120° F (49° C). Have the film processed promptly.

The best trips to the national parks are those made with the most planning. This book will serve as a guide to the mammals you may hope to see in the parks, and where the best places to observe them are. When you have decided which parks you want to visit, write to the superintendent at the address shown and ask for general information. You will promptly receive a packet of maps, information about accommodations and campsites, availability of services, reading lists, and much useful information.

As your planning proceeds, you can then use this book to plan day-by-day expeditions to particular parts of the park in search of specific mammals, and you can also read about the lives of the different species you may encounter. Carry the book with you to the park as an aid in identifying and locating mammals. Spend some time at the visitor center to learn more about the mammals and to receive the latest information about where the best places to see them are. I hope that your visits to the parks will be as enjoyable as mine have been.

Note: *In the mammal accounts, the most substantial reference to a mammal appears in boldface type.*

△ Plate 1a   Caribou. See pp. 280–81, 289–91.

▽ Plate 1b   Grizzly Bears. See pp. 240–41, 243–45.

△ *Plate 2a*   *Fox Squirrel. See pp. 206–7, 209.*

△ *Plate 2b*   *Western Chipmunk. See pp. 187–88.*

△ *Plate 3a*   *Mountain Goat. See pp. 294–96.*

△ *Plate 3b*   *Bison. See pp. 293–94.*

△ Plate 4a   *Bobcat. See pp. 273, 275–76, 277.*

▽ Plate 4b   *Mountain Lion. See pp. 273–75, 276.*

△ *Plate 5a*   *Black-tailed Jack Rabbit. See pp. 179–80.*

△ *Plate 5b*   *Red Fox. See pp. 235–37.*

△ Plate 6a   *Yellow-bellied Marmot. See pp. 191–92.*

△ Plate 6b   *Black-tailed Prairie Dog. See pp. 203–5, 206.*

△ *Plate 7a*   *Mule Deer. See pp. 284–86.*

△ *Plate 7b*   *Pronghorn. See pp. 291–93.*

△ Plate 8a   *California Ground Squirrel. See pp. 194, 198–99.*

▽ Plate 8b   *Harbor Seals. See p. 271.*

Mammals of the National Parks

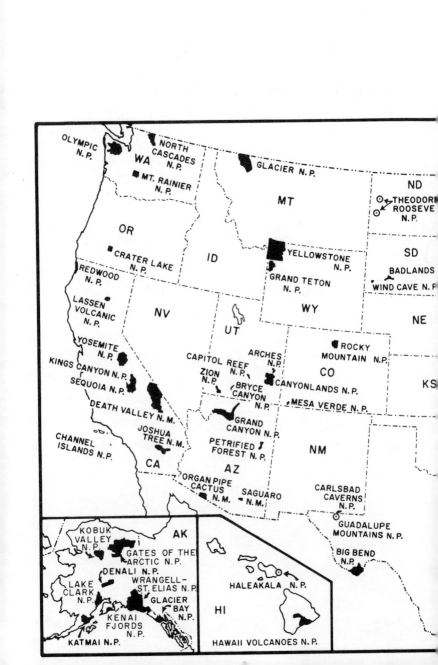

Part 1

PARKS AND MONUMENTS

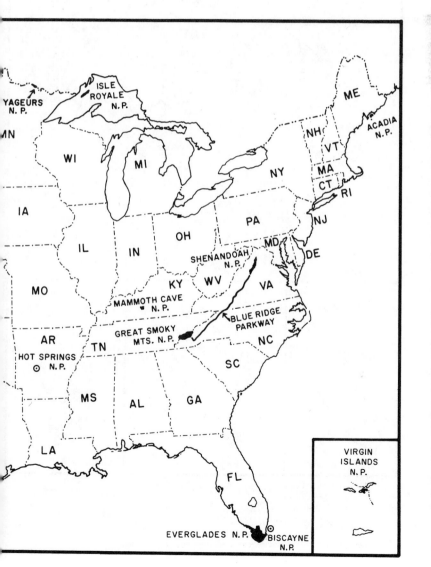

ACADIA NATIONAL PARK

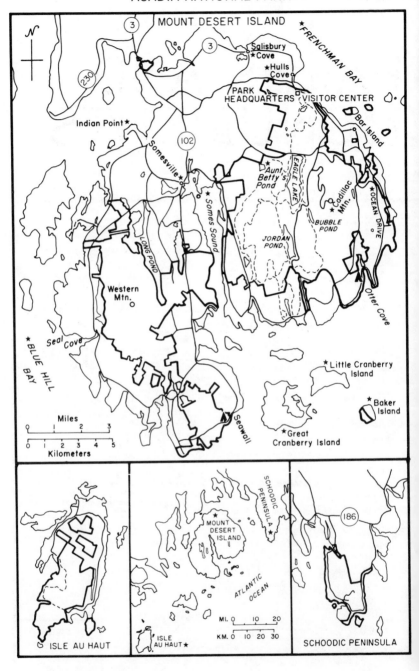

Acadia National Park

Route 1, Box 1
Bar Harbor, Maine 04609

Smallest of the national parks on the eastern seaboard, Acadia's 39,097 acres (15,417 hectares) feature Mount Cadillac, which is the highest elevation on the eastern seaboard (1,530 feet; 466 meters), and the dramatic rocky coast of northern New England. Most of the park lies on Mount Desert Island, but it is fragmented and also includes portions of the Schoodic Peninsula to the east and of Isle au Haut and Baker Island to the south, as well as some islands in Frenchman Bay. Discovered in 1604 by Champlain, Mount Desert Island is rich in human and geologic history. The rocks and only true fjord on the U.S. Atlantic Coast exemplify ice-age geology, and the vegetation features spruce, fir, and balsam fir, hardwood birch vales, and beech, tamarack, and black spruce in the bogs.

The land mammals of Acadia are not regularly seen, but there are opportunities to see marine species such as Gray and Harbor Seals, Minke, Pilot, and Fin Whales, and Harbor Porpoises. Although the park is open in winter, the visitor center is closed then, and commercial boat tours are likewise not in operation. July and August are the peak months for visitors, and many also come to see the magnificent fall foliage in late September and early October.

Red Squirrels occur throughout the park, from mountain tops to the shore, and are usually encountered along trails and are often heard chattering. **Gray Squirrels** are much less common, and they live mainly in the mature hardwood forests, where they are not often seen. Populations of Gray Squirrels at Salisbury Cove, Hull's Cove, and Somesville are descendants of animals released in 1922. **Eastern Chipmunks** are common from mountain tops to the bays and lakes of the park, and they are usually seen around campsites.

Although there are hundreds of **White-tailed Deer** throughout the park, they are usually seen only at dawn or dusk, mainly along roadsides and in clearings in the autumn. The only other native, large, hoofed animal that might be encountered is the **Moose**. It is rare, however, and not resident in the park, all sightings being of transient animals that have wandered through and have not remained. Another large mammal, the **Black Bear**, also is a rare resident, but bears have been seen in the park.

Raccoons are common, especially around campgrounds at night. **Striped Skunks** in Acadia seem not to be campground pests, but are seen regularly at night along the roads.

Beavers are common in the ponds and are not hard to observe at dawn or dusk at places such as New Mill Meadow, opposite the Precipice, and Aunt Betty's Pond. The park Beavers are descendants of two pairs reintroduced in 1920. Ponds with cattail swamps, such as Aunt Betty's Pond, are places to observe **Muskrats** at dusk or at night.

The only member of the rabbit family in Acadia is the **Snowshoe Hare**. It is sometimes seen along roadsides at dawn and dusk, and in cedar swamps. This hare's main predator, the **Red Fox**, is common in the park, but it is seldom seen by visitors.

Other land mammals in Acadia are rarely seen, and some may not be resident. The **Bobcat** has been observed and **Fisher** tracks have been seen on Day Mountain, but neither species may actually reside in Acadia. **River Otters, Mink,** and **Long-tailed Weasels** are residents, but they are seldom seen, and the same is true of **Porcupines. Woodchucks** inhabit some clearings on Cadillac Mountain and are in fields along Bay View Drive, but are not common and are not often seen.

Harbor Seals inhabit the waters and rocky ledges around Acadia, and they are usually seen from the commercial boat cruises. Some are generally hauled out on rocky ledges at Indian Point (on private land) in the spring and summer, and people on the Baker Island Cruise usually see some. Frenchman Bay and Blue Hill Bay are better places to find Harbor Seals, and some of the outlying islands have sizable populations. **Gray Seals** are less common, but they are sometimes seen on the Bass Harbor Cruise.

Harbor Porpoises may be seen in any bay, especially when the mackerel are running. Frenchman Bay, Blue Hill Bay, the mouth of Somes Sound, and around the Cranberry Islands are among the better places in which to find them.

Swimming close to the park's coastal waters at times are **Minke Whales** and **Common Pilot Whales**. Deep-water species, such as **Fin Whales** and **Hump-backed Whales**, are extremely rare close to Acadia.

CHECKLIST

___ Snowshoe Hare	___ Red Fox
___ Eastern Chipmunk	___ Black Bear
___ Woodchuck	___ Raccoon
___ Gray Squirrel (I?)	___ Fisher
___ Red Squirrel	___ Long-tailed Weasel
___ Beaver (RI)	___ Mink
___ Muskrat	___ Striped Skunk
___ Porcupine	___ River Otter
___ Common Pilot Whale	___ Harbor Seal
___ Harbor Porpoise	___ Gray Seal
___ Fin Whale	___ Bobcat
___ Minke Whale	___ White-tailed Deer
___ Hump-backed Whale	___ Moose

Arches National Park

446 South Main Street
Moab, Utah 84532

Featuring delicate red sandstone arches and other intriguing examples of wind and water erosion, Arches National Park's 73,379 acres (29,696 hectares) lie entirely within the pinyon-juniper plant community. The arid environment (annual rainfall 8 inches; 203 millimeters) and narrow altitudinal range (about 4,000–5,600 feet; 1,219–1,707 meters), as well as high summer and low winter temperatures, limit the diversity of species that inhabit the park.

Paved roads lead to many of the major geological features, and there are some unpaved and four-wheel-drive vehicle roads to remoter areas. With relatively few hiking trails, Arches National Park is open to much cross-country hiking. Rabbits and hares, ground squirrels, and Mule Deer are the most frequently seen mammals.

The only large mammal resident in the park is also one of the more often seen species. The **Mule Deer** lives throughout the park and may be observed almost anywhere, especially in the vicinity of Devil's Garden. Both **Pronghorns** and **Mountain Sheep** have been part of the park's fauna, but neither may be present any more. Pronghorns have not been seen here since 1941, and the last observation of Mountain Sheep was near Double O Arch in 1974.

During the day, **White-tailed Antelope Squirrels** are often seen scampering about, displaying their conspicuous tails. They have even been seen at the visitor center, as well as at many other places throughout the park, including Courthouse Towers and Park Avenue. **Colorado Chipmunks**, too, are abundant, preferring rockier areas than those used by the antelope squirrels, and are regularly seen. The only other member of the squirrel family now in Arches is the **Rock Squirrel**. It is not very common, and most observations of it have been near the visitor center and on the rocks near the Delicate Arch Trail, Park Avenue, and the Double O Arch Trail. The White-tailed Prairie Dog may have once been part of the park's fauna, but is probably extirpated now, the last sighting having been in 1966 in Salt Valley.

The large **Black-tailed Jack Rabbit** is abundant in the park and, although primarily nocturnal, is usually seen at dawn or dusk along the sides of roads, and even during the day near Devil's Garden, in the Salt Valley Wash, and around Wolfe Ranch. The other member of this family in Arches is the **Desert Cottontail**. It, too, is abundant and is often seen throughout the park, especially at dawn or dusk. Park Avenue is a good place to find Desert Cottontails.

Although **Porcupines** are numerous in the park, they are not seen in proportion to their abundance. Most sightings of Porcupines have been around Courthouse Wash, Balanced Rock, and Park Avenue. Similarly,

ARCHES NATIONAL PARK

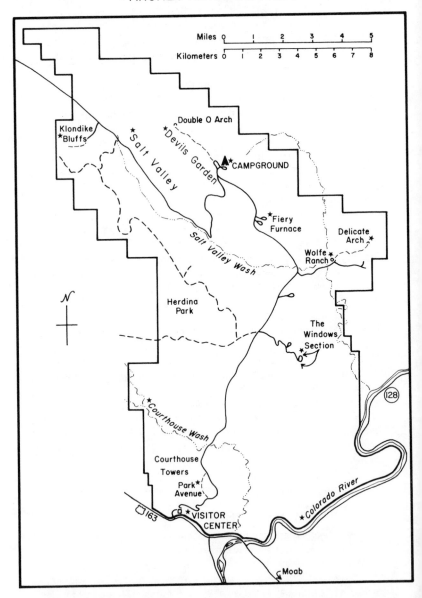

Desert Woodrats are common, but are seldom seen because they are nocturnal. Their debris nests are often found in rock crevices. Another species, the **Bushy-tailed Woodrat**, also lives in the park, but is regarded as uncommon.

Gray Foxes are common in Arches, but all observations of them have been at night, when they may be encountered almost anywhere. Another fox, usually considered a rare species, may be becoming fairly common in the park. This is the **Swift Fox**, which has been observed a number of times at night, starting in 1978, mainly in the Fiery Furnace area at the Salt Valley overlook, at Courthouse Wash, and at the Delicate Arch viewpoint. Whether the observed species is actually the Swift Fox or the closely related **Kit Fox** needs confirmation. **Coyotes** are probably not rare, but there are few sightings of them in summer, and mostly they have been seen at night in autumn throughout the park and, in particular, near Delicate Arch and at Courthouse Towers.

Although **Bobcats** are seldom seen, they are probably fairly common in the park and, judging from tracks and occasional nighttime observations, may be abundant around Delicate Arch. The only other member of the cat family in the park is the **Mountain Lion**. Mountain Lions are rare, but their tracks have been seen at Delicate Arch and Navajo Arch, and the animals themselves have been seen at these places. However, there is some doubt as to whether these Mountain Lions are residents of the park or merely transients. The same question has been raised for the **Black Bears**, which are thought, from the few observations of them, to be transients moving through the park from the La Sal Mountains to the east.

A number of small carnivores also inhabit Arches, none of them common. Perhaps the most abundant, but rarely seen, is the **Badger**, the diggings of which have been seen in several places, including Salt Valley. Both **Spotted** and **Striped Skunks** are quite rare, and the same is true of the **Raccoon**, for which there is only a single record, from Courthouse Wash. More common is the Raccoon's relative, the **Ringtail**, which has been observed at several places, such as among the rocks on the switchbacks above the visitor center, but only at night.

Beavers are known to inhabit the Colorado River along the southern border of the park, and they sometimes wander up to be seen at Courthouse Wash and Freshwater Canyon. There is a single, old record of a **Muskrat** in the residence area.

CHECKLIST

___ Desert Cottontail

___ Black-tailed Jack Rabbit

___ Colorado Chipmunk

___ White-tailed Antelope
    Squirrel

___ Rock Squirrel

___ White-tailed Prairie Dog
    (E?)

___ Beaver

___ Desert Woodrat

(continued)

(continued)

___ Bushy-tailed Woodrat   ___ Badger
___ Muskrat   ___ Spotted Skunk
___ Porcupine   ___ Striped Skunk
___ Coyote   ___ Mountain Lion
___ Swift Fox (or Kit Fox)   ___ Bobcat
___ Gray Fox   ___ Mule Deer
___ Black Bear   ___ Pronghorn (E?)
___ Ringtail   ___ Mountain Sheep (E?)
___ Raccoon

Badlands National Park

P.O. Box 6
Interior, South Dakota 57750

In the Badlands, prairie grassland overlies hundreds of feet of mud and silt sediments—mainly of the Oligocene geological epoch—which were deposited starting about 35 million years ago. Erosion by water in this 244,300 acre (98,866 hectare) park has cut into the prairie, exposing bare cross sections of the fossil-rich ancient sediments, which gives the park bizarre beauty and its name. Two recent additions to the park, the South Unit and the Palmer Creek Badlands, are undeveloped and relatively inaccessible.

With a typical prairie fauna, the park is best known for its Black-tailed Prairie Dogs, Bison, and Pronghorns, all of which are usually visible. It is also one of the two national parks that still may have some rare Black-footed Ferrets, and also has the rare Swift Foxes. The best mammal-viewing areas are along the Badlands Loop from the Cedar Pass region to the Pinnacles Entrance and the Sage Creek Wilderness area, and the best times are from spring to fall.

Black-tailed Prairie Dog colonies are numerous in Badlands, but most colonies are not visible from the roads. A large prairie dog town, mainly on the north side of the road, is about 5 miles (8 kilometers) west of the Pinnacles Entrance and is an excellent place to see the animals during the day. There are also many towns in the southern sectors of the park, as well as in the Sage Creek Wilderness area.

Another common species is the **Desert Cottontail**, found throughout the park. This pale-colored rabbit is readily seen around the Cedar Pass Visitor

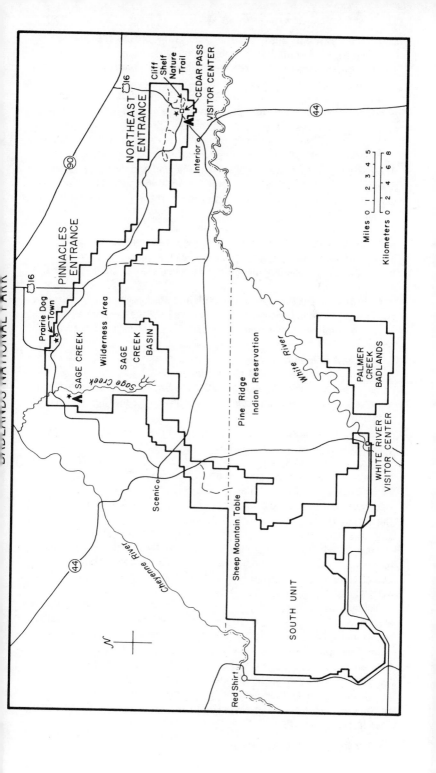

BADLANDS NATIONAL PARK

Center, at dawn and dusk, especially on lawns behind the cabins. The much larger **White-tailed Jack Rabbit** undergoes cyclical population increases, but at present is seldom seen in the park. When abundant, these hares are seen mostly along the roads at night.

Bison were extirpated from the Badlands area before 1900, but were reintroduced in the park in 1963. They have thrived and now number several hundred. Bison are usually in the Sage Creek Wilderness area, but most of the time some can be seen from the road north of the Sage Creek Campground itself, around the campground, and near the prairie dog town. Another species that was wiped out in the Badlands was the **Mountain Sheep**. These, too, were reintroduced in 1964, and persist in small numbers; there are occasional sightings of them in the badlands around the Pinnacles Entrance.

Pronghorns also disappeared from the park early in the twentieth century, but have repopulated the area on their own from surrounding lands. They are now common in the grasslands of the park, and they can often be seen along the Badlands Loop from the northeast entrance to the Pinnacles Ranger Station.

Mule Deer occur throughout the park and are fairly common. At night, they are occasionally observed along the roads. **White-tailed Deer** are rare, but are sometimes encountered near the Pinnacles Entrance and around Sage Creek.

The **Least Chipmunk** is common in the park and is usually encountered at lookouts along the roads and on nature trails such as Cliff Shelf. Another common rodent is the **Thirteen-lined Ground Squirrel**, found in grasslands. It is often seen in prairie dog towns, as well as in the Sage Creek Wilderness area and near Cliff Shelf.

Coyotes occur throughout the park and are occasionally encountered. Because they prey on prairie dogs, Coyotes are sometimes found in the vicinity of the colonies, especially at dawn and dusk. In the evening, they can sometimes be heard howling and, in winter, are often seen near the Cedar Pass Visitor Center. **Gray Wolves** undoubtedly once ranged through Badlands, but have long been extinct in the area, and the same is true of the **Grizzly Bear**.

Prairie dog colonies also attract **Badgers**, but they are seen only occasionally and then more commonly in the southern sectors of the park. Another denizen of prairie dog towns is the **Black-footed Ferret**. This extremely rare species may still occur in the park, but there have been no authentic sightings of it since about 1970.

Bobcats are considered common, but are seldom seen in Badlands, except for a few observations of them near the Cedar Pass Visitor Center, the Pinnacles area, and around Cliff Shelf. Another seldom seen species is the **Porcupine**, which is sometimes encountered in Sage Creek Wilderness area, but may also enter the park in the South Unit, through which the White River passes. The **Mountain Lion** has sometimes been reported in the Sage Creek Wilderness area, but the status of this large cat in Badlands is unknown. The **Striped Skunk** is also uncommon and rarely seen in the park.

A number of other species may occasionally occur within the boundaries of Badlands. These are found along the White River, and include **Beaver, Mink, River Otter, Raccoon**, and **Fox Squirrel**, as well as the **Swift Fox**, which has been seen at night on the road to the White River Campground and near Cedar Pass. The **Red Fox** also occurs in Badlands, but is rare, and the same is true of the **Long-tailed Weasel**. **Elk** once regularly passed through the Badlands while migrating, but are not resident now. There have been a few rare stragglers noted in the park in some years.

CHECKLIST

___ Desert Cottontail	___ Long-tailed Weasel
___ White-tailed Jack Rabbit	___ Black-footed Ferret
___ Least Chipmunk	___ Mink (?)
___ Thirteen-lined Ground Squirrel	___ Badger
___ Black-tailed Prairie Dog	___ Striped Skunk
___ Fox Squirrel (?)	___ River Otter (?)
___ Beaver (?)	___ Mountain Lion
___ Porcupine	___ Bobcat
___ Coyote	___ Elk
___ Gray Wolf (E)	___ Mule Deer
___ Red Fox	___ White-tailed Deer
___ Swift Fox	___ Pronghorn
___ Grizzly Bear (E)	___ Bison (RI)
___ Raccoon	___ Mountain Sheep (RI)

Big Bend National Park

Big Bend National Park, Texas 79834

One of the larger national parks, with 708,118 acres (286,568 hectares), Big Bend features rugged mountains rising out of desert, bordered by the broad Rio Grande River. The Chisos Mountains, with Emory Peak (7,835 feet; 2,388 meters) as their highest point, form the center of the park and are characterized by Douglas fir, ponderosa pine, and Arizona cypress at the higher elevations, and pinyon pine, juniper, and oaks on the slopes. At low elevations are plants characteristic of the Chihuahua Desert, including creosote bush, ocotillo, yucca, and many kinds of cactus. At the lowest elevations (1,700 feet; 518 meters) along the Rio Grande are river-bottom plant communities, with mesquite, cane, willow, and cottonwood.

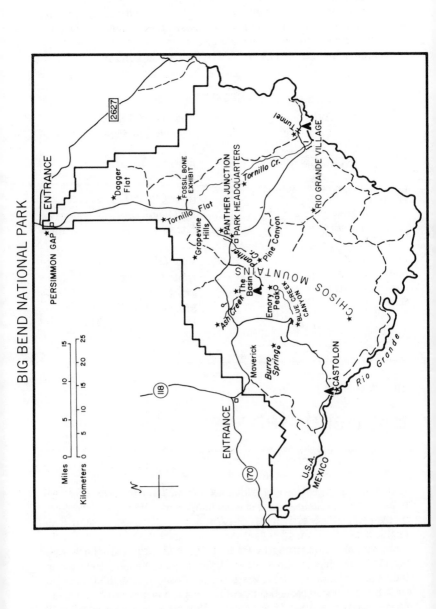

BIG BEND NATIONAL PARK

Big Bend has a variety of mammals, mainly either desert forms or species with wide distribution, but very few that cannot be found in other national parks. It is the best park for seeing Collared Peccaries and is a good place to see deer and ground squirrels. The park is accessible throughout the year; it is warm in the lowlands and chilly at night in the mountains in winter. It is extremely hot in the desert in summer, but comfortable then in the mountains.

Mule Deer are common in the park, especially at elevations of 2,500–5,000 feet (762–1524 meters), but occur even down to the Rio Grande flood plain. They generally can be seen along the roads early in the morning and at dusk. **White-tailed Deer** are numerous above 4,000 feet (1,219 meters) elevation in the mountains, especially in the oak-pinyon zone.

Another common hoofed animal is the **Collared Peccary** or **Javelina**, a piglike mammal that inhabits the lowlands and foothills of the park. Peccaries are often found near water, especially in the morning and evening. During the day, they may shelter in canyons, and those around Ash Creek are good places to look for them, as well as along the Grapevine Hills Road, and at the Old Ranch site near the junction of the road from The Basin or Panther Junction to Santa Elena Canyon. Peccaries are also found in The Basin, where a musky, somewhat skunklike odor may be a sign of their presence.

Two other hoofed mammals have been recorded from the park. **Pronghorns** are rare. They may occur in the western part of the park and have also been seen between Persimmon Gap and Panther Junction. **Mountain Sheep** once inhabited Big Bend, and there are old records of them from Tornillo Flat, Dagger Flat, the Fossil Bone Exhibit, the Tunnel, Panther Junction, and Blue Creek Canyon. However, because there have been no recent sightings, they are thought to be extinct in the park. Observations of either Pronghorns or Mountain Sheep should be reported to park rangers.

Rabbits and hares are abundant and often seen in Big Bend. **Black-tailed Jack Rabbits** are common at the lower elevations, diminishing in abundance to about 5,400 feet (1,646 meters). They are frequently seen along the roadsides late in the evening and early in the morning, especially below Panther Junction and toward the Fossil Bone Exhibit, and sometimes are out, even in summer, as late as midmorning. The large-eared, but small-sized, **Desert Cottontail** is also abundant at low elevations, up to about 4,700 feet (1,433 meters). It is also often seen along the roads or darting across them late in the evening and early in the morning. Much less common and not regularly seen is the **Eastern Cottontail**, which lives in brush and edges of forest at elevations from 4,700 feet (1,433 meters) and higher.

Coyotes are found throughout the Big Bend lowlands and up to 5,500 feet (1,676 meters) and are common. They are widespread and may be seen at almost any time, day or night. On the flats north of Panther Junction early in the morning is a good place and time to find them. **Gray Wolves** are not resident in Big Bend, although some may venture into the park from Mexico at times. There have been no recent reports of them. Foxes are not common. The **Gray Fox**, although occurring throughout the park, is rarely seen, except around the residential area in The Basin, where it is sometimes sighted. **Kit Foxes**, lowland inhabitants, are very rare, and a recent survey failed to locate any of them in the park.

Of the four kinds of skunk known from the Big Bend, only the **Striped Skunk**, found throughout the park, is common. The smaller **Spotted Skunk** is also widespread, but is much less often seen and is considered rare. The large, white-backed **Hog-nosed Skunk** is uncommon, but is found at elevations as high as 7,500 feet (2,285 meters) and has been seen on the road to The Basin. The rarest of all the Big Bend skunks is the **Hooded Skunk**, generally a low-country denizen, for which the single park record is at Tornillo Creek. There are two other members of the weasel family in the park. **Badgers** inhabit the lowlands up to the foothills of the Chisos Mountains. They are not common in Big Bend and are seldom seen. Their diggings may be encountered more often than the animals themselves. **Long-tailed Weasels** are sparse in the park, but there have been a number of sightings of them during the daytime in The Basin Area.

Although there are four kinds of ground squirrel in Big Bend, none of them is common. The most frequently seen is the **Texas Antelope Squirrel**, which is found at elevations as high as 6,100 feet (1,859 meters) and is most regularly seen in the Chisos Mountains, but seems uncommon elsewhere. **Rock Squirrels** are also found in rocky places, especially in the Chisos Mountains, but are not common. They are likely to be encountered on the road to The Basin. The **Spotted Ground Squirrel** is a denizen of the lower elevations, mainly 2,000–4,000 feet (610–1,219 meters), and is uncommonly seen on the road between Rio Grande Village and Panther Junction. The rarest of the Big Bend ground squirrels is the **Mexican Ground Squirrel**, which has been found only at Rio Grande Village.

Mountain Lions are resident in the park, and although they are not regularly sighted, there are enough observations to indicate that they are permanent, breeding inhabitants. Most of the sightings have been from the road between Panther Junction and The Basin—also the area of most traffic in the park. Of the two other wild cats known from Big Bend, the **Bobcat** is found throughout the park, but sightings are infrequent, Burro Springs being the best locale. Occasionally an **Ocelot** has been recorded from Rio Grande Village, presumably a transient from Mexico.

Raccoons are rarely seen in Big Bend and are mainly inhabitants of the wet areas along the Rio Grande. In excessively wet years, however, they are known to move into The Basin. The long-nosed relative of the raccoon, the **Coati**, is a transient in the park, passing through at times, but not maintaining a permanent population. When present, Coatis are found in the lowlands. **Ringtails** are fairly common and are known to occur throughout the park. Because of their nocturnal habits, however, they are seldom seen. **Black Bears** are not resident in the park but occasionally do occur as transients. They are thought to come out of the Sierra del Carmen, across the Rio Grande in Mexico.

Beavers are present along the Rio Grande, but are not common. Their largely nocturnal habits, the muddy water, and the fact that they do not build houses but inhabit dens in the river banks, make them a rare sight.

Two other mammals may occur in Big Bend. Although there are no recent records of **Virginia Opossums** in the park, there are old records from Pine Canyon and Pimate Spring. **Porcupines** have not actually been found in Big Bend, but they have been reported from just north of the park, and they may eventually be found within the boundaries. Sign or observations of Porcupines or Virginia Opossums should be reported to park rangers.

CHECKLIST

___ Virginia Opossum (?)
___ Eastern Cottontail
___ Desert Cottontail
___ Black-tailed Jack Rabbit
___ Texas Antelope Squirrel
___ Mexican Ground Squirrel
___ Spotted Ground Squirrel
___ Rock Squirrel
___ Beaver
___ Southern Plains Woodrat
___ White-throated Woodrat
___ Mexican Woodrat
___ Porcupine
___ Coyote
___ Gray Wolf (E?)
___ Kit Fox (E?)
___ Gray Fox
___ Black Bear

___ Ringtail
___ Raccoon
___ Coati
___ Long-tailed Weasel
___ Badger
___ Spotted Skunk
___ Striped Skunk
___ Hooded Skunk
___ Hog-nosed Skunk
___ Mountain Lion
___ Ocelot
___ Bobcat
___ Collared Peccary
___ Mule Deer
___ White-tailed Deer
___ Pronghorn
___ Mountain Sheep (E?)

Biscayne National Park

P.O. Box 1369
Homestead, Florida 33030

Bordered by the Atlantic Ocean on the east and the Florida mainland on the west, Biscayne National Park consists of a series of about twenty-five north-south islands, the largest of which is Elliott Key. Most of the park's 175,000 acres (70,821 hectares) consist of reefs and water, and the park is not accessible by automobile, nor are there any commercial boat tours to the islands at the present time. Biscayne was proclaimed a park in 1980.

The fauna of the park is not completely known. **Marsh Rabbits** are on some of the keys, including Elliott, and are occasionally seen. The most visible mammal, however, is a nonnative species, the Mexican **Red-bellied Squirrel**. This rodent was released on Elliott Key in 1938 or 1939, probably from stock that originated from Tamaulipas, Mexico. It is now abundant on the island, and its leaf nests are visible in the trees. Although the typical coloration of this squirrel is a gray back and bright reddish belly and sides, melanism is common, and about three out of four of these squirrels are in the black color phase. **Raccoons** are also present on some of the islands and are often seen, especially around campgrounds.

In the waters around the keys, **Manatees** are sometimes seen, and **Bottle-nosed Dolphins** are not at all rare. A number of other typically Floridian species are known to live on the adjacent mainland or islets within the park's borders. Among these are **Bobcats, River Otters**, and **Virginia Opossums**.

CHECKLIST

___ Virginia Opossum	___ River Otter
___ Marsh Rabbit	___ Bobcat
___ Red-bellied Squirrel (I)	___ Bottle-nosed Dolphin
___ Raccoon	___ Manatee

Blue Ridge Parkway

700 Northwestern Bank Building
Asheville, North Carolina 28801

Connecting Shenandoah National Park in Virginia with Great Smoky Mountains National Park in North Carolina, the Blue Ridge Parkway's 469 miles (755 kilometers) follow the crest of the southern Appalachian Mountains. The actual National Park Service land, which encompasses 81,569 acres (33,010 hectares), with an average altitude of about 3,000 feet (914 meters), is a narrow strip that winds through the forested mountain wilderness.

For most people, the main attractions of the parkway are the magnificent vegetation and pioneer history of the area. Relatively few mammals live within the actual bounds of the narrow parkway, but many species can be seen crossing the road or at recreation sites. The mammals to be seen here—early and late in the day are the best times—include most, if not all, of those that are common to Shenandoah or Great Smoky Mountains parks. **Eastern Chipmunks, Red Squirrels, Gray Squirrels,** and **Woodchucks** are usually seen during the day, and **Eastern Cottontails, White-tailed Deer, Virginia Opossums, Raccoons,** and **Spotted** and **Striped Skunks** are often observed at night, dawn, or dusk.

For a checklist of mammals, see those for Shenandoah or the Great Smoky Mountains. In addition, **Elk** were reintroduced near Peaks of Otter in 1917, and, although they survived for some years, they now seem to be gone from the area once more.

Bryce Canyon National Park

Bryce Canyon, Utah 84717

Dropping sharply away from the high Paunsaugunt Plateau, Bryce Canyon is a wonderland of limestones, shales, sandstones, and conglomerates sculpted into colorful and bizarre formations by erosion. Long and narrow, Bryce's 35,835 acres (14,502 hectares) are bisected by a paved road on the plateau most of the length of the park, and there are numerous hiking and horse trails down into the eroded areas. Because of Bryce's altitudinal range—6,600–9,100 feet (2,012–2,774 meters)—three plant communities are represented. At the lower elevations, pinyon pine, juniper, and sagebrush predominate, as well as Gambel oak. Over most of the plateau, ponderosa pine, limber pine, juniper, and manzanita make up the woodlands, but, at the higher elevations to the south near Rainbow Point, Douglas fir becomes the dominant tree, along with white fir and blue spruce.

Although mainly visited in the summer, Bryce is open throughout the year, and roads to the main viewing points are kept open. The most visible animals in the park in summer are chipmunks, Golden-mantled Ground Squirrels, and Mule Deer, as well Utah Prairie Dogs, which have been reestablished in the park.

Of the three species of chipmunk in Bryce, the most abundant and frequently seen is the **Uinta Chipmunk**, which abounds on the plateau and is the common species along the rim. Down in the canyon, where the Uinta also ranges, are both **Least** and **Cliff Chipmunks**, neither of which is abundant. As usual with chipmunks, field identification of a particular species is difficult, if not impossible. The other striped rodent in the park is the **Golden-mantled Ground Squirrel**. It is most abundant along the rim, at automobile turnouts along the road, and especially around campgrounds and cabins. The large, grayish-brown **Rock Squirrel** is not common on the

BRYCE CANYON NATIONAL PARK

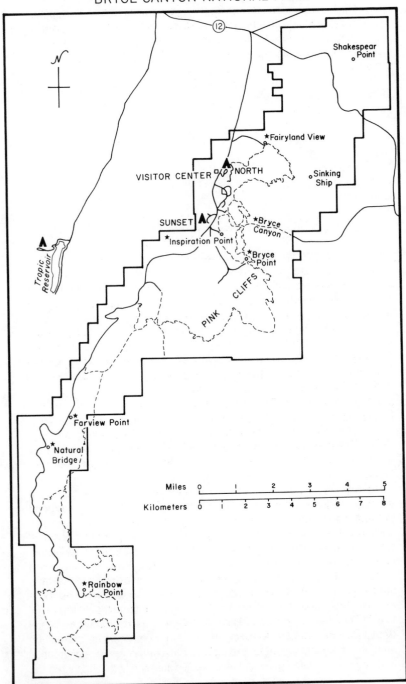

plateau, but is often seen in the broken country below the rim, especially along the trails. Another rock dweller on the plateau and edge of the rim is the large **Yellow-bellied Marmot**, often seen sunning on rocks. Marmots usually can be found around the stables housing the horses used on trail trips. The only tree squirrel in Bryce is the **Red Squirrel**, which is mainly an inhabitant of the spruce-fir forests at the park's southern end.

Mule Deer are common on the plateau and are usually seen, at dawn or dusk, in meadows and other areas near the visitor center, campground, and lodge. The only other hoofed animal in the park is the **Elk**, which may not remain in the park throughout the year and is seen only in winter, when some from neighboring areas may venture into the flats. **Pronghorns** were extirpated from the park area by 1936.

Gray Foxes are the most frequently seen carnivores, usually observed at night along roadsides or at campgrounds. **Coyotes** live in the lower parts of the park and are not commonly seen in Bryce. The last **Gray Wolf** in the park was killed in 1928.

Although both **Striped** and **Spotted Skunks** are present in the park, only the former is common, mainly at lower elevations. The Spotted Skunk, which prefers broken country among rocks, is rather uncommon there, although it is known to occur below the rim and at the southern end of Bryce. **Badgers** are seldom seen, although they may be common in some low elevation meadows. **Long-tailed Weasels** are hardly ever seen, but they are recorded from the park. The **Wolverine** was extirpated by 1897.

The **Utah Prairie Dog** once inhabited Bryce in large numbers, but was exterminated. In 1974, these white-tailed rodents were reintroduced into the park, where they have survived and bred. There are prairie dog colonies in meadows southwest of the visitor center, west of the grocery store, and north of Sunset Campground; the animals are easily observed.

Porcupines are fairly common wherever there are conifers, but scarred trees are seen more often than the animals themselves; however, Porcupines occasionally invade campgrounds in search of salty materials.

The **Desert Woodrat** is common at the lower elevations, where its stick nests are found in broken-rock areas. **Beavers** lived along streams in the park, but may no longer be present.

Bobcats are common in Bryce, but they are rarely seen. Also present in the park, and also a rare sight, are **Mountain Lions**, which are known to live in the southern parts. **Black Bears** are occasionally seen in the same area, but observations are few; **Grizzly Bears** have been extinct in the park since 1916. The **Ringtail** is another seldom-seen species, which lives in rocky areas below the plateau rim.

Both the **Desert Cottontail** and the **Nuttall's Cottontail** occur in the park, but neither is common. The Desert Cottontail lives at the lower elevations near the eastern boundary; Nuttall's Cottontail inhabits high-elevation fir forest. **Black-tailed Jack Rabbits** are found at elevations as high as 9,100 feet (2,758 meters), but **White-tailed Jack Rabbits** are rare and have been seen only from October to February.

CHECKLIST

___ Nuttall's Cottontail	___ Coyote
___ Desert Cottontail	___ Gray Wolf
___ Black-tailed Jack Rabbit	___ Gray Fox
___ White-tailed Jack Rabbit	___ Black Bear
___ Least Chipmunk	___ Grizzly Bear (E)
___ Cliff Chipmunk	___ Ringtail
___ Uinta Chipmunk	___ Long-tailed Weasel
___ Yellow-bellied Marmot	___ Wolverine (E)
___ Rock Squirrel	___ Badger
___ Golden-mantled Ground	___ Spotted Skunk
Squirrel	___ Striped Skunk
___ Utah Prairie Dog	___ Mountain Lion
___ Red Squirrel	___ Bobcat
___ Beaver (E?)	___ Elk
___ Desert Woodrat	___ Mule Deer
___ Porcupine	___ Pronghorn (E)

Canyonlands National Park

**446 South Main Street
Moab, Utah 84532**

Consisting of high plateaus dissected by the Green and Colorado rivers, Canyonlands National Park's 337,530 acres (136,611 hectares) exhibit colorful and intricate examples of rugged geological erosion. Relatively little of the park is accessible by two-wheel-drive automobile, and although there are a number of four-wheel-drive roads, the park is best enjoyed by hikers. Concessionaires in nearby towns offer vehicle rentals and tours, as well as horseback trips and river expeditions. The park is primarily desert country (annual precipitation 5–9 inches; 127–229 millimeters), but there are isolated pockets of Douglas fir and ponderosa pine, and the rivers also provide habitat for aquatic species. Temperature extremes may range from $-20°$ F. ($-29°$ C.) in winter to more than $110°$ F. ($43°$ C.) in summer. Daily ranges of 30° F. (17° C.) may occur.

The fauna of Canyonlands is still being surveyed, and there may be species that are unrecorded from within the park's borders. Cottontails, jack rabbits, Coyotes, chipmunks, and ground squirrels are common on the plateaus, and Beavers and Gray Foxes are abundant along the rivers. Among the most frequently seen rodents in the park is the **Colorado**

CANYONLANDS NATIONAL PARK

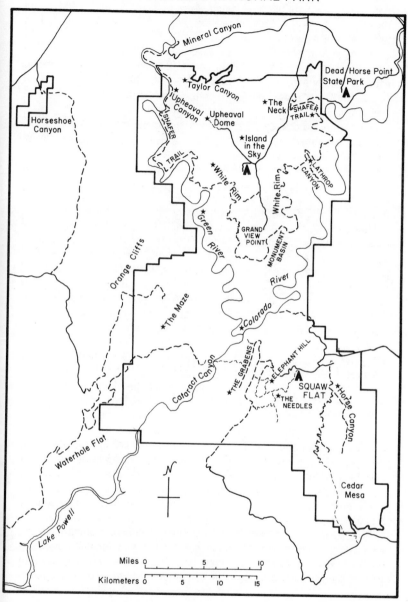

Chipmunk, which is abundant in brush lands, broken-rock areas and in the pinyon-juniper woodlands. Another abundant and conspicuous species is the **White-tailed Antelope squirrel**, which is found throughout Canyonlands wherever there are rocks near sandy grasslands. The distinctive white underside of the tail readily distinguishes this species from the chipmunk.

Although **Rock Squirrels** are found at all elevations in Canyonlands, wherever there is rock rubble, they do not seem to be especially abundant and thus cannot be said to be seen regularly. Around Squaw Flat Campsite in the Needles District is one of the better places to find them.

On the plateaus of Canyonlands, **Black-tailed Jack Rabbits** are common and, although mainly nocturnal, are often seen by hikers during the day, especially early and late. They also occur along the Green and Colorado rivers. The **Desert Cottontail** is a common species that occurs throughout the park.

Coyotes are common in Canyonlands and range throughout the park. They are sometimes seen along the road in Island in the Sky. **Gray Foxes** are common, especially along the rivers and their tributaries, and they have also been seen on the plateau near the ranger station at the Island in the Sky entrance, and at Squaw Flat in the Needles District. Two other foxes also occur in the park, although they are less abundant. Both the **Red Fox** and the **Kit Fox** have been seen in the Island in the Sky sector.

Mule Deer are found throughout Canyonlands, on river bottoms as well as on the plateaus. They are fairly common, but are not seen often. **Pronghorns** no longer occur in the park, although they have been reintroduced not far to the east of the Needles District, outside the park boundaries. **Mountain Sheep** occur in the park; the current population is believed to number about three hundred. They are found along the rivers and also have been observed on White Rim, which is probably the best area to observe them, especially in the north near the Shafer Trail.

Four species of woodrat are known to inhabit the park; one or more species may be found almost anywhere. These are the **White-throated Woodrat**, the **Mexican Woodrat**, the **Desert Woodrat**, and the **Bushy-tailed Woodrat**. The most abundant and widespread of these is the Desert Woodrat, which is found in the Island in the Sky and the Maze districts in woodlands, the rock rubble of the rim, and down on the river flats. The least abundant is the Bushy-tailed Woodrat, although it is found in both the Needles and the Island in the Sky districts and is also thought to inhabit the Maze area. Both the White-throated and Mexican Woodrats have been found only in the Needles District, and both species have been found living near Cave Spring. All of these woodrats are nocturnal and are rarely seen by visitors, although their brush-debris nests may usually be found.

Both **Beavers** and **Muskrats** occur in the Colorado and Green rivers, as well as in permanent water tributaries. They are not extremely abundant, however, and, in common with their occurrence elsewhere, are not often seen in Canyonlands. Beaver bank dens and other sign are often encountered.

Sign of **Porcupines** is common, especially in the pinyon pine areas of Canyonlands. The Porcupines themselves are seldom seen. They occur throughout the park, and sign of their activity—debarked trees—have been noted in all three major park districts: the Maze, Needles, and Island in the Sky.

Badgers seem to be common along the river bottoms, but as usual with this species, they are not often seen. There are ample diggings, however, to indicate their presence. Both **Striped** and **Spotted Skunks** occur on the plateaus of the park, although the former is not as common as the latter, which has been seen at night around Grandview Point and probably also occurs in the canyons near the rivers. **Long-tailed Weasels** may occur in Canyonlands, but there is no record of them. **Ringtails** probably occur throughout the park, but are nocturnal and secretive, and are thus rarely seen.

Bobcats are common in Canyonlands and, although not usually seen by visitors, have been observed at various places throughout the park, especially in the Island in the Sky District. Much less common, and rarely seen, is the **Mountain Lion**, which roams throughout the park and for whose presence there is good evidence in Upheaval Canyon, as well as in the Needles District.

Both **Gunnison's Prairie Dog** and the **White-tailed Prairie Dog** are known to have lived near the borders of Canyonlands, but there are no records of them from within the park, and a recent mammal survey of the region failed to find them.

Another species that has not yet been recorded from the park is **Abert's Squirrel**. It is primarily a rodent that inhabits ponderosa pine forests, and it is thought that, if present at all, it could live in the extreme southeastern portion of the Needles District. Any sighting of this large squirrrel should be reported to park rangers.

CHECKLIST

___ Desert Cottontail	___ Porcupine
___ Black-tailed Jack Rabbit	___ Coyote
___ Colorado Chipmunk	___ Red Fox
___ White-tailed Antelope Squirrel	___ Kit Fox
___ Rock Squirrel	___ Gray Fox
___ White-tailed Prairie Dog (?)	___ Ringtail
___ Gunnison's Prairie Dog (?)	___ Long-tailed Weasel (?)
___ Abert's Squirrel (?)	___ Badger
___ Beaver	___ Spotted Skunk
___ White-throated Woodrat	___ Striped Skunk
___ Desert Woodrat	___ Mountain Lion
___ Mexican Woodrat	___ Bobcat
___ Bushy-tailed Woodrat	___ Mule Deer
___ Muskrat	___ Pronghorn (E)
	___ Mountain Sheep

CAPITOL REEF NATIONAL PARK

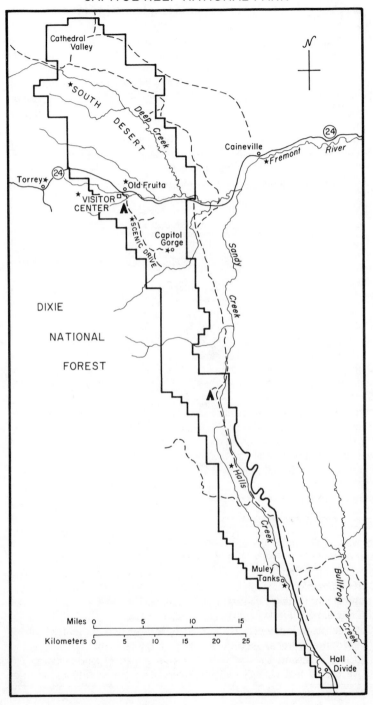

Capitol Reef National Park

Torrey, Utah 84775

This 72-mile-long (115-kilometer) park averages less than 10 miles (16 kilometers) in width. It features exposed geological formations deposited as long as 250 million years ago. The park is essentially desert (it receives less than 7 inches—178 millimeters—of annual precipitation), but the Fremont River parallels its only paved road (State Highway 24) and provides a narrow belt of luxuriant vegetation and habitat. A few four-wheel-drive roads give access to the northern and central parts of Capitol Reef, but most places can be reached only by hiking. Fruit orchards established by the pioneers are maintained as a historical element of the park, and the crops attract many birds and some mammals. Much of Capitol Reef's 241,904 acres (97,874 hectares) lie at 5,000–7,000 feet (1,524–2,137 meters) elevation. Pinyons and junipers are the predominant trees in dry areas; cottonwoods and willows favor the river edge.

There are a number of mammals that are known to occur near Capitol Reef, but which have not yet been recorded from within its boundaries, partly owing to the narrow configuration of the park. Other species may have been eliminated when the area was first farmed, but there are others that are readily observed because of the maintenance of the orchards. Among these are marmots, jack rabbits, cottontails, skunks, and deer.

The only large mammal that is regularly seen in Capitol Reef is the **Mule Deer**. It occurs throughout the park and is usually seen at dawn or dusk, commonly in the orchards at Fruita. At present there are no **Mountain Sheep** in the park, although there is ample evidence that they occurred here and were hunted by Indians in the past. The last sighting of Mountain Sheep here was in 1948. The **Pronghorn** is another species that was extirpated from Capitol Reef, and although it has been reintroduced to the west of the park near Loa, it has not yet been reported in the park. It is not certain that **Bison** were originally in the park area, although this seems likely, but at present the few sightings of them are from reintroduced stock that spread from nearby areas, and they are not resident in Capitol Reef.

In the lower parts of Capitol Reef, the **White-tailed Antelope Squirrel** is a commonly seen species, especially in the southern extent of the park, as well as around Fruita and even high on the reef. The only chipmunk certain to be in the park is the **Colorado Chipmunk**, which is abundant in pinyon-juniper areas, dry and rocky canyons, and around Fruita. **Least** and **Cliff Chipmunks** are not actually recorded from Capitol Reef, but may occur in the higher northern district. **Rock Squirrels** are abundant, especially in the orchards at Fruita, and in the rocky canyons and river flood plains. More restricted in the park is the **Golden-mantled Ground Squirrel**, which is reported only from the northern district and is relatively common on the adjacent Thousand Lake Mountain. **Yellow-bellied Marmots** live mainly above 6,800 feet (2,073 meters), where they can sometimes be seen sunning themselves on rocks, but they also inhabit the orchards at Fruita, where they are abundant in summer and readily seen.

Beavers inhabit the permanent rivers of the park and live in the Fremont River, as well as Hall's Creek. Beaver sign can be found at the Hickman Bridge Trail at the Fremont River, between Spring Canyon and Grand Wash. **Muskrats** also utilize the permanent water, and their nests can be seen at the campground at the Fremont River. **Mink** are also known to live on the riverbanks, but are seldom seen. **River Otters** are thought to be extinct here. **Raccoons**, which also remain near water, are not yet recorded from within the park's borders, but their tracks have been seen outside the park on Hall's Creek in the south and Sulphur Creek to the east. Whether these Raccoons are from released animals or are native is uncertain.

Black-tailed Jack Rabbits are widespread in the park, but are not often encountered, except at night, especially in the orchards at Fruita, where they are numerous. **White-tailed Jack Rabbits** may occur in the northern sector's higher elevations, but records are few, and the specimen that was found dead at the historic schoolhouse is suspect. **Snowshoe Hares** are rare and probably limited to the northern part of the park and perhaps on the western border. **Desert Cottontails** are common throughout the park, particularly in the lower elevations, and are often seen along the roads at dawn or dusk near Fruita.

Two species of fox are known from Capitol Reef. The commoner is the **Gray Fox**, which is widespread and is abundant around Fruita, where it often can be seen at night. **Red Foxes** are rare, but have been seen in the northern sector and also near Red Slide in the south. **Kit Foxes** are thought to occur near the park and may inhabit the southern portions. **Coyotes**, although widespread throughout Capitol Reef, are not particularly common and are not often seen. **Gray Wolves**, which once inhabited this park, are believed to be extinct now. **Black Bears** are rare, although they have occasionally been seen near Grand Wash and Chimney Rock, but may not be resident within the park. **Grizzly Bears** are extinct in the area.

Of the smaller carnivores, **Ringtails** are numerous, but are seldom seen because they are wholly nocturnal. Some live in the rocks near the Petroglyphs and at Hickman Bridge. In similar areas, there are **Spotted Skunks**, and they and Ringtails are both known to be numerous at times around the park service residential area and the campground at Fruita. **Striped Skunks** are common near Fruita and are more frequently seen than Spotted Skunks. Striped and Spotted Skunks also are both recorded from the southern part of the park. **Badgers** are fairly common, but not often seen, especially in the southern part of Capitol Reef. **Ermine** live in the northern part of the park and have also been seen on Chimney Rock Trail, but **Long-tailed Weasels**, which probably live along the Fremont River, have not actually been recorded there.

Both **Bobcats** and **Mountain Lions** occur in the park and are not rare, although observations of them are. Mountain Lions probably roam throughout the park and have been observed on the Scenic Drive and in the park service residential area, and tracks have been seen on the Cohab Trail. **Bobcats** are also widespread and have been seen at Hickman Bridge and Chimney Rock.

The **Desert Woodrat** is the common species in the central part of the park and is the one that builds its nests in the Fremont River valley. The **White-throated Woodrat** is known to inhabit the extreme southern part of the park, especially Hall's Creek. The **Bushy-tailed Woodrat** may exist in the higher, northern district, but has not actually been recorded there.

Porcupines are abundant in Capitol Reef, but are not often seen, although their debarking of trees is evident. They are sometimes pests around the orchards and the visitor center. No tree squirrels are definitely recorded from the park, but **Red Squirrels** have been seen close to the northeastern boundary of Capitol Reef. **Utah Prairie Dogs** exist in small populations to the west of the park and, in 1979, were reintroduced at Jones Bench at the north end of the park. Most prairie dogs in this part of Utah were poisoned in the 1930s.

CHECKLIST

___ Desert Cottontail	___ Coyote
___ Snowshoe Hare (?)	___ Gray Wolf (E)
___ Black-tailed Jack Rabbit	___ Red Fox
___ White-tailed Jack Rabbit (?)	___ Gray Fox
___ Least Chipmunk (?)	___ Kit Fox (?)
___ Cliff Chipmunk (?)	___ Black Bear
___ Colorado Chipmunk	___ Grizzly Bear (E)
___ White-tailed Antelope Squirrel	___ Raccoon (?)
___ Rock Squirrel	___ Ermine
___ Golden-mantled Ground Squirrel	___ Mink
	___ River Otter (E)
___ Yellow-bellied Marmot	___ Badger
___ Utah Prairie Dog (RI)	___ Spotted Skunk
___ Red Squirrel (?)	___ Striped Skunk
___ Beaver	___ Mountain Lion
___ Desert Woodrat	___ Bobcat
___ Bushy-tailed Woodrat	___ Mule Deer
___ White-throated Woodrat	___ Pronghorn (E?)
___ Muskrat	___ Bison (RI)
___ Porcupine	___ Mountain Sheep (E)

CARLSBAD CAVERNS NATIONAL PARK

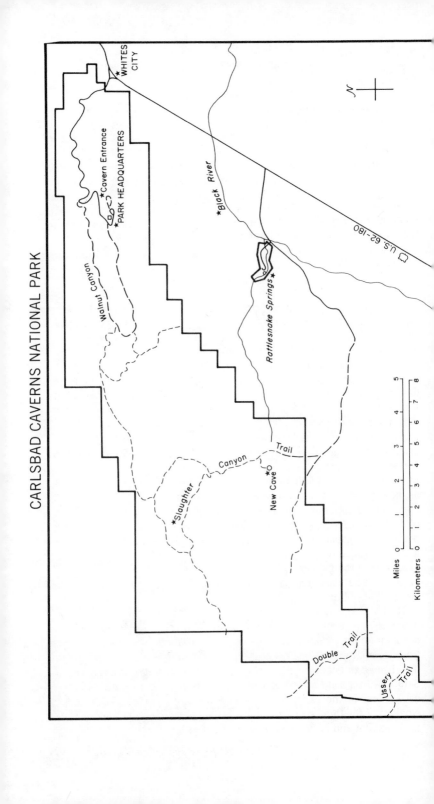

Carlsbad Caverns National Park

3225 National Parks Highway
Carlsbad, New Mexico 98220

Although the major feature of Carlsbad National Park is its huge cavern, one of seventy in the 46,753 acre (18,916 hectare) park, Carlsbad offers much in the way of wildlife. Situated just northeast of Guadalupe Mountains National Park in Texas, Carlsbad Caverns National Park has rugged, rocky ridges with cacti and other arid-land plants at lower elevations (3,600 feet; 1,097 meters), and juniper, pine, madrone, and even Douglas fir in the higher places (up to 6,350 feet; 1,935 meters).

The park is open throughout the year, and the famed caverns, formed by water dissolving 250-million-year-old limestones, are accessible daily. The most exciting animal phenomenon of the park, the bat flight, only occurs from late spring until frost in October or November. There are, however, many opportunities to see other mammals at Carlsbad throughout the year.

Although more than a dozen species of bat have been recorded from the park, only one is abundant, the **Brazilian Free-tailed Bat**. The population of this species is migratory and spends the winter in Mexico. In spring the bats return to Carlsbad to form a nursery colony, which has numbered in the millions, and, from May through September, they provide one of the most dramatic displays of wildlife readily available to the layperson. Starting at dusk (the park service posts the expected time of the flight at the visitor center), bats spew forth from the main cavern mouth by the hundreds and thousands and are easily seen from the amphitheater above it. Sometimes they come out like puffs of smoke and seem to disappear in the air as they fly out of sight; at other times, they come forth in long, wispy trails that seem to waft out over the valley to the east. Because the exact time of departure cannot be predicted, observers should come earlier than the posted time, especially if there has been rain the night before. Less dramatic, but equally fascinating, is a visit to the cavern mouth before daybreak, when the bats are returning in smaller groups than when they departed. The bat roost is more than 1,000 feet (303 meters) from the cavern mouth, but not on the usual visitor's path.

The bats, their droppings, and predatory insects also attract mammalian predators. There are three species of skunk in the park. The **Striped Skunk** occurs from the mouth of Walnut Canyon to the cavern itself and is sometimes seen at night along the road to Whites City. Mainly an insect eater, it is known to enter the main cavern in search of food. The **Hog-nosed Skunk**, with a solid white back and all-white tail, is common around the cavern and in the canyons of the foothills. The **Spotted Skunk** is the smallest of the three species in the park. It prefers the broken, rocky cliffs as its habitat and is common where there are a lot of crevices. It is known to venture far into the cavern, and it is the most carnivorous and weasellike of the three species. Although common, it seems more alert and is faster than the other species and is less often observed.

Another carnivore that is common around the cavern mouth and is also known to go far into the cave is the **Ringtail**. It is sometimes seen by visitors to the cavern and also lives among the rocks and in the canyons throughout the park. The Ringtail's relative, the **Raccoon**, is widespread at Carlsbad, from the lowlands of the Black River to the mouth of the cavern itself. Raccoons are often seen along the road at night.

Long-tailed Weasels live near the cavern entrance and are known to hunt inside, but, although not rare, are seldom seen. **Coyotes**, on the other hand, never seem abundant right around the cavern, but they are present in the park and often heard howling at night, but are not frequently seen. The **Gray Wolf** was extirpated in the Carlsbad area early in the twentieth century.

Gray Foxes are common in the canyons and ridges of the park. They are frequently seen along the road at night, especially around the residential area across the parking lot from the visitor center. The small, desert-dwelling **Kit Fox** lives in the low, arid valleys of the park, but is extremely rare.

Desert Cottontails inhabit the park from the low flats to the limestone ridges around the cavern, but despite their abundance, they are not often seen. The best situation in which to observe them is to drive the road from the cavern to Whites City slowly at night, or very early in the morning. The large and conspicuous **Black-tailed Jack Rabbit** is common on the low flats below the cavern, but is also numerous on the limestone ridges. Although Black-tailed Jack Rabbits are essentially nocturnal and most usually seen in the early hours of the morning along the roadsides, it is not rare to see them during the day around the cavern itself.

A diurnal rodent, usually mistaken for and called a chipmunk, the **Texas Antelope Squirrel** inhabits the rocky foothills at the base of the escarpment. Another diurnal squirrel is the **Rock Squirrel**, a large ground-dweller that, as its name implies, inhabits the rocky canyon walls. Rock Squirrels are abundant around the cavern entrance, often sunning on rocks, and are frequently seen by visitors.

Mule Deer are common throughout the park. They are often seen around the cavern entrance and by hikers in Slaughter Canyon and along the Walnut Canyon road. Whether **Elk** ever occurred within the borders of what is now the park is unknown, but some of the Elk that were reintroduced into Guadalupe Mountains National Park have now moved into the western part of Carlsbad Caverns Park. **Pronghorns** were once fairly common, even within sight of the cavern entrance, but are no longer present there. They still occur, although rarely, in the broad valleys to the east and west of the park. **Mountain Sheep**, which lived in Slaughter Canyon and adjacent canyons as late as the 1920s, have been extirpated. No wild **Bison** have been seen in the area for nearly a century.

Mountain Lions once were found at the cavern entrance, but have been scarce in the area for many years. However, they have been observed within the park borders in recent years. Although the environment seems ideal for **Bobcats**, they are rarely seen and are thought to be uncommon in the park.

Two species of woodrat inhabit the park, one in the lowland valleys and the other in the caves and on the canyon walls. The latter, the **White-throated Woodrat**, is common around the cavern entrance and adjacent rocks, cliffs, and shallow caves, but is not usually seen by visitors because it is nocturnal. The other species, the **Southern Plains Woodrat**, is abundant in the open and arid valleys of the park, especially where there are cacti. The houses of Southern Plains Woodrats are made of plant debris and are often seen under cacti, agaves, or yuccas, whose sharp spines provide additional protection. These animals are wholly nocturnal and therefore rarely seen.

Despite the absence of trees, **Porcupines** are common in the park, but are not often seen because of their nocturnal habits. **Black Bears** are very rare in the park. Some may wander through from other locales, such as the Guadalupe Mountains, and they are known to have inhabited Slaughter Canyon. The **Grizzly Bear**, which once may have inhabited the park, has been extirpated in the region.

In the low flats along the Black River, **Muskrats** and **Beavers** once occurred. The Beavers have been extirpated, and it is questionable whether any Muskrats remain. **Badgers** are another denizen of the valleys and flats (rather than the limestone ridges) and, although common, are seldom seen. The once-thriving colony of **Black-tailed Prairie Dogs** in the Black River Valley, about two miles (3.2 kilometers) south of the cavern, seems to have been wiped out, for there have been no recent reports of them.

CHECKLIST

___ Brazilian Free-tailed Bat
___ Desert Cottontail
___ Black-tailed Jack Rabbit
___ Texas Antelope Squirrel
___ Rock Squirrel
___ Black-tailed Prairie Dog (E?)
___ Beaver (E)
___ Southern Plains Woodrat
___ White-throated Woodrat
___ Muskrat (E?)
___ Porcupine
___ Coyote
___ Gray Wolf (E)
___ Kit Fox
___ Gray Fox

___ Black Bear (?)
___ Grizzly Bear (E)
___ Ringtail
___ Raccoon
___ Long-tailed Weasel
___ Badger
___ Spotted Skunk
___ Striped Skunk
___ Hog-nosed Skunk
___ Mountain Lion
___ Bobcat
___ Elk (RI)
___ Mule Deer
___ Pronghorn (E?)
___ Bison (E)
___ Mountain Sheep (E)

CHANNEL ISLANDS NATIONAL PARK

Channel Islands National Park

1699 Anchors Way Drive
Ventura, California 93003

Designated in 1980, Channel Islands National Park eventually will include five of the eight Channel Islands. At present, the National Park Service has full jurisdiction over Anacapa and Santa Barbara islands and manages, on a day-to-day basis, San Miguel Island, which is actually administered by the U.S. Navy. Most of Santa Cruz Island and Santa Rosa Island are now privately owned, but eventually will be incorporated into Channel Islands National Park. Access to the park is only by boat, and commercial boat service is available. Permits are required for camping on Anacapa and Santa Barbara, and any access to San Miguel requires a permit from the park headquarters. Visits to other islands require the permission of the owners. Waters within one nautical mile (1.85 kilometers) of the islands are also within the park.

As might be expected, the land mammals on these islands are few, but the waters and shores are rich in marine species. Harbor Seals and California Sea Lions are regularly seen on or around all the park's islands; Northern Elephant Seals and Northern Fur Seals are resident on some islands, and Gray Whales are commonly observed from commercial boat trips from January to March.

The only native land mammal that might be readily observed is the **Insular Gray Fox**, a form of Gray Fox that may be encountered during the day on San Miguel Island, but is not on Santa Barbara or Anacapa. It does live on Santa Rosa and Santa Cruz. Insular Gray Foxes are numerous and regularly seen during daylight hours. The **European Rabbit**, a nonnative species, is widespread on Santa Barbara and is usually seen by visitors. Another nonnative species, the **Black Rat**, is also sometimes seen during the day on Anacapa. **Spotted Skunks** inhabit Santa Rosa and Santa Cruz.

In the waters around all the islands, **Harbor Seals** are usually seen swimming, floating, and sometimes hauled out on the shores and rocks. The only other true seal in these waters is the **Northern Elephant Seal**, a huge species that is regularly seen at Santa Barbara and San Miguel islands. These seals are most numerous in winter (their breeding season), but some are present throughout the year.

Of the sea lions and fur seals, the **California Sea Lion** is the commonest and is seen around all of the islands. Visitors to Anacapa can usually find some California Sea Lions on the southern fringes of East Anacapa, and there are large colonies on Santa Barbara and San Miguel.

Northern Fur Seals are common breeding residents of the shores of San Miguel, and they can be found at Point Bennett and Castle Rocks throughout the year. Among the thousands of California Sea Lions at Point Bennett on San Miguel, there are usually a few dozen **Northern Sea Lions**, which are hard to discern. Also quite rare, but present every summer, are several young male **Guadalupe Fur Seals**.

Sea Otters once inhabited these waters, but are no longer present.

Gray Whales are the most frequently seen whales in these waters, but only from the end of December until the end of March, when they pass through the area during their annual migration. At these times, the commercial boat tours to Anacapa encounter them regularly, and passing whales can also be observed from the islands. The **Short-finned Pilot Whale** is another species that is sometimes, even regularly, seen in the channel between Anacapa and the mainland.

None of the other whales and dolphins that inhabit these waters are seen with regularity, but **Fin Whales, Common Dolphins, Pacific White-sided Dolphins, Bottle-nosed Dolphins**, and **Dall's Porpoises** are more often seen than the rarer **Hump-backed Whales, Grampuses, False Killer Whales, Killer Whales, Northern Right-whale Dolphins, Blue Whales**, and **Minke Whales**.

CHECKLIST

___ *European Rabbit (I)*
___ *Black Rat (I)*
___ *Bottle-nosed Dolphin*
___ *Northern Right-whale Dolphin*
___ *Pacific White-sided Dolphin*
___ *Killer Whale*
___ *Grampus*
___ *False Killer Whale*
___ *Short-finned Pilot Whale*
___ *Dall's Porpoise*
___ *Gray Whale*
___ *Fin Whale*

___ *Minke Whale*
___ *Blue Whale*
___ *Hump-backed Whale*
___ *Insular Gray Fox*
___ *Northern Fur Seal*
___ *Guadalupe Fur Seal*
___ *Northern Sea Lion*
___ *California Sea Lion*
___ *(Spotted Skunk)*
___ *Sea Otter (E)*
___ *Harbor Seal*
___ *Northern Elephant Seal*

Crater Lake National Park

P.O. Box 7
Crater Lake, Oregon 97604

A volcanic explosion some 6,600 years ago caused the top of Mount Mazama to collapse, and the remnant water-filled crater, technically a caldera, is the feature of 183,180 acre (74,131 hectare) Crater Lake National Park. Surrounded by peaks as high as 8,926 feet (2,721 meters) and sloping away to 5,000 feet (1,524 meters) or lower altitudes at the borders, Crater Lake's surface is at an elevation of 6,176 feet (1,882

CRATER LAKE NATIONAL PARK

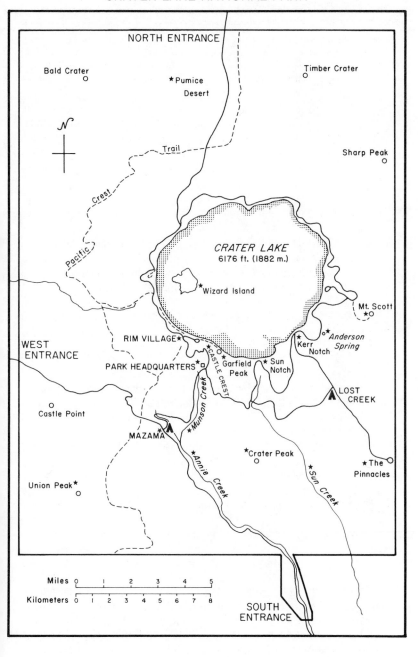

meters), and the lake reaches a depth of 1,932 feet (589 meters). It is the seventh deepest lake in the world and the deepest in the United States.

The vegetation of the park is typical of the Cascade Mountains, with white and lodgepole pine, mountain hemlock, and firs at the higher elevations, and ponderosa and sugar pines on the lower slopes. The wildlife is also a characteristically Cascade fauna, and this is a good park in which to see both subspecies of western deer—the Mule Deer and the Black-tailed Deer—as well as several species of chipmunk. It is one of the few parks in which Red Foxes and Long-tailed Weasels are commonly seen. Crater Lake is open throughout the year, with an average of 50 feet (15 meters) of snowfall in winter. The best mammal-viewing time is summer, and many species can be seen close to Rim Village.

Rodents are the most frequently seen mammals at Crater Lake. Probably the commonest mammal in the park is the **Golden-mantled Ground Squirrel**, which is abundant at the view turnouts along the roads and around campgrounds; it is regularly seen at Rim Village. Present at the same sites are two kinds of chipmunk that occur throughout the park and that are abundant. The slightly larger and darker one is the **Townsend's Chipmunk**; the brighter colored and smaller one is the **Yellow-pine Chipmunk**, which may be the commoner of the two in open areas in the forests. Either or both of these chipmunks are readily seen at lake overlooks and around campgrounds and buildings.

The **Douglas Squirrel** is the common tree squirrel of the park and is widespread. It, too, sometimes forages at turnouts, and its characteristic chattering is often heard. The **Western Gray Squirrel**, primarily an inhabitant of ponderosa pine forest, seems to have disappeared from the park, as none has been seen there since 1940. Observations of this species should be reported to a park ranger. Another seldom-seen member of the squirrel family is the **California Ground Squirrel**. Its main habitat is outside the park boundaries, but it occurs in the southern and western parts of the park, although not in areas of dense forest.

Two large rodents are common at Crater Lake. The **Yellow-bellied Marmot** is found on rocky sites and talus slopes at higher elevations. Marmots are frequently seen at roadside observation points or sunning themselves on rock surfaces such as Castle Crest or Garfield Peak. **Porcupines** are most common in the lodgepole pine zone of the park, although they are widespread throughout. They are not often seen in summer, but are commonly observed in winter. The largest of the North American rodents, the **Beaver**, which is known to have inhabited streams at lower elevations, seems no longer to be present in the park. The last documented occurrence of Beavers was in 1946 at Annie Creek.

At Crater Lake, a visitor has a good chance to see numerous kinds of carnivore. Among the commonest is the **Red Fox**, which is found throughout the park. Visitors at Rim Village have seen Red Foxes fairly often, and they have been, at times, especially common near Mazama Campground and the park headquarters. The **Gray Fox** is rare in Crater Lake, however, sometimes occurring in ponderosa pine forest, but mainly inhabiting the southern and western areas bordering the park, which are warmer. It is seldom seen. The only other member of the dog family at Crater Lake is the **Coyote**. It is widespread, but not abundant, and is not often seen. At night, Coyotes can sometimes be heard howling, especially near Union Peak.

The **Long-tailed Weasel**, a species that is seldom seen elsewhere, is common throughout Crater Lake and has been often sighted during the day, especially in Munson Valley and at Rim Village. The **Ermine**, a short-tailed weasel, is also widespread, but is nocturnal and not usually seen. A larger member of the weasel family, the **Marten**, is common at elevations higher than 5,500 feet (1,676 meters) in the park and is occasionally seen by visitors. Marten tracks in snow are often seen in winter. Much rarer is the **Fisher**, once thought to be extinct in the park, but now known to be present. Observations of Fishers should be reported to park rangers.

Black Bears are common in Crater Lake, especially around meadows and streams, and sometimes around campgrounds. The best areas in which to see them seem to be Mazama Campground, and Munson, Annie, and Sun creeks.

At higher elevations in rocky areas, especially talus slopes, **Pikas** are common. Although Pikas are hard to spot, their ventriloqual whistles are often heard, and patient observers can usually see them at Cleetwood Cove, near the Pinnacles, Garfield Peak, Sun Notch, Kerr Notch, Castle Crest, and even in the crater on Wizard Island. Other relatives of the Pika—rabbits and hares—are less often seen in the park. The **Snowshoe Hare**, mainly a forest species, is seldom seen; in winter its tracks in the snow indicate its presence and abundance. **Nuttall's Cottontail** is not common and is mainly confined to the lower elevations of ponderosa pine forest, where it is seldom seen.

Deer are often seen in the park. At Crater Lake, two subspecies of Mule Deer are present; the **Mule Deer** and the **Black-tailed Deer**. Mule Deer are mainly in the eastern part of the park, although their ranges may overlap those of the Black-tailed Deer at the higher elevations. Mule Deer are best seen in drier, eastern places, such as around Mount Scott, whereas the Black-tailed Deer is the subspecies more commonly seen along the major roads and trails of the park, especially in moist meadows of the western part. Both subspecies move to lower elevations in winter. The **White-tailed Deer** is extremely rare in Crater Lake Park, occurring only in the western part.

Elk are present in the park only in summer and then just in the western and southern parts. In winter, they migrate outside of the park. In summer, the best place to see Elk is near Union Peak, but they are also (less commonly) seen near Crater Peak. These Elk are descendants of ones brought from Yellowstone National Park to replenish the native stock. **Pronghorns** are not resident in Crater Lake, but they sometimes pass through the park and have been observed in the Pumice Desert and the burned area in the northeastern part of the park.

Although seldom seen because they are nocturnal, **Bushy-tailed Wood-rats** are fairly common in rocky areas and pine forest. Their presence is evident from their large stick-and-debris nests, sometimes built in old buildings. **Mountain Beaver** populations in Crater Lake have diminished in recent years, and, in any event, the animals were seldom seen. There are burrows at Anderson Spring near Mount Scott, and at a few other moist meadows and stream sides.

Crater Lake is also the home of a number of other mammals that may be encountered only rarely. **Mountain Lions** are probably not resident in the park, but they have supposedly been seen at times in the Pinnacles area and also near park headquarters. **Bobcats** are found at the lower and middle elevations of the park but, except for sightings of them along roads at night, are seldom seen. **Wolverines** are transients, rather than residents, at the higher elevations. A sign of their presence is usually tracks in the snow. **Badgers** are common, although rarely seen. Their prodigious diggings show where they occur, especially in the Annie Creek area and in the meadows of the Pumice Desert.

The only species of skunk in the park is the **Striped Skunk**, but it is rare, mainly occurring along the eastern and southern boundaries. Striped Skunks have also been seen around park headquarters. **Raccoons** are not often seen at Crater Lake, but there are some in the ponderosa forests, especially near streams. Two species that are known to have inhabited the park, but that are believed to be no longer present, are the **Mink** and the **River Otter**. Both species lived near rivers and streams, especially at lower elevations. **Muskrats** may not have been native to this part of Oregon, and occasional observations of Muskrats in or near the park are thought to be of animals introduced in the upper Klamath marshes nearby.

CHECKLIST

___ Pika
___ Nuttall's Cottontail
___ Snowshoe Hare
___ Mountain Beaver
___ Yellow-pine Chipmunk
___ Townsend's Chipmunk
___ Yellow-bellied Marmot
___ California Ground Squirrel
___ Golden-mantled Ground
    Squirrel
___ Western Gray Squirrel (E?)
___ Douglas Squirrel
___ Beaver (E)
___ Bushy-tailed Woodrat
___ Muskrat (I?)
___ Porcupine
___ Coyote
___ Red Fox

___ Gray Fox
___ Black Bear
___ Marten
___ Fisher
___ Ermine
___ Long-tailed Weasel
___ Mink (E?)
___ Wolverine
___ Badger
___ Striped Skunk
___ River Otter (E?)
___ Mountain Lion
___ Bobcat
___ Elk (RI)
___ Mule Deer
___ White-tailed Deer
___ Pronghorn

Death Valley National Monument

Death Valley, California 92328

Death Valley National Monument is a place of extremes; it boasts the lowest elevation in the United States—282 feet (86 meters) below sea level—and heights to 11,049 feet (3,368 meters). Death Valley has recorded the highest temperature in the Western Hemisphere, 134° F. (56.7° C.), in July 1913, and ground temperatures up to 201° F (93.9° C.), also in July, but in 1972. In 1929 there was no precipitation at all; the maximum precipitation recorded in any one year (1913) was 4.5 inches (113 millimeters). Covering an area of 2,067,795 acres (836,816 hectares), mainly in California but with a small portion in Nevada, Death Valley is one of the largest of the parks and monuments in the contiguous United States.

Despite high temperatures and aridity, Death Valley supports a diverse mammalian fauna in the depths of the valley, as well as in the relatively cooler surrounding mountains. Most species avoid the heat by their nocturnal or subterranean habits, but there are some that are above ground and active in the heat of the day. Death Valley is open throughout the year and is visited most frequently during winter. Summer is hot, of course, but it is also a good time to visit, especially since there are few visitors at that time. The lives of many of the mammals are tied to mesquite thickets, which are the best places in which to look for them. The Panamint Chipmunk is unique to this area and is found in no other park or monument.

Even during the midday heat, ground squirrels are active in Death Valley. The **White-tailed Antelope Squirrel** tolerates heat well and is the most commonly seen mammal, especially along roadsides in the northern part of the monument. It inhabits mesquite thickets on the valley floor, as well as mountains up to 6,000 feet (1,829 meters). The **Round-tailed Ground Squirrel** is strictly a low-desert species, mainly inhabiting mesquite thickets around Furnace Creek and adjacent sandy areas. These ground squirrels may estivate from late summer onward, not reappearing until late winter. The **California Ground Squirrel** is present in the monument only on Hunter Mountain and Cottonwood areas, in the northwestern sector.

Another diurnal species is the **Panamint Chipmunk**, which is known only from the Panamint and Grapevine mountains at the eastern side of the monument. It inhabits the pinyon-juniper zone and may also occur around Scotty's Castle. In the same mountains, and also in the Cottonwood Mountains, are **Porcupines**. Although Porcupines are seldom seen, bark stripped from trees is evidence of their presence.

Two species of cottontail live in Death Valley. The **Desert Cottontail** is found on the valley floor, evidently only near mesquite thickets. **Nuttall's Cottontail** is characteristic of the mountains. Cottontails in Death Valley

DEATH VALLEY NATIONAL MONUMENT

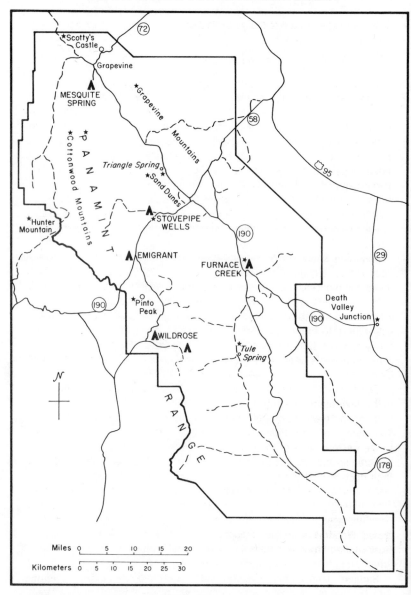

are often active until midmorning, and hikers approaching mesquite areas are sure to see some. Cottontails are also known to graze on the lawns at Scotty's Castle. **Black-tailed Jack Rabbits** are present in the monument, from the lowest elevations up into the mountains, but are not numerous. They seem more nocturnal than the cottontails and thus are less often seen.

Coyotes are not scarce in Death Valley and are known to occur from the low salt flats up into the mountains. Coyotes are frequently found around mesquite thickets because there are many rodents and rabbits there, but are not often seen, except along roads at night; they have also been observed around Scotty's Castle. Another member of the dog family is common, but rarely seen. This is the **Kit Fox**, found mainly in the valley, where rodent populations are substantial. Kit Foxes have been seen around many of the campsites, especially at Sand Dunes, Furnace Creek, Stove Pipe Wells, Triangle Spring, and Scotty's Castle. The other fox in the monument, the **Gray Fox**, lives in the mountains, especially on the east side of the Grapevine Mountains and west of Death Valley Junction. The Gray Fox may also occur at Scotty's Castle.

Mountain Sheep range throughout the monument, rarely on the valley floor and usually up in the mountains, but they are not often seen. They prefer the inaccessible ridges and canyons of the surrounding mountains, but they do tend to remain within easy walking distance of water. The only other native hoofed animal in the monument is the **Mule Deer**, but it does not venture to the valley floor and is found in the mountains, especially the Panamints, Cottonwoods, and Grapevines, mainly in the pinyon-juniper zone. Both Mule Deer and Mountain Sheep have been seen near Scotty's Castle.

Two nonnative hoofed animals—**Feral Donkeys** and **Feral Horses**—live in the monument. The burros (dating from the 1880s) and free-living horses both inhabit the Cottonwood Mountains, and the horses are also on Hunter Mountain and Pinto Peak. Burros are often seen along roads, especially in the western parts of the monument.

Three species of woodrat are known of in Death Valley, but because of their nocturnal habits are not usually seen. The **Desert Woodrat** ranges at elevations of −280 feet (−85 meters), in the salt marshes, to as high as 7,500 feet (2,286 meters), in the adjacent mountains. The stick houses beneath mesquite bushes are usually made by this species. The **Dusky-footed Woodrat** is known to live only near Furnace Creek, whereas the **Bushy-tailed Woodrat** inhabits only the high country of the northern Panamint Mountains in the pinyon-juniper zone.

Badgers live from below sea level at Furnace Creek and Tule Spring up into the mountains, wherever there are enough rodents for their food. Although Badger diggings may be seen, seldom are the animals observed. Both the **Spotted Skunk** and **Ringtail** inhabit the monument, but not the low parts of Death Valley. They prefer rocky areas on the mountainsides and, because of their nocturnal habits, are rarely seen.

Bobcats are seldom seen in Death Valley, although they are known to live around Furnace Creek and Triangle Spring, as well as up into the surrounding mountains. They prefer areas of thick cover with good populations of small mammals and are nocturnal. **Mountain Lions** are rare in the monument, but they have been recorded mainly in the mountains and have been seen near Scotty's Castle.

CHECKLIST

___ Nuttall's Cottontail
___ Desert Cottontail
___ Black-tailed Jack Rabbit
___ Panamint Chipmunk
___ White-tailed Antelope
    Squirrel
___ California Ground Squirrel
___ Round-tailed Ground
    Squirrel
___ Desert Woodrat
___ Dusky-footed Woodrat
___ Bush-tailed Woodrat
___ Porcupine

___ Coyote
___ Kit Fox
___ Gray Fox
___ Ringtail
___ Badger
___ Spotted Skunk
___ Mountain Lion
___ Bobcat
___ Horse (I)
___ Donkey (I)
___ Mule Deer
___ Mountain Sheep

Denali National Park

P.O. Box 9
McKinley Park, Alaska 99755

Lying mainly to the north of the Alaska Range, with North America's highest peak, 20,320 foot (6,194 meters) Mount McKinley, Denali's 5,696,000 acres (2,305,114 hectares) feature tundra and spruce taiga, as well as glaciers, gravel-banked rivers, and small lakes. The main tourist season is June, July, and August (although the park is open year-round), when the weather is often rainy and cool, and there is less than a 40 percent chance of seeing Mount McKinley itself. But with up to twenty-four hours of daylight, there are profuse blooms of wildlfowers, and Denali's open expanses provide virtual certainty of observing Grizzly Bears, Moose, Caribou, Dall's Sheep, and Red Foxes, as well as many smaller species, such as Hoary Marmots and Arctic Ground Squirrels. Because private vehicles are not permitted on the park roads and accommodations are limited, advance preparation is necessary for a visit.

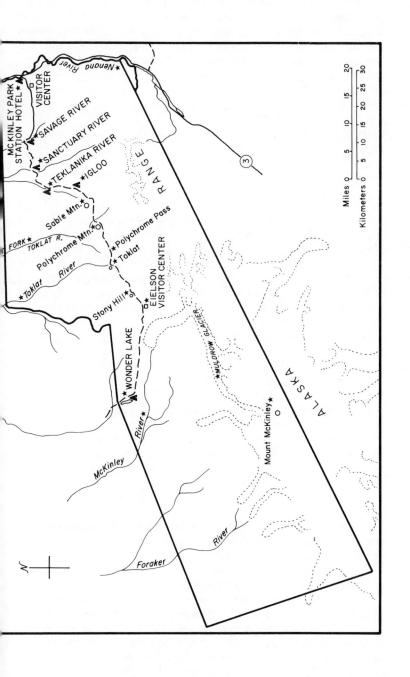

The huge and often pale-colored **Grizzly Bears** of Denali Park are often seen wandering across the tundra. Around Sable Pass and the Toklat River are good places to find them, and near Stony Hill is another. **Black Bears** are much rarer in the park and live mainly in wooded country in the east and north sides. Along the Nenana River is the best place to look for them.

Moose are common in the park and may be seen almost anywhere. There are usually some between the Savage and Sanctuary rivers, along Igloo Creek, in the willows at Sable and Polychrome passes, near Wonder Lake, and even around the hotel. The only other antlered animals in the park are **Caribou**. They wander out of the park but, in midsummer, usually move through it toward the Teklanika and Sanctuary rivers. In June and July, Sable Pass, Muldrow Glacier, Stony Hill, East Fork, Polychrome Pass and Flats, the Toklat area, and near the Eielson Visitor Center are all promising sites, and in August a few Caribou usually can be found near Wonder Lake.

Dall's Sheep are a major attraction of Denali, and these white sheep can usually be seen in summer between the Nenana River and Muldrow Glacier, especially on the mountain slopes south of the road between Igloo Mountain and Sable Mountain. Another all-white hoofed animal that has been seen in the park is the **Mountain Goat**, which has been observed at Igloo Mountain and Mile 3 on the road. Mountain Goats are not residents of the park, however, and the few sightings may be of stragglers from the nearest goat population, which is 60 miles (97 kilometers) away in the Talkeetna Mountains.

From April to October, the most frequently seen small mammal in Denali is the **Arctic Ground Squirrel**. This tundra dweller can usually be seen at lookouts on the road, such as at Stony Hill Lookout, as well as at Wonder Lake, Teklanika Campground, along the East Fork, and around the hotel and Eielson Visitor Center.

Less abundant are the large **Hoary Marmots**, which usually inhabit rocky areas. Some can usually be seen, especially in sunny weather, near Polychrome Pass and the Savage River Bridge. Another rock-slide dweller is the **Collared Pika**, which is often heard whistling before it is seen. Collared Pikas are generally numerous among the rocks above the Savage River Bridge, at Polychrome Pass, and near Eielson Visitor Center.

Red Squirrels which inhabit spruce woods along rivers, can usually be found between the Teklanika, Savage, Sanctuary, and Toklat rivers, as well as around the hotel. **Porcupines** also live in spruce forests, but are not often seen, except when they come around campgrounds, such as the one at the Teklanika River.

Red Foxes are abundant in the park and are found throughout, from tundra to spruce forest. They are active during the day and are commonly seen along the road, especially at Sable Pass, the Toklat River, and Polychrome Pass, as well as around the Savage River Campground. The Red Foxes in this park exhibit many of the color phases known for this species; they may be black, silver, orange, or yellow—even if they are from the same litter. The white-tipped tail will identify them, however, as well as distinguish them from the other two members of the dog family in Mount McKinley Park; the **Gray Wolf** and the **Coyote**.

Gray Wolves are relatively abundant, but they are rarely seen by visitors, and then only by chance. Most sightings have been near Mile 52 on the road, and in the valley near Stony Creek. Much rarer than Gray Wolves or Red Foxes are Coyotes. They are seldom seen, although they sometimes are heard howling in the Nenana River area.

Snowshoe Hares undergo cycles of abundance, but, even when the population is low, there are usually some around the hotel, Igloo Creek, and in the brushy country along the eastern and northern boundaries of the park. The **Lynx**, which preys mainly on Snowshoe Hares, also shows cyclical population fluctuations. In general, Lynx are seldom seen, but Mile 60 on the road, and the Tokat and Savage river valleys, are the most likely places to find them.

Beavers inhabit many lakes and ponds, such as Wonder Lake, Horseshoe Lake, and ones near Muldrow Glacier and the Nenana River— where their dams and lodges can be seen. Although mainly nocturnal, Beavers are sometimes seen swimming as early as midafternoon during the long summer days. **Muskrats** are neither as abundant nor widespread as Beavers. There are usually some in eastern ponds in the Horseshoe Lake area, and near Wonder Lake, in the west.

None of the five members of the weasel family in the park is seen very often. **Martens** live mainly in the spruce woods on the northern and western park borders, and evidence of their presence is usually tracks in the snow in winter. **Wolverines** range throughout the park, but it is a matter of chance to see this large weasel. Most sightings of them have been in the low passes around Igloo Creek and Sable Pass. The **Mink** may be more abundant around lakes and rivers than observations indicate. The **Least Weasel** is uncommon in this park; there are only a few records of this mammal. **Ermine** are present and are sometimes seen in spruce forest areas. **River Otters** have been reported at Wonder Lake, Riley Creek, and at the Savage River, and they also occur along the Nenana River, but are not often seen.

CHECKLIST

___ Collared Pika	___ Grizzly Bear
___ Snowshoe Hare	___ Marten
___ Hoary Marmot	___ Ermine
___ Arctic Ground Squirrel	___ Least Weasel
___ Red Squirrel	___ Mink
___ Beaver	___ Wolverine
___ Muskrat	___ River Otter
___ Porcupine	___ Lynx
___ Coyote	___ Moose
___ Gray Wolf	___ Caribou
___ Red Fox	___ Mountain Goat
___ Black Bear	___ Dall's Sheep

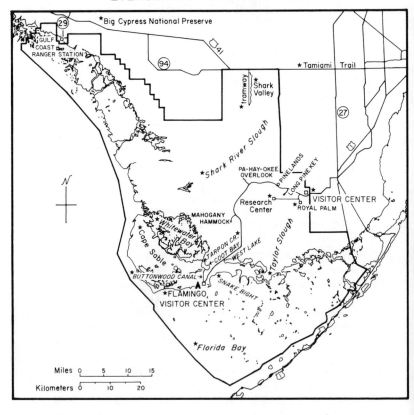

Everglades National Park

P.O. Box 279
Homestead, Florida 33030

At the southern tip of the Florida peninsula, with a maximum elevation of 20 feet (6 meters) and an area of 1.4 million acres (566,790 hectares), Everglades National Park is the largest, wettest, southernmost, and most tropical of the continental eastern parks. Over the low-lying limestones flows a vast, wide river of fresh water, which glides slowly to the sea and nurtures a great variety of life, including a "river of grass," the largest saw-grass marsh in the world. Slight elevations in this area are above the water flow and support pine woods and tropical hardwoods. In the waters, both fresh and salt, are aquatic mammals.

The park is not especially good for mammal-watching, although it is superb for bird-watching. It does have Manatees, the Round-tailed Musk-rats, and the Marsh Rabbits, which are mammals found only in southern Florida parks. This is perhaps the best park in which to see Mountain Lions, River Otters, Bobcats, and Bottle-nosed Dolphins. The best mammal-watching areas are at Flamingo, Shark Valley, and around the visitor center; the best months are January to May.

The **Marsh Rabbit** is the most readily seen mammal in the Everglades, being found throughout the park, seldom far from the shelter of low vegetation. These rabbits are usually seen along roadsides early in the morning, especially near Flamingo. In the summer, they are often on the parking lot at Shark Valley. The **Eastern Cottontail** is virtually absent from the Everglades, prefering drier sites, but it has been seen at the eastern edge, on the lawn of the visitor center.

The **Raccoon** is also an abundant species in the park and may be encountered anywhere. Although Raccoons are basically nocturnal, they are sometimes seen in the daytime near the campground at Flamingo, as well as at night, in the trees at the parking lot there. They are also often seen at Shark Valley in the daytime during the dry season (December to May).

The **River Otter** is a species that lives in many national parks but is rarely seen, however, it is observed comparatively often in the Everglades. It is found in habitats near fresh and salt water. River Otters have been usually seen near roadside canals and other residues of water in the dry period from January through May, especially in Shark Valley.

Although it is common in the park, the **Bobcat** is not often seen. It is, however, more often viewed by visitors to this park than in any other where it occurs. Bobcats have been seen in many habitats, crossing the highway, and also swimming across rivers and canals. At Shark Valley, they are sometimes seen on the hammocks, and they have been sighted in the mangrove zones with some frequency. The only other cat in the park is the **Mountain Lion**, and this large species seems more abundant in the cypress swamps and western parts of the park and rare in the typical Everglades. Many of the observations of Mountain Lions have been around Cape Sable and Flamingo, and even along Snake Bight Trail. At times, one Mountain Lion has been resident near Flamingo, where there is perhaps the best opportunity of seeing this rare species anywhere in its North American range. There are also valid records of Mountain Lions seen in farmlands near the research center.

The **Virginia Opossum** lives in just about every one of the various terrestrial habitats in Everglades Park and is probably most abundant in tropical hammocks and mangrove swamps. Primarily nocturnal, these opossums are rarely seen in the daytime. It is regrettable that their presence is often indicated by the corpse of one on the road early in the morning.

The **White-tailed Deer** is common throughout the park and is frequently seen. Among the best places to see this species is along the road near the research center on the eastern side of the park, and from the tramway at Shark Valley.

A species that is hard to see, and one that varies in abundance in the park, is the **Round-tailed Muskrat**. This is an aquatic rodent that is seldom seen, but its small, round houses made of grasses are sometimes found. There are colonies of Round-tailed Muskrats in the Shark Valley Slough, and, in the lower part of the park, they can be found at Taylor Slough, not far from the visitor center.

Although the small **Spotted Skunk** lives in southern Florida, none has been recorded from the park. It seems to prefer somewhat drier habitats. The larger **Striped Skunk** is widespread in the park, but it is nowhere abundant and is seldom seen. While the presence of both these species seems frequently to be proclaimed by their odor, one of the hammock trees, the stopper tree, gives off a skunklike odor, and skunks may be less abundant here than might be thought.

Two other members of the weasel family, the **Long-tailed Weasel** and the **Mink**, are both considered rare in the park. The Long-tailed Weasel is extremely rare, but there are a few sightings of Mink at Shark Valley and Cape Sable, as well as around Flamingo.

Compared with many other parks, Everglades has few members of the squirrel family. The **Gray Squirrel** is not rare outside the park on the eastern and northwestern edges, but within the park it occurs only around Royal Palm Hammock, and it is thought to be a newcomer to the fauna. The **Fox Squirrel** is also found more commonly outside the park, in the east and west, but is a regular part of the fauna, primarily seen in the southern or western parts of the park, especially in the pine lands. Fox Squirrels were probably more abundant in the past and are now considered relatively rare.

The **Gray Fox** is not common in the park, but is probably more abundant in the pine lands than elsewhere. It is very seldom seen, and then usually at night. The **Red Fox** is not native to the area, but in recent years some Red Foxes have been seen near the park, probably from stock that escaped from captivity.

The **Black Bear** was once common in Florida and lived in the Everglades. Destruction of its habitat and illegal killing have so reduced its numbers in Florida that the last remnants of the species in the state are in the Big Cypress Preserve to the north of the park. Within the park itself, Black Bears inhabited many habitats, mainly in the western part, although they probably were never very common. In recent years, the sightings of bears have become fewer and fewer, and it is now thought that bears are no longer resident in the park.

The **Manatee**, sometimes called the Sea Cow, is a large mammal that is wholly aquatic. It prefers shallow waters, both fresh and salt, where it can feed on aquatic vegetation. Manatees are rare in Florida Bay, but in the more sheltered Hell's Bay and Whitewater Bay there are resident populations whose exact locations are generally are known to park rangers. Manatees are shy animals, easily disturbed by human beings, and park visitors who wish to see them should be prepared to be quiet and patient, and to view them by boat, preferably nonpowered. In some places, Manatees now may be active at night, rather than during the day, as in the much-traveled Tarpon Creek.

In Florida Bay, **Bottle-nosed Dolphins** are often seen, sometimes from the breezeway at the Flamingo Visitor Center. These animals, familiar from television popularization, travel in small groups and often come close, and harmlessly, to boats. They sometimes venture into Whitewater Bay via the Buttonwood Canal. Other small whales are not regularly seen in Florida Bay, although some, including **Common Pilot Whales**, may sometimes enter the bay for short periods of time.

A few other species of mammals are sometimes seen in the Everglades, but they are not native ones. **Nine-banded Armadillos**, which were introduced to Florida many years ago, have been seen on the Tamiami Trail near the park, as well as in the park itself. A small, long-bodied tropical cat, the **Jaguarundi**, escaped from captivity and now persisting in Florida, has been seen in the park on rare occasions. **Feral Hogs**, which are descendants of domestic animals, have also been recorded at times in the park.

CHECKLIST

___ Virginia Opossum	___ Raccoon
___ Nine-banded Armadillo (I)	___ Long-tailed Weasel
___ Marsh Rabbit	___ Mink
___ Eastern Cottontail	___ Spotted Skunk (?)
___ Gray Squirrel	___ Striped Skunk
___ Fox Squirrel	___ River Otter
___ Round-tailed Muskrat	___ Mountain Lion
___ Bottle-nosed Dolphin	___ Jaguarundi (I)
___ Common Pilot Whale	___ Bobcat
___ Gray Fox	___ Manatee
___ Red Fox	___ Feral Hog (I)
___ Black Bear (E?)	___ White-tailed Deer

Glacier Bay National Park

P.O. Box 1089
Juneau, Alaska 99801

Located in the southeast corner of Alaska and chiefly bordered by Canada, Glacier Bay National Park is a huge (2,805,269 acres; 1,135,265 hectares) area of mountains, glaciers, fjords, and newly emerged lands. The park, which is topped by 15,320 foot (4,670 meter) Mount Fairweather, has attractions that are available only to those who arrive by boat

GLACIER BAY NATIONAL PARK

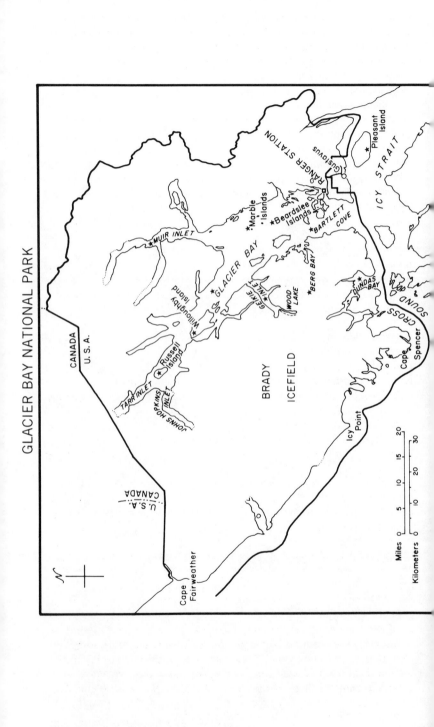

or airplane, because there are no roads to or through the park and the only trails are around the ranger station at Bartlett Cove, where the only formal campground is maintained. Two hundred fifty years ago, virtually all of what is now Glacier Bay was covered by glaciers reaching to the sea. These tidewater glaciers have now receded to more than 45 miles (72 kilometers) from the mouth of Glacier Bay.

Although there are not many species of mammals at Glacier Bay, there are good opportunities to see whales and porpoises, seals, mountain goats, moose, and both black and brown bears. Most of these can best be seen from boat trips, available at Bartlett Cove. The main visitor season is mid-May to mid-September, and the weather is often rainy and cool. Advance preparations and reservations are essential for a visit to this park.

Hump-backed Whales migrate into the park in June and usually remain there during the summer. They are the commonest large whales in the area. Unfortunately, the activity of tourist boats in the bay may be discouraging the whales' presence. The larger **Fin Whale** does not often venture far up into the bay; it is rarely encountered on the outer coast of the park. Among the other whales that are regularly seen in Glacier Bay are **Killer Whales**. These are generally seen in small pods, but are not especially common. The **Minke Whale**, a small baleen whale, is sometimes seen in the middle of Glacier Bay, but is also not common. It is more likely to be found, as is the **Gray Whale**, on the outer coast.

Harbor Porpoises are common throughout the park's waters, especially in the southern parts of Glacier Bay and in Icy Strait. Another species, **Dall's Porpoise**, is not common in the main channels of the park itself, and is more often seen in the Inland Passage and Icy Strait.

Of the other marine mammals of the region, **Harbor Seals** are frequently seen throughout the park. In midsummer, they can always be found in the upper bay, usually resting on icebergs at the foot of tidal glaciers, where they also have their pups. **Northern Sea Lions** are uncommon in the park, living mainly at the southern margin at Cross Sound. **Northern Fur Seals** are rarely found in the of the park, but there are a few old records of them on the outer coast. **Sea Otters**, which once occurred in the area and were extirpated, have been reintroduced along the outer coast. Sea Otters do not yet seem to be permanent residents of the area again, however, despite occasional sightings of them.

Red Squirrels are the only tree squirrels in the park and, because they inhabit coniferous forest, are abundant only in the southern parts. They are commonly seen and heard around the lodge at Bartlett Cove, and they have also been seen in the Muir Inlet area. **Hoary Marmots** inhabit rocky, alpine regions, and they are common on the outer coast sections of the park. They also are seen locally in the lower Glacier Bay and Muir Inlet areas. **Porcupines** are not found in the northern parts of the park, but are occasionally seen in the forested portions of lower Glacier Bay, especially near Bartlett Cove and the Beardslee Islands. **Snowshoe Hares** may inhabit the forests of lower Glacier Bay, but they are rarely seen.

Both **Brown** and **Black Bears** inhabit the park. The larger Brown Bear is widespread, but more frequently seen in the northern parts, especially near Russell Island and on the outer coast. The Black Bear also may be seen anywhere in the park, but it is most frequently observed near forested regions. The **Glacier Bear** (a color phase of the Black Bear) is a rarity, but is known from Dundas Bay and may also exist at Berg Bay.

Coyotes wander throughout the park and, although they may be encountered almost anywhere, are more often seen in the mature forest areas around the lower bay, as well as at Muir Inlet. **Gray Wolves** also live anywhere in the park, but they are seldom seen. **Red Foxes** are not common and seem to live only in the southern and outer coast parts of the park.

River Otters are common in both fresh- and salt-water habitats of the lower parts of Glacier Bay. When birds are nesting on the Marble Islands, River Otters may take up residence there and are sometimes seen foraging for nestlings. **Mink** inhabit fresh-water streams, as well as edges of salt-water bays, but are rarely seen, although they may be abundant.

Mountain Goats prefer the higher meadows during the summer, but often can be spotted on rocky cliffs along the bay at such places as Geike Inlet, on the rocks above the interglacial forest stumps. **Moose** are usually found either near spruce forests or in areas of developing vegetation where there are willows. **Mule Deer** of the Black-tailed subspecies were introduced on Willoughby Island in Glacier Bay. They have been seen swimming ashore from there and perhaps will take up residence on the mainland. They are native on Pleasant Island and other islands, just to the south of the park, and are common there.

Martens are common in the forested regions where there are mature trees. They are occasionally seen during the day in the woods around Bartlett Cove. The diminutive **Ermine** is also common in the park, both in the upper and lower parts, and is seen occasionally. Much rarer, and only in the lower bay region, is the **Least Weasel**. The largest member of the weasel family, the **Wolverine**, roams throughout the park, but is seldom seen. **Lynx** are also widespread in Glacier Bay, but are considered rare there and are hardly ever observed.

CHECKLIST

___ Snowshoe Hare (?)	___ Minke Whale
___ Hoary Marmot	___ Hump-backed Whale
___ Red Squirrel	___ Coyote
___ Porcupine	___ Gray Wolf
___ Killer Whale	___ Red Fox
___ Harbor Porpoise	___ Black Bear
___ Dall's Porpoise	___ Brown Bear
___ Gray Whale	___ Northern Fur Seal
___ Fin Whale	___ Northern Sea Lion

(continued)

___ Marten	___ Sea Otter (RI)
___ Ermine	___ Harbor Seal
___ Least Weasel	___ Lynx
___ Mink	___ Mule Deer (I)
___ Wolverine	___ Moose
___ River Otter	___ Mountain Goat

Glacier National Park

West Glacier, Montana 59936

Bordering Canada and contiguous with that country's Waterton Lakes National Park, Glacier National Park is a major portion of the unique Waterton-Glacier International Peace Park. Within Glacier's 1,013,598 acres (410,193 hectares) is a serene wonderland of rugged mountains, glaciers and glacier-scarred valleys, vast forests, glistening lakes, and grasslands. Traversed by the 50-mile (80 kilometer) Going-to-the-Sun Road, the park has few other paved roads, but there are more than 700 miles (1,127 kilometers) of hiking trails and horse trails. Although the size and number of glaciers have diminished greatly in the past century, there are still dozens of small glaciers accessible to hikers in the park. The Continental Divide bisects the park, and many peaks lie along it, the highest being Mount Cleveland, which is 10,466 feet (3,190 meters).

Glacier Park is noted for its wildlife, and there are good opportunities to see Mountain Sheep, Columbian Ground Squirrels, Moose, and Mountain Goats here. Although the park is open throughout the year, most of its roads are closed by snow from September to June, and summer is the prime visitor season.

Columbian Ground Squirrels abound in Glacier Park from low to high elevations. Although for 8 months of the year they hibernate, in summer they are seen scurrying about from dawn to dusk, mainly in open country and around many campsites and building sites. A much rarer species that occurs only in the lower grasslands on the eastern side of the mountains, such as near St. Mary, is the **Thirteen-lined Ground Squirrel. Golden-mantled Ground Squirrels** are less common than the Columbian, but are usually seen in rocky areas along Going-to-the-Sun Road, as well as at many campsites in open forest throughout the park. Restricted to the Waterton Valley in the extreme northern part of Glacier Park is **Richardson's Ground Squirrel**, which, although similar to the Columbian, does not have the rust-colored nose or feet of the latter.

GLACIER NATIONAL PARK

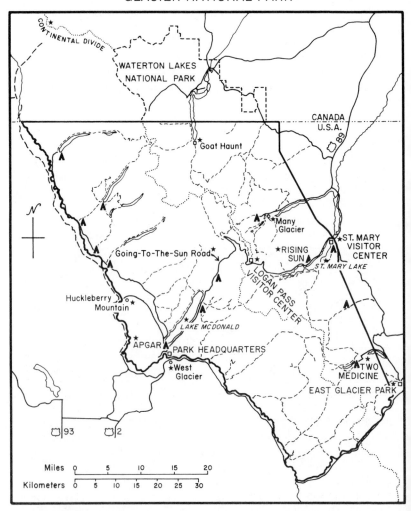

Three species of chipmunk, virtually indistinguishable in the field, inhabit Glacier. The **Yellow-pine Chipmunk** is more abundant in the brush areas of the lower elevations in the park, as well as in rocky places and open forest zones. In the higher spruce and fir forests, the **Red-tailed Chipmunk** is the common species around rock outcrops; and, ranging up into the alpine meadows and talus areas, is the third species, the **Least Chipmunk**. As with all the ground squirrels in the park, chipmunks hibernate for much of the winter.

The only tree squirrel in Glacier is the **Red Squirrel**, a noisy, highly visible inhabitant of the coniferous forests. Large caches (called middens) of conifer cones are additional evidence of the presence of Red Squirrels. The largest members of the squirrel family here are the **Hoary Marmots**. These bulky (to 15 pounds; 7 kilograms), grizzled rodents inhabit high-elevation talus slopes and other rocky areas near meadows, and their loud whistles are often heard. They frequently can be seen as they sun themselves on rocks.

Both **Mule Deer** and **White-tailed Deer** inhabit this park, and the former may be seen, in summer, almost anywhere. Males, especially, inhabit the high meadows, such as those near Logan Pass, but in fall large herds of Mule Deer descend to spend the winter in the lower, wooded valleys. White-tailed Deer live mainly in the western lower stream valleys, as well as in areas of second-growth vegetation, which occur in burned areas or abandonded Beaver meadows. White-tailed Deer are less abundant than Mule Deer and are believed to be decreasing in number because of natural changes in the park's vegetation. The largest of the park's hoofed animals are **Moose**. They may range to the higher elevations, but generally are found, in summer, near lakes and ponds, where they feed on aquatic vegetation. There are only a few hundred Moose in the park. One of the better places to find them is along McDonald Creek Valley, southwest of Avalanche Creek Campground, on Going-to-the-Sun Road—a place called ''Moose Country.''

Mountain Goats are a major attraction at Glacier, and there are several places where there is a good chance of seeing them. One of these is at Logan Pass on Mount Oberlin or Clements Mountain; another, especially in summer, is at the extreme southern end of the park, 3 miles (4.8 kilometers) east of the Walton Ranger Station on U.S. Highway 2, where there is a salt lick. Both sites are accessible to motorists. Hikers can often find Mountain Goats near Sperry Chalet and Gunsight Pass. There are between 1,000 and 2,000 of these white goats in the park. Less numerous are the **Mountain Sheep**. There are fewer than 1,000 of these animals. Although Mountain Sheep are often in the high meadows in summer they can usually be seen from the road just to the east of Many Glacier Hotel, and hikers on the Highline Trail often see them. In winter, Mountain Sheep live at lower elevations, unlike Mountain Goats. **Elk** are numerous in the park, and in winter, when they have come to the lower elevations, they can readily be observed from U.S. Highway 2 near West Glacier; the northeastern herds migrate up the Belly River Valley into Waterton Lakes National Park. In summer, Elk are less readily seen, usually traveling in smaller groups and living mainly in the denser spruce and fir forests, but sometimes are seen near some of the northwestern lakes. **Bison**, which once occurred in the northeastern part of the park, are extinct; a small reintroduced herd is maintained in Waterton Lakes Park. **Caribou** also once lived in the area, but none has been seen since the 1930s.

There are good numbers of both **Grizzly** and **Black Bears** in Glacier National Park. The latter, which are not always black in color, mainly inhabit the park's spruce and fir forests, and especially berry-bush areas on the western side near Camas Creek Road. They are occasionally dangerous pests around campgrounds. The larger Grizzly Bear is less numerous, but more wide-ranging, than the Black Bear. Grizzlies also show great variation in color, but their large size, humped shoulders, and dished face usually distinguish them. Like the Black Bears, Grizzlies are attracted to berries in the fall, the time in which both species put on fat for their winter dormancy (not hibernation). Park visitors should thus be extremely cautious. When in bear country, hikers should take precautions not only to avoid bears, but also to make noise, in order to establish their presence.

The numerous watercourses of Glacier Park provide habitat for several aquatic species, predominant among which is the **Beaver**. Dams and lodges built by Beavers can be found along many of the lower streams on both the eastern and western slopes of the park. Along the road to Many Glacier is a good place to see Beaver works, and many of the park's meadows are silted former Beaver ponds. Less abundant, surprisingly, are **Muskrats**, which also inhabit ponds and streams. Although Muskrats, like Beavers, are largely nocturnal and not often seen, their small, rounded nests can usually be found along the Flathead River and Bear Creek, on the southern border of the park, and also in the marshes along McDonald Creek. **River Otters** inhabit some of the park's lakes and rivers, especially in the southern and western parts, and are sometimes seen along the Flathead River. Inhabiting the same areas is the **Mink**, fairly common here, and sometimes seen along the North Fork of the Flathead River.

Although no rabbits are known from Glacier, two species of hare inhabit the park. The **Snowshoe Hare** is fairly abundant at times but, because it prefers dense forest and brush, is seldom seen. Its populations fluctuate, and in some years there are very few Snowshoe Hares, as evinced by the paucity of their tracks in winter snow. Extremely rare, and then seen only in the eastern grasslands of the park, is the large **White-tailed Jack Rabbit**. More readily seen is the diminutive relative of these hares—the **Pika**. It inhabits talus slopes and boulder fields near alpine meadows, where its high-pitched cries are usually the first evidence of its presence. Although Pikas blend in well with their rocky background, patient hikers can readily observe them.

Coyotes are numerous in Glacier Park and range throughout it up to the timberline. They are seen mainly in the eastern grasslands, but are often heard in both east and west, howling at night. **Gray Wolves** also occur in the park, although, as a result of trapping, none were seen for many years before 1940. They are rarely seen or heard, and the few observations of Gray Wolves have been at the periphery of the park, mainly in the north. **Red Foxes**, although recorded from both eastern and western sides of the mountains, are seldom seen, and more likely to occur in the low, eastern grasslands.

The weasel family is well represented in Glacier Park, even if its members are not frequently seen. Hikers in the forests on the western side occasionally see **Martens**, but the **Fisher**, also a coniferous forest dweller, seems to be very rare, as is true of the largest of the weasel family, the **Wolverine**. Cracker Lake is one place where there have been several observations of

Wolverines. Three species of weasel—the **Ermine**, the **Least Weasel**, and the **Long-tailed Weasel**—inhabit the park, but they are not often seen. **Badgers** live mainly in the lower grasslands and sometimes are seen around St. Mary. **Striped Skunks** live in the grasslands and lower forests on both sides of the park, but they are not usually seen. In some years, however, they have been so abundant around campgrounds and other settlements that the park service has had to remove them.

The largest of the American cats, the **Mountain Lion**, still exists in Glacier Park, where it inhabits coniferous forest. The population is small, and sighting one of these animals is a rare occurrence. Both the **Bobcat** and **Lynx** are more numerous, but also are seldom observed. The Lynx lives primarily in coniferous forest, judging from tracks in the snow, whereas the Bobcat prefers the grassier and brushier areas.

Porcupines occur throughout Glacier and in some years are very numerous. Hikers may encounter them on trails, especially in the forested areas, and they sometimes invade camps, attracted by salty objects. Also widespread in the park, especially in rocky areas, are **Bushy-tailed Woodrats**. They are seldom seen, but their nests of sticks are often found in rock crevices and abandoned buildings. **Raccoons** are newcomers to the park, unknown here before 1970, when some appeared in the eastern lowlands. Those that invaded the western parts are believed to be descendants from Raccoons introduced many miles to the south of the park, and they are becoming more numerous.

CHECKLIST

___ Pika

___ Snowshoe Hare

___ White-tailed Jack Rabbit

___ Least Chipmunk

___ Yellow-pine Chipmunk

___ Red-tailed Chipmunk

___ Hoary Marmot

___ Richardson's Ground Squirrel

___ Columbian Ground Squirel

___ Thirteen-lined Ground Squirrel

___ Golden-mantled Ground Squirrel

___ Red Squirrel

___ Beaver

___ Bushy-tailed Woodrat

___ Muskrat

___ Porcupine

___ Coyote

___ Gray Wolf

___ Red Fox

___ Black Bear

___ Grizzly Bear

___ Raccoon (!?)

___ Marten

___ Fisher

___ Ermine

___ Least Weasel

___ Long-tailed Weasel

___ Mink

___ Wolverine

___ Badger

___ Striped Skunk

(continued)

Grand Canyon National Park

P.O. Box 129
Grand Canyon, Arizona 86023

Truly one of the wonders of the world, the Grand Canyon of the Colorado River is a great eroded cleft in the earth, averaging 9 miles (14 kilometers) wide and one mile (1.6 kilometers) deep. At the bottom, the river winds 277 miles (446 kilometers) beneath the mesas, plateaus, and canyons that its force has carved, and it exposes a geological record that dates back nearly two billion years. Within the park's 1,218,375 acres (493,064 hectares) are habitats that vary from the hot desert conditions of the Inner Gorge to the cool, moist, spruce-fir forests of the North Rim. There is a wide range of altitudes—from 2,400 feet (732 meters) at the river, to 8,000 feet (2,438 meters) on the North Rim. Also diverse are the ways in which visitors experience the park. On both the South and North Rims, paved roads lead to viewpoints overlooking the canyon. There are relatively few hiking trails into the canyon, and the climb is rigorous. Horse and mule trips into the canyon are popular, and boat rides on the river became so frequent that it became necessary to limit them to preserve the fragile canyon environments. The South Rim is open throughout the year. The North Rim is closed from mid-October to mid-May.

The Colorado River and the canyon form a barrier for some species of mammal: Nuttall's Cottontails, Red Squirrels, and Golden-mantled Ground Squirrels are only on the north side; Desert Cottontails, Spotted Ground Squirrels, and Gunnison's Prairie Dogs are only on the south side. Raccoons, River Otters, and Beavers are limited to the Inner Gorge area; Spotted Skunks and Mountain Sheep are limited to the canyon proper, including the Inner Gorge, but not the rims.

In the summer, chipmunks are constantly encountered on both the North and South Rims of the canyon. On the south side of the river, only the **Cliff Chipmunk** is present, and it inhabits not only the pinyon-juniper zones

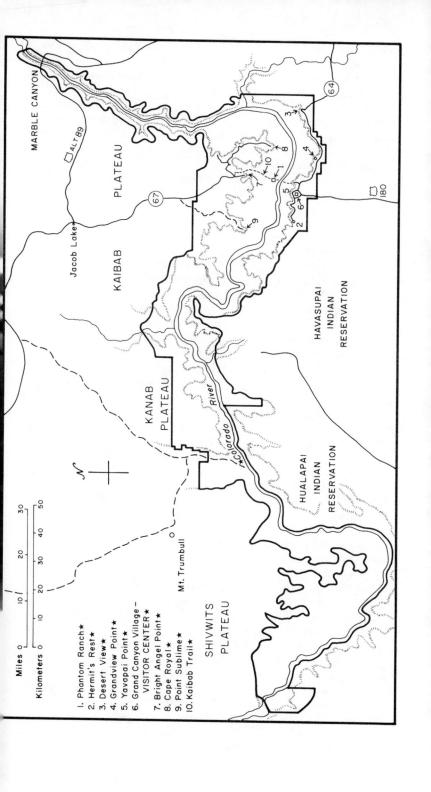

Miles 0 | 10 | 20 | 30
Kilometers 0 | 10 | 20 | 30 | 40 | 50

N

1. Phantom Ranch★
2. Hermit's Rest★
3. Desert View★
4. Grandview Point★
5. Yavapai Point
6. Grand Canyon Village – VISITOR CENTER★
7. Bright Angel Point★
8. Cape Royal★
9. Point Sublime★
10. Kaibab Trail★

MARBLE CANYON

ALT 89

Jacob Lake

KAIBAB PLATEAU

67

KANAB PLATEAU

SHIVWITS PLATEAU

Mt. Trumbull

Colorado River

HAVASUPAI INDIAN RESERVATION

HUALAPAI INDIAN RESERVATION

64

180

on the rim, but lives in the canyon down to the level of Indian Gardens and Havasu Canyon. It is also found in pinyon-juniper areas on the north side of the river. In addition to Cliff Chipmunks, there are **Least** and **Uinta Chipmunks** on the north side of the canyon. The Least Chipmunk is not as often seen as the Uinta Chipmunk, which is the common species around North Rim campgrounds. In addition, on the north side, there are **Golden-mantled Ground Squirrels**, sometimes confused with chipmunks. They are abundant and conspicuous on the North Rim, where they beg for food and roam campsites and picnic sites in search of food left by people. These chipmunks and the Gold-mantled Ground Squirrel hibernate in winter.

Another conspicuous and abundant ground squirrel is the **Rock Squirrel**, a large species that is usually encountered at view points on the South Rim. Rock Squirrels are present on the north side of the canyon, as well as in its depths, wherever there are the exposed rocky areas they prefer. More restricted in its habitat preference is the **White-tailed Antelope Squirrel**, which prefers open, arid areas, especially midst sagebrush, cactus, and yucca. White-tailed Antelope Squirrels are present on the Tonto Plateau and in the more westerly parts of the park. They also live on the north side of the river in lower elevations in the eastern sector, as near Nankoweap Canyon. In these warmer environments, they may not hibernate in winter. **Harris' Antelope Squirrel** occurs at the extreme western border of the park, at the edge of Lake Mead's south shore. The **Spotted Ground Squirrel**, another desert species, prefers sand soil and is found only at the south boundary of the park in a few places, such as around Bass Camp and Pasture Wash.

Of the large mammals in Grand Canyon Park, the most readily seen are **Mule Deer**. In summer, they are frequently seen around Grand Canyon Village and all along the South Rim, and also around the North Rim visitor accommodations. In winter, bad weather may force them to descend from the rims into warmer areas. Dawn and dusk are the best times to see deer. **Elk** are not resident in the park, but live in forests to the south of its borders. They do venture into the park, however, sometimes even to the South Rim, and recently there have been sightings of Elk several times each year. **Pronghorns** are not common in the park, but there have been occasional sightings of them near the South Entrance. Pronghorns also were introduced on the Tonto Plateau in 1924, but have not been seen there since the 1950s.

For those willing to descend to the bottom of Grand Canyon, there are good opportunities to see some small carnivores. Around the campsite at Phantom Ranch, **Spotted Skunks** are abundant at night. Because they prefer rocky areas near water, the canyon bottom is an ideal location for them, and they are found up to about 4,500 feet (1,372 meters) on both sides of the river. The same habitat is preferred by **Ringtails**, and Phantom Ranch is an excellent place to find them at night. Ringtails range up to the rims, and, on the South side, they often inhabit buildings, including the hotels. In some places, such as Havasu Canyon, **Raccoons** are abundant, but are rather limited in their park distribution to a few canyon bottoms on both sides of the river. **Striped Skunks** are common on the South Rim and are sometimes encountered at night near Grand Canyon Village and campgrounds, but seem not to descend into the canyon. Whether there are Striped Skunks on the North Rim is still not known.

The Colorado River and its tributaries are habitat for aquatic species, and **Beavers** are common. They are sometimes seen during the day by boaters, but most of their activity is at night. Although lodges and dams can be found in some of the Colorado tributaries, most of the Grand Canyon Beavers live in bank dens, and the most usual sign of their presence is tree cuttings. Probably because there is no suitable aquatic vegetation for their food, **Muskrats** are not known to occur in the park's rivers or creeks. **River Otters** seem never to have been numerous in the Colorado River, and some people have speculated that the few observations of them were of transient otters, rather than residents. Now, because of the cold water temperature of the park's river—which is a consequence of the Glen Canyon Dam and the subsequent reduction of the fish fauna—there are doubts that otters could persist in the park permanently.

On the South Rim, in the ponderosa pine forest, the common species of squirrel is **Abert's Squirrel**. Unlike chipmunks, some ground squirrels, and Gray Squirrels, Abert's Squirrels do not become accustomed to human beings and, because their diet is mainly pine seeds, do not especially frequent campsites. Because they have a fairly large home range, they are distributed thinly. Nevertheless, they are often seen around Canyon Village and places in the ponderosa forest to the east of it. On the North Rim, Abert's Squirrel is called the **Kaibab Squirrel** and has, at times, been regarded as a separate species. The Kaibab Squirrel has a black belly and a tail that is white above and below; the Abert's Squirrel of the South Rim has a white belly and underside of the tail, but the top of the tail is gray. On the North Rim, these squirrels are less frequently seen than on the south side. Also on the North Rim are **Red Squirrels**, which usually inhabit the spruce and fir forests and, less commonly, the ponderosa pines. Hikers along the rim in the spruce-fir forest hear and see them fairly often. Also common in the conifers on both North and South Rims are **Porcupines**. They are often seen along the roads, although they usually spend most of their time in trees.

The only cottontail on the South Rim is the **Desert Cottontail**, which is common in brushy areas and is usually seen at dawn or dusk. The North Rim species, **Nuttall's Cottontail**, is much less often seen, perhaps because the great depths of winter snow in that area makes survival difficult. **Black-tailed Jack Rabbits** occur on both sides of the river. They prefer open country, and are thus commoner on the South Rim than the North. They are often seen along side roads of the South Rim at dawn and dusk. There are reports that suggest the **White-tailed Jack Rabbits** may occur on the North Rim.

Coyotes range throughout the park, from the river edge to the forested rims. They are less common in the Inner Gorge than at higher elevations and may be encountered, usually early or late in the day, on either rim.

They are sometimes heard howling at night. In the past, **Gray Wolves** probably occurred on both North and South Rims, but predatory-animal control operations eliminated any resident populations. Occasional sightings of large canids led to the hope that wolves still venture into Grand Canyon Park from remnant outlying populations; skeptics believe that these observations are of Coyotes or free-roaming domestic dogs. **Gray Foxes** are as widespread in the park as Coyotes and are probably more abundant. They are seen fairly often on the rims, as well as along trails in the canyon. **Kit Foxes** have been reported recently near the western edge of the North Rim, but observations of **Red Foxes** in the park are questionable, because the nearest valid records are of some 200 miles (322 kilometers) east of the park.

Black Bears are probably not resident in the park, but sometimes they wander in from adjacent areas. There are rare observations of them on the South Rim, and, although they may have occurred on the North Rim in the past, there are no recent reports of them there. If **Grizzly Bears** ever occurred in the park, the North Rim seems the most likely area, but they have been extinct there for more than a century. The largest carnivores known to reside in the park now are **Mountain Lions**. These large cats are probably more abundant and more often seen in Grand Canyon Park than in any other national park; there are a number of sightings nearly every year, some of them from the most frequently visited areas, such as around the visitor center on the South Rim. On the North Rim, Mountain Lions are also quite numerous, if less often seen; a half-century ago, hundreds were killed on the Kaibab Plateau as part of a program to eliminate predators there. In the past, transient **Jaguars** may have reached Grand Canyon from the south, but there have been no reports of them since the early twentieth century. **Bobcats** in the Grand Canyon are numerous and are not uncommonly seen. They roam throughout the park, and their tracks have been found along most trails and at Indian Gardens and Phantom Ranch, as well as on the North Rim. The best location for seeing a Bobcat, however, seems to be near the Desert View Entrance, where these animals have been seen in the daytime.

Six species of woodrat live in Grand Canyon Park. **Stephen's Woodrat**, the **White-throated Woodrat**, the **Arizona Woodrat**, and the **Mexican Woodrat** live only on the south side of the river. The first lives among rocks in the pinyon-juniper areas; the second lives mainly among rocks in the ponderosa pine zone. The third species lives in the drier areas at the western edge of the park, whereas the last seems also to prefer a drier habitat, such as sagebrush and cactus. A fifth species, the **Desert Woodrat**, lives on both sides of the Colorado River. On the south side, it lives mainly on the Tonto Plateau, where the White-throated Woodrat also lives; on the north side, it occupies the broken rocks at cliff bases as well as pinyon-juniper areas. It is extremely abundant along the Colorado River itself. The last species, the **Bushy-tailed Woodrat**, lives only on the north side of the river, where it builds its nests in rock crevices. Wholly nocturnal, none of these woodrats are usually seen, although their conspicuous nests are often found.

Gunnison's Prairie Dogs have lived in open areas at the southern edge of the park, but the existence of the species here is marginal, and the persistence of prairie dogs in the park is questionable. **Badgers**, which prey on prairie dogs and ground squirrels, are not rare, but are more abundant in the western parts of the North Rim. They are, however, occasionally seen around Grand Canyon Village. Diggings in the pinyon-juniper areas along the South Rim are ample signs of their presence. The **Least Weasel** is another predator that is not commonly seen. Least Weasels occur on both rims, mainly in the wooded areas, and have sometimes been seen during the day on the north side. **Hog-nosed Skunks** are unrecorded in the park, but they may occur on the river edge, near the eastern border.

Within the canyon, and commoner on the south side than the north, are **Mountain Sheep**. They prefer broken, rocky country, and it is rare to see them on either of the rims. In summer, they are often in the Inner Gorge, and they drink from the river; usually they are in side canyons of the Tonto Plateau. The number of Mountain Sheep here is believed to be decreasing, possibly because of competition with feral Donkeys. These Donkeys, which escaped from, or were released by, prospectors, outnumber the Mountain Sheep and seem more prolific. Not only do they utilize forage and water that the native sheep seem to need but they also damage archaeological sites and hasten trail erosion. Currently, attempts to control the Donkey population are under way.

CHECKLIST

___ Nuttall's Cottontail
___ Desert Cottontail
___ Black-tailed Jack Rabbit
___ White-tailed Jack Rabbit (?)
___ Cliff Chipmunk
___ Least Chipmunk
___ Uinta Chipmunk
___ Harris' Antelope Squirrel
___ White-tailed Antelope Squirrel
___ Rock Squirrel
___ Spotted Ground Squirrel
___ Golden-mantled Ground Squirrel
___ Abert's Squirrel (Kaibab Squirrel)
___ Red Squirrel
___ Gunnison's Prairie Dog (E?)
___ Beaver

___ White-throated Woodrat
___ Bushy-tailed Woodrat
___ Desert Woodrat
___ Mexican Woodrat
___ Arizona Woodrat
___ Stephen's Woodrat
___ Porcupine
___ Coyote
___ Gray Wolf (E)
___ Red Fox (?)
___ Kit Fox
___ Gray Fox
___ Black Bear
___ Grizzly Bear (E)
___ Ringtail
___ Raccoon
___ Least Weasel
___ Badger
___ Spotted Skunk

(continued)

(continued)

___ Striped Skunk
___ Hog-nosed Skunk (?)
___ River Otter
___ Jaguar (E)
___ Mountain Lion
___ Bobcat

___ Donkey (I)
___ Elk
___ Mule Deer
___ Pronghorn
___ Mountain Sheep

Grand Teton National Park

P.O. Box 67
Moose, Wyoming 83012

The bare-rocked, snow-bedecked rugged Teton Range forms the western wall that is the backdrop to the cool lakes, rivers, and forests of Grand Teton National Park. Grand Teton Peak (13,770 feet; 4,197 meters) dominates the range, which includes more than a dozen mountains higher than 11,000 feet (3,353 meters), whose snowfields drain into Jackson, Jenny, Leigh, and other lakes, and ultimately into the Snake River, which runs through the eastern part of the park. Rich in "mountain man" trapping history, Grand Teton's 310,418 acres (125,623 hectares) offer camping, boating, fishing, hiking, horseback riding, swimming, and mountaineering. The park is bisected by the John D. Rockefeller, Jr., Memorial Parkway. Summer days are warm and nights are cool; there is an average of 16 feet (5 meters) of snow in the mountains during winter, which allows for an extensive winter program.

Grand Teton is one of the better parks for observing mammals, and visitors have good opportunities to see a number of species, including Moose, Pronghorn, and various ground squirrels.

Busily putting on fat in the summer, **Uinta Ground Squirrels** scurry about low sagebrush areas, where they are easily observed. Cunningham Cabin and Jackson Lake Lodge are two sure places to find them. Three species of chipmunk that are largely indistinguishable from one another—the **Least Chipmunk**, the **Yellow-pine Chipmunk**, and the **Uinta Chipmunk**—are found throughout the park. They are usually seen around campsites and viewpoints. Another species that lives in higher altitudes, usually near forests, is the **Golden-mantled Ground Squirrel**. Like the chipmunks, these squirrels are conspicuous foraging near campgrounds and parking areas, and Inspiration Point is a good place to find them. The largest of the Teton ground squirrels is the **Yellow-bellied Marmot**, which inhabits rocky hill-

GRAND TETON NATIONAL PARK

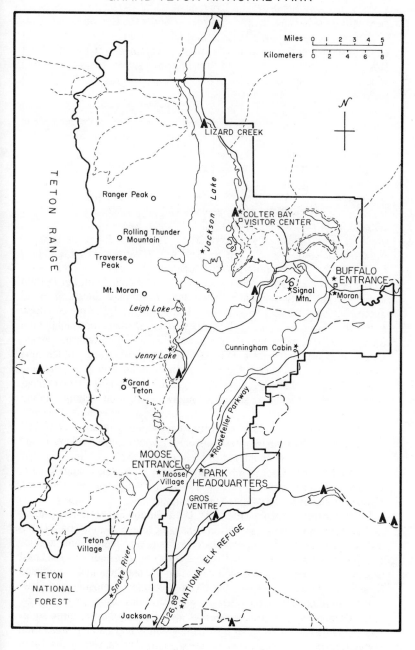

Miles 0 1 2 3 4 5
Kilometers 0 2 4 6 8

N

LIZARD CREEK

TETON RANGE

Ranger Peak ○

Jackson Lake

COLTER BAY
VISITOR CENTER

Rolling Thunder ○
Mountain

Traverse ○
Peak

Signal
Mtn.

BUFFALO
ENTRANCE

Mt. Moran ○

Leigh Lake

Moran

Jenny Lake

Cunningham Cabin ★

★ Grand
Teton ○

Rockefeller Parkway

MOOSE
ENTRANCE
★ Moose
Village

★ PARK
HEADQUARTERS

GROS
VENTRE

Teton ○
Village

Snake River

TETON
NATIONAL
FOREST

NATIONAL ELK REFUGE

Jackson 26,89

★ NATIONAL ELK REFUGE

sides from the lower elevations to above timberline and is often seen sunning on rocks. Some examples of melanistic (black) marmots have been seen in the park. The only tree squirrel present is the **Red Squirrel**, abundant in the conifers, and often heard scolding hikers. Its large caches of cones are frequently found in the woods. **Porcupines** range throughout Grand Teton and are common in the coniferous forest, as well as in the willows along the watercourses, and it is not uncommon for hikers to see them.

There are several hundred **Moose** present throughout the year, and, in summer, there are usually some around Jackson Lake Lodge and in the Snake River bottoms. In winter, more Moose come into the area from the higher forests and can be found along the Gros Ventre River, at the southeastern border of the park, as well as along streams north of the Buffalo Entrance. **Pronghorns** are not very numerous, but there are usually some in the sagebrush flats. In summer, **Elk** are seldom seen in the Tetons, because they inhabit the high meadows east and northeast of the park. Some remain near Jenny Lake and Burnt Ridge, where they may be seen at dawn or dusk. In winter, however, thousands of Elk move into the National Elk Refuge, adjacent to the park's southeastern border, and they usually can be seen in Antelope Flats and along the Gros Ventre River.

The hundred or so **Mountain Sheep** of Grand Teton spend most of the summer high in the mountains, often outside the park. In winter, some descend and are occasionally seen in Jackson Hole. The contrary is true of the **Mule Deer**, also not especially numerous here, which summer in the park and are occasionally seen in meadows or sagebrush flats, but which usually migrate southward out of the park in winter. A few remain near Jackson at East Gros Ventre Butte. There are only two or three dozen **Bison** in the park, mainly in the vicinity of Signal Mountain and the Snake River. It is questionable whether any **White-tailed Deer** are still resident in the park.

In the talus slopes and rock piles at higher elevations, **Pikas** are common and are usually heard and seen by hikers. In summer, their haystacks are evident. **Snowshoe Hares** inhabit the coniferous forests and, although they are abundant at times, are not regularly seen. **White-tailed Jack Rabbits** live in the lower grasslands and sagebrush, but are only occasionally encountered. **Bushy-tailed Woodrats** are not especially common and are rarely seen; their nests are usually in rocky cliffs and often in abandoned cabins.

The rivers and lakes of this park provide ample habitat for aquatic species. Predominant among these is the **Beaver**, whose dams, lodges, and cuttings are common along the Snake River and its tributaries. Also common, and sometimes seen at dusk on quiet nights, are Muskrats, which inhabit most ponds and small lakes where there is adequate aquatic vegetation. Both **Mink** and **River Otters** are streamside dwellers and, although not abundant, are occasionally seen swimming or foraging along the shorelines.

Coyotes are widespread in the Tetons and fairly abundant. They are often seen in meadows and on Antelope Flats, in the bottoms of the Snake River, and, in winter, in the Elk Refuge at Jackson Hole. **Wolves** are believed to be extinct in the area, but there are rare unconfirmed observations that are thought to be of wolves. **Red Foxes** are rare in this park. Although **Black Bears** are not rare and may be encountered almost anywhere in the park, sightings are not especially frequent. **Grizzly Bears** are very rare in the Tetons, and there is some doubt whether any remain now.

The most frequently seen small carnivore is the **Marten**. It mainly inhabits coniferous forest zones and is not uncommonly observed by hikers. Both **Short-tailed** and **Long-tailed Weasels** live in the park, and although the latter (judging from tracks in the snow) is common, neither is often seen. **Badgers** are numerous in the sagebrush flats and meadows, but are seldom seen; there is much evidence of their diggings around ground squirrel burrows, such as at the Cunningham Homestead. **Striped Skunks** which are not common in the Tetons, mainly live along the cottonwoods near streams. **Wolverines**, which roam the high forests near timberline, are rare.

None of the members of the cat family are regularly seen in the Tetons. The **Lynx**, which lives mainly in the coniferous forest, is rare. **Bobcats** are more abundant and roam a variety of habitats, but are not often seen. **Mountain Lions** may also occur almost anywhere in the park, but are rarely seen.

CHECKLIST

___ Pika
___ Snowshoe Hare
___ White-tailed Jack Rabbit
___ Least Chipmunk
___ Yellow-pine Chipmunk
___ Uinta Chipmunk
___ Yellow-bellied Marmot
___ Uinta Ground Squirrel
___ Golden-mantled Ground
    Squirrel
___ Red Squirrel
___ Beaver
___ Bushy-tailed Woodrat
___ Muskrat
___ Porcupine
___ Coyote
___ Gray Wolf (E?)
___ Red Fox
___ Black Bear

___ Grizzly Bear (E?)
___ Marten
___ Ermine
___ Long-tailed Weasel
___ Mink
___ Wolverine
___ Badger
___ Striped Skunk
___ River Otter
___ Mountain Lion
___ Lynx
___ Bobcat
___ Elk
___ Mule Deer
___ White-tailed Deer (E?)
___ Moose
___ Pronghorn
___ Bison
___ Mountain Sheep

GREAT SMOKY MOUNTAINS NATIONAL PARK

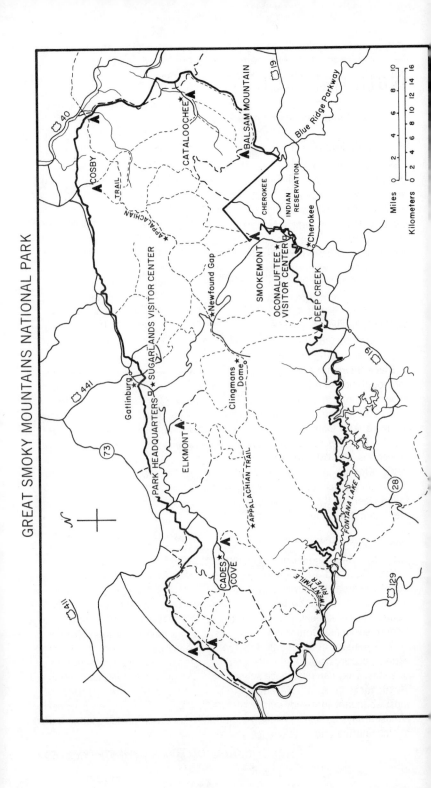

Great Smoky Mountains National Park

Gatlinburg, Tennessee 37738

Straddling the Tennessee—North Carolina border, Great Smoky Mountains National Park's 517,368 acres (209,374 hectares) include one of the highest mountains in eastern United States, Clingman's Dome (6,643 feet; 2,025 meters), and fifteen other peaks taller than 6,000 feet (1,829 meters). There is more than 80 inches (2,032 millimeters) of rainfall in some parts of the park. A virgin hardwood forest dominates the flora, and spruce and firs thrive at higher elevations in the eastern sector. Some of the mountain summits are treeless "balds," grass or heath areas that may have originated when Indians felled the trees here.

The wildlife of the park is typical of the Appalachian Mountains, of which the Great Smoky Mountains are the southern end. Of the eastern parks, this is the best in which to see Black Bears, Woodchucks, and White-tailed Deer. Cades Cove is the best place to see mammals. The park is open throughout the year, is especially crowded on weekends, and is of spectacular beauty in the spring and fall.

Black Bears are a major attaction in Great Smoky Mountains Park, and the population numbers in the hundreds. They may be encountered almost anywhere, and in some areas their density is higher than two per square mile (259 hectares).In some places in the park, bears are a nuisance and a hazard, and extensive studies of the Smoky Mountains bears are under way, so that it is not uncommon to see a bear with colored tags in its ears or wearing a collar with a radio transmitter. Black Bears usually sleep in dens, often up in hollow trees, from December to March.

The only other commonly seen large mammal in the park is the **White-tailed Deer**. In the 1930s they were almost extinct in the area, but now are found throughout the park. Deer are least common in the eastern spruce-fir areas, and the most likely place to see them is at Cades Cove, where some can usually be observed throughout the day. **Elk** were extirpated by 1849, and **Bison** fifty years earlier.

Gray Squirrels are common in the oak and beech woods and are frequently seen at the lower elevations of the park. **Red Squirrels** are most numerous in the spruce-fir forests and are often heard and seen. The **Fox Squirrel** has also been recorded in the park, but is considered rare and is seldom seen. Fox Squirrels are not known at elevations above 4,000 feet (1,219 meters) in the mountains. **Eastern Chipmunks** abound at all elevations, but are commoner in the deciduous forest areas than in the spruce-fir zones. They can usually be seen around campgrounds and picnic areas. **Woodchucks** occur all the way up to 6,300 feet (1,920 meters), but are most abundant in meadowlands and roadsides at lower places. They can almost always be seen, in summer, at Cades Cove and Newfound Gap, and at Oconaluftee and Cataloochee.

Virginia Opossums are often found near old fields and open woods near streams and are common throughout the park, more so at lower elevations. They can usually be seen around Cades Cove, Oconaluftee, and Cataloochee. Around these and other campgrounds, **Striped Skunks** are active at night and sometimes are pests. They range in elevations up to 5,200 feet (1,585 meters) but are more often found in the lower open fields and woodlands. The **Spotted Skunk** is less common and seems to live mainly below 2,800 feet (853 meters). **Raccoons**, too, are commoner at lower elevations, although they are found throughout the park. They can usually be seen along roads at night, especially around Cades Cove. **Long-tailed Weasels**, although widespread in the park, are rarely observed.

Great Smoky Mountains is one of the few parks in which **Bobcats** are seen occasionally along roadsides at night; the Roaring Fork Motor Nature Trail is one of the better places. The **Gray Fox** is another nocturnal species that is sometimes seen along the roads, especially at the lower elevations. Although they have been seen at all elevations in the park, **Red Foxes** are rare, and it is possible that they are descendants of animals that were transplanted from other areas many years ago.

Rabbits live throughout the park, but are more abundant at lower elevations, near fields, thickets, and open woodlands. The **Eastern Cottontail** is probably the species most often seen, but the **New England Cottontail** is known to live in dense woods, from Low Gap (33 feet; 10 meters), to as high as 2,100 feet (640 meters), near Elkmont. The two species are virtually indistinguishable from each another.

In low-elevation streams at the periphery of the park, **Muskrats** may be fairly common, but are seldom seen. The same is true of the **Mink**, but another aquatic species, the **River Otter**, has been gone from the park since 1927. **Beavers** were also extirpated here, but now have been found again at some places of lower elevations, such as Hazel Creek.

Mountain Lions were wiped out of the Great Smoky area by 1920, but there now seems to be evidence that these large, secretive cats may once more roam the mountains; sightings are quite rare.

The **European Wild Boar**, a nonnative species, is abundant in the park, numbering in the hundreds. These wild pigs are mainly nocturnal and avoid humans. In large numbers, they have caused great destruction to the park environment, and there is much evidence of their rooting and wallowing. They seem to be concentrated more in the western part of the park, especially between Cades Cove and Fontana Lake.

CHECKLIST

___ Virginia Opossum	___ Woodchuck
___ Eastern Cottontail	___ Gray Squirrel
___ New England Cottontail	___ Fox Squirrel
___ Eastern Chipmunk	___ Red Squirrel

(continued)

___ Beaver

___ Muskrat

___ Gray Wolf (E)

___ Red Fox (I?)

___ Gray Fox

___ Black Bear

___ Raccoon

___ Long-tailed Weasel

___ Mink

___ Spotted Skunk

___ Striped Skunk

___ River Otter

___ Mountain Lion (?)

___ Bobcat

___ Wild Boar (I)

___ Elk (E)

___ White-tailed Deer

___ Bison (E)

Guadalupe Mountains National Park

3225 National Parks Highway
Carlsbad, New Mexico 88220

Guadalupe Mountains is a small park (76,293 acres; 30,875 hectares) in Texas which is just a few miles southwest of Carlsbad Caverns National Park, New Mexico. The mountains—including Guadalupe Peak, which, at 8,749 feet (2,667 meters), is the highest point in Texas—are populated with mammals not unlike those in the Rocky Mountains, whereas the lower elevations are inhabited by a fauna with relations to the Chihuahuan Desert of northern Mexico.

There are few roads in Guadalupe Mountains, but there are many fine hiking trails, and the scenery is dramatic. In this park there are good opportunities to see Gray Fox and Hog-nosed Skunk around the campground in Pine Springs Canyon and, for those willing to climb above 5,900 feet (1,800 meters), a chance to see the Gray-footed Chipmunk, a species found in no other national park. The best areas to see mammals throughout the year are around McKittrick Canyon, Upper Dog Canyon, and at the campgrounds.

Below 6,300 feet (1,920 meters), rabbits and hares are the most commonly seen mammals. The largest species, the **Black-tailed Jack Rabbit**, is often encountered along roads at night or at dawn or dusk. Similarly, in the same areas and hours, the small, pale **Desert Cottontail** is abundant and frequently visible. Much sparser in population is the darker **Eastern Cottontail**, one of the rarest species in Guadalupe Mountains. A small population of this rabbit, perhaps fifty animals, is known to inhabit The Bowl (8,000 feet; 2,377 meters). Eastern Cottontails are seldom seen.

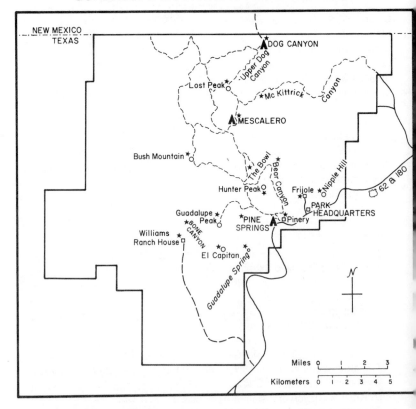

Mule Deer are abundant, widespread, and frequently seen throughout the park, especially in brushy or forested parts. There are usually some near McKittrick Canyon. **White-tailed Deer**, which once may have inhabited the higher forests, have been extirpated. The native Guadalupe elk, a distinctive large and pale subspecies called **Merriam's Elk**, was wiped out before 1900. Another subspecies, the **Rocky Mountain Elk**, was introduced in 1928 and is thriving. Elk are common and widespread throughout higher parts of the park. Upper Dog Canyon, Frijole, Bush Mountain, The Bowl, and McKittrick Canyon are among the better localities in which to see them.

The **Rock Squirrel**, a large species, is abundant in rocky areas from 5,000 feet (1,524 meters) to 7,000 feet (2,134 meters) and is often seen sunning on rocks. This species has been seen, among other places, at Frijole, Williams Ranch House, at the head of McKittrick Canyon, and in Upper Dog Canyon. Another diurnal species, an inhabitant of rock areas and low desert up to 6,300 feet (1,920 meters), is the **Texas Antelope Squirrel**. It is often seen dashing across roads during the day, usually mistaken for a chipmunk. One distinguishing characteristic of this squirrel is that, as it runs, it carries its tail flat over its back, rather than straight up, as do chipmunks. The only chipmunk in the park is the **Gray-footed Chipmunk**, a high-country creature, not found lower than 5,900 feet (1,800 meters). It is abundant around The Bowl, McKittrick Canyon, Upper Dog Canyon, Bear Canyon, Guadalupe Canyon, Hunter Peak, Bone Canyon, and can also be seen on Bush Mountain and near Mescalero Campground. The **Spotted Ground Squirrel** is an uncommon species in the park, occurring at lower elevations on the western side, where it is not abundant. Spotted Ground Squirrels have also been seen near the junction of the road to Williams Ranch with U.S. Highways 62 and 180, and at Guadalupe Spring.

The **Raccoon** prefers areas of woodland near water and is fairly common in the eastern and northern parts of the park. There are always some Raccoons around the Pine Springs Campground and Frijole, and at night they can be seen foraging for scraps left by campers. The Raccoon's relative, the **Ringtail**, probably is common and may occur in rocky places in many parts of the park; however, this species is wholly nocturnal, shy, and thus seldom seen. Ringtails have been recorded in Frijole, McKittrick Canyon, Upper Dog Canyon, and The Bowl.

Three species of skunk inhabit Guadalupe. The large, white-backed "rooter," the **Hog-nosed Skunk**, distinguished by its solid white back, bare nose, and all-white tail, occurs throughout the park and is most easily seen at night around Frijole and the campgrounds. Also common and widespread is the **Striped Skunk**, which brazenly forages for food at campgrounds. The diminutive **Spotted Skunk** probably lives among the rocks and cliffs of the park, but the only records of it are from Pine Springs and Williams Ranch House, and it must be considered rare in Guadalupe.

Two other members of the weasel family live in Guadalupe. The **Badger** is a low-elevation resident and is uncommon. It does live near the mouth of McKittrick Canyon and sign of its presence—diggings—have been seen at Guadalupe Spring near the base of El Capitan. The **Long-tailed Weasel** probably occurs in the park, but has been seen only near McKittrick Canyon and other localities just outside the park's borders. It may, because of its distinctive facial mask, be mistaken for the extremely rare **Black-footed Ferret**, for which there are no valid records at Guadalupe.

The **Gray Fox** is a widespread and common species in the park and may be encountered at night almost anywhere, especially around the Pine Springs Campground. The small desert species, the **Kit Fox**, also inhabits the park, but is much less common. It has been seen mainly at night near the old road south of Pinery, close to the residential area. The **Coyote** is relatively abundant and widespread in the park and is sometimes seen around McKittrick Canyon. At night, especially in winter, Coyotes can be heard calling. Another member of the dog family, the **Gray Wolf**, once inhabited the park area, but was wiped out by cattlemen.

Three species of woodrat live in the park, all of them common but seldom seen. The **White-throated Woodrat** occurs throughout the area around the base of the mountains; most of the debris nests seen by visitors are of this species. The **Mexican Woodrat** is restricted to the higher elevations above

4,921 feet (1,500 meters) and builds its houses mainly in rock crevices and sometimes in abandoned buildings. The **Southern Plains Woodrat** occurs only at the lower desert elevations in the southwestern part of the park. Although the woodrats themselves are seldom seen, their conspicuous stick and cactus houses are evidence of their presence.

Porcupines range throughout the park, from the desert to the mountain tops and, although common, are not regularly seen. McKittrick Canyon and Frijole are two areas where they might best be encountered.

Although they occur throughout Guadalupe, **Bobcats** are very seldom seen, even though they may be common. Most observations of them have been at McKittrick Canyon, Dog Canyon, or near the Pine Spring Campground. Some **Mountain Lions** also inhabit the park, but they are neither common nor often seen. Most of the observations of Mountain Lions have been near McKittrick Canyon, The Bowl, and Dog Canyon.

Black Bears are now rare in the park and are seldom seen. They once were abundant in the higher parts of the mountains. There may not be a resident population of Black Bears in Guadalupe now; observations and sign of them in The Bowl and Upper Dog Canyon may represent transients. The **Grizzly Bear** was once part of the fauna of the park, but was extirpated.

Another species that once inhabited the park is the **Black-tailed Prairie Dog**. This species was poisoned by cattlemen. Dog Canyon got its name from the prairie dogs that once lived there. One attempt to reintroduce prairie dogs near Nipple Hill was unsuccessful. **Pronghorns** inhabited the desert grasslands near the park, but the last of them disappeared around 1960. There are indications that some Pronghorns may be venturing back into the park from adjacent lands. Observations of Pronghorns should be reported to park rangers. **Bison**, which inhabited what are now the park's grasslands, had been killed off by the middle of the nineteenth century.

Mountain Sheep, which once lived in the Guadalupes, are extinct. There are, however, five other sheeplike animals occurring either in the vicinity or in the park itself, and these are frequently mistaken for Mountain Sheep. These are the North African wild animal called the **Barbary Sheep**, the Mediterranean wild sheep called the **Mouflon**, the domestic breed of sheep called the **Ramboulet**, and two kinds of domestic goats, the **Spanish Goat** and the **Angora Goat**.

The **Collared Peccary** in Guadalupe is from stock that was introduced there. Some have been seen around the mouth of McKittrick Canyon, Smith Springs, and Frijole.

CHECKLIST

___ Desert Cottontail	___ Gray-footed Chipmunk
___ Eastern Cottontail	___ Texas Antelope Squirrel
___ Black-tailed Jack Rabbit	___ Spotted Ground Squirrel
	(continued)

(continued)

___ Rock Squirrel
___ Black-tailed Prairie Dog
___ White-throated Woodrat
___ Mexican Woodrat
___ Southern Plains Woodrat
___ Porcupine
___ Coyote
___ Gray Wolf (E)
___ Gray Fox
___ Kit Fox
___ Black Bear
___ Grizzly Bear (E)
___ Ringtail
___ Raccoon

___ Long-tailed Weasel
___ Badger
___ Spotted Skunk
___ Striped Skunk
___ Hog-nosed Skunk
___ Mountain Lion
___ Bobcat
___ Collared Peccary (I)
___ Elk (RI)
___ Mule Deer
___ White-tailed Deer (E)
___ Pronghorn (E)
___ Bison (E)
___ Mountain Sheep (E)

Haleakala National Park

P.O. Box 369
Makawao, Hawaii 96768

Reaching from the top of Mount Haleakala, a 10,023-foot (3,055-meter) dormant volcano, to the coast of the island of Maui, Haleakala National Park encompasses 28,072 acres (11,360 hectares). It features the 7.5 mile (12 kilometers) by 2.25 mile (3.6 kilometers) eroded crater of Haleakala, which is almost 2,500 feet (762 meters) deep, and has fine examples of volcanic geology and erosion. Morning is the best time to visit the crater rim because clouds often obscure it in the afternoon. It is accessible by road and has a visitor center.

There are no native mammals in Haleakala National Park, but there are a number of kinds of introduced species. Goats are present and browse in the crater, and feral pigs live in the moist, lush parts of the park. In addition, there are nonnative rats and mice, and mongooses also abound. Each of these nonnative species has caused grave damage to native plants and animals. The park is rich in native birds, but also has a number of introduced kinds, including pheasants. Haleakala is open throughout the year.

Hawaii Volcanoes National Park

Hawaii National Park, Hawaii 96718

Mauna Loa, the world's largest active volcano, is the dominating feature of Hawaii Volcanoes National Park. The park rises 13,667 feet (4,166 meters) above sea level and more than 30,000 feet (9,144 meters) above the floor of the ocean. Its 229,177 acres (92,746 hectares) lie on the southeastern section of the largest of the Hawaiian Islands, Hawaii itself. Although Mauna Loa is larger, another volcano in the park, Kilauea, is more active. Because Mauna Loa is not accessible by car, Kilauea is the one more often viewed by tourists. Although Kilauea is only 4,000 feet (1,219 meters) above sea level, its dramatic activity and lava displays are a major attraction. The park service maintains a tape-recorded automatic telephone service to inform inquirers of current activity of either Mauna Loa or Kilauea.

There are no native mammals in the park, but introduced goats, pigs, and mongooses can be seen. The park has a rich flora and bird fauna, and it is open throughout the year.

Hot Springs National Park

P.O. Box 1860
Hot Springs, Arkansas 71901

This small park (5,801 acres; 2,348 hectares) in central Arkansas is little more than a city park devoted mainly to hot mineral waters (143° F.; 61.7° C.), reputedly medicinal, which flow forth at a rate of almost a million gallons (3,785,412 liters) a day from forty-seven natural springs. The five oak-hickory clad mountains, which are part of the Ouachita Mountains and which make up most of the park, have numerous roads an trails. The park is composed of three units, Sugarloaf Mountain, West Mountain, and the Hot Springs—North-Indian mountain sections, with the city of Hot Springs in between these units.

As might be expected in such a small area, the mammals are mainly those that are small and tolerant of human presence. During the day, **Gray Squirrels** are commonly seen along trails and are usually fed by residents. The **Fox Squirrel** occurs in the park area, but it is uncommon. **Eastern Chipmunks** are the other frequently seen denizens of the park, occurring throughout the wooded areas. Early in the morning and late in the day, as well as at night, the **Eastern Cottontail** abounds; it may be the most com-

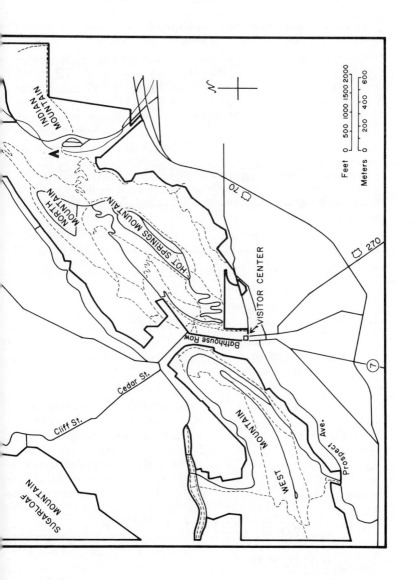

INDIAN MOUNTAIN

NORTH MOUNTAIN

HOT SPRINGS MOUNTAIN

70

VISITOR CENTER

270

Bathhouse Row

Cedar St.

Cliff St.

SUGARLOAF MOUNTAIN

WEST MOUNTAIN

Prospect Ave.

7

Feet 0 500 1000 1500 2000

Meters 0 200 400 600

mon mammal in the park. **Swamp Rabbits** are thought to have once inhabited this region, but it is unlikely that they still live here.

In the evening and at night, **Virginia Opossums** roam the wooded parts of the park, often along trails and roads, and **Raccoons** are frequently seen, too, as they forage around picnic sites and trash cans. Another mammal that takes advantage of garbage is the **Striped Skunk**, a common nighttime raider of trash bins. If **Spotted Skunks** remain in the park, they are extremely rare.

Plentiful in the park, and frequently seen at night, are **Nine-banded Armadillos**. Hot Springs National Park is the only park where this species is native (they also live in Everglades, but are from introduced stock). **Gray Foxes** are sometimes seen along the roads at night, and they are the commoner of the two species of fox that live here. **Red Foxes**, perhaps only a single resident pair, are rarely seen in the park.

A few **White-tailed Deer** live in the woods, but are not often seen; the same is true of **Woodchucks**. Several other species may still persist in the park, but there are no recent records of them. Among these are **Eastern Woodrats, Mink, Beavers, Bobcats**, and **Coyotes**. **Red Wolves** and **Black Bears** once inhabited this part of Arkansas, but were extirpated from the park area long ago.

CHECKLIST

___ Virginia Opossum	___ Red Wolf (E)
___ Nine-banded Armadillo	___ Red Fox
___ Eastern Cottontail	___ Gray Fox
___ Swamp Rabbit (?)	___ Black Bear (E)
___ Eastern Chipmunk	___ Raccoon
___ Gray Squirrel	___ Mink (?)
___ Fox Squirrel (?)	___ Spotted Skunk
___ Woodchuck	___ Striped Skunk
___ Beaver (?)	___ Bobcat (?)
___ Eastern Woodrat (?)	___ White-tailed Deer
___ Coyote (?)	

Isle Royale National Park

87 North Ripley Street
Houghton, Michigan 49931

Closer to both Canada and Minnesota than to mainland Michigan, of which state it is a part, Isle Royale National Park is an island some 45 miles (72 kilometers) long and less than 9 miles (14 kilometers) wide in

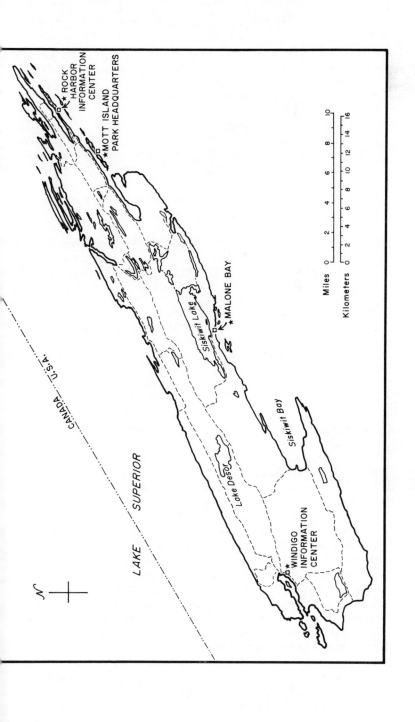

Lake Superior. Accessible only by boat or seaplane, the park's 571,796 acres (231,473 hectares) have more than 170 miles (273 kilometers) of foot trails through coniferous and deciduous woods, through hills and valleys, and past ponds and lakes. The highest point on Isle Royale is Mount Desor, 1,394 feet (425 meters) high. There are many opportunities for boating in the park's waters, which extend 4.5 miles (7 kilometers) from the island's borders.

The mammals of Isle Royale are few, mainly species that are good swimmers, capable of crossing on ice in years when Lake Superior froze, or that may have rafted accidentally to the island. Although Isle Royale is famed for its Gray Wolves, visitors have few chances of seeing them, because they avoid human beings. The other main attraction, Moose, is readily observed, especially around Rock Harbor. There are also good opportunities to see Red Foxes and Red Squirrels on Isle Royale. It is open from mid-May to mid-October, and reservations for transportation from the mainland and accommodations should be made well in advance of the intended dates of visitation.

Although **Moose** may have been on the island in the mid-nineteenth century, they were gone early in the twentieth, either because of hunting or ecological changes on the island. Just when the ancestors of the present Moose population arrived on the island is unknown; they evidently were not present in 1904, and it is thought that they may have crossed from Canada in the winter of 1912, when Lake Superior froze. Others believe that the Moose swam to Isle Royale from the mainland about this time. In the 1920s there was a growing population of Moose, and by 1930 Moose numbered more than one thousand. There was a die-off in 1933, and the population declined further when a forest fire reduced the food supply. In 1936 there were four hundred to five hundred Moose, and the population began increasing again until Gray Wolves arrived. These predators kept the Moose population at about six hundred for many years. In 1980, however, it was evident that the Gray Wolves were increasing and the Moose decreasing, partly from the former's predation and partly as the ecological changes on the island reduced the latter's food supply. **Caribou** existed on the island early in the twentieth century, but they disappeared by about 1920 because of overhunting, the changing environment, or both. An attempt to introduce **White-tailed Deer** on the island was unsuccessful.

The famed Gray Wolves of Isle Royale were not part of the original fauna of the park. They are believed to have invaded the island by crossing the ice from the mainland in the winter of 1949. Since then, these Gray Wolves have been steadily studied by scientists, and much valuable information has been produced. The number of wolves, stable at twenty to twenty-five for many years, is now known to fluctuate, and has reached a high of nearly sixty, divided into five packs, which are distributed throughout the park. The summer visitor has little chance of seeing a wolf (about one in a thousand visitors sees a wolf), because wolves avoid people. Visitors camped away from noise often hear wolves howling, which they may do at any time of day or night throughout the year. As soon as visitors leave the park in the fall, wolves begin to use the hiking trails. Much of the research on wolves is conducted in winter, when visibility is better and the animals can be tracked in the snow.

Before the Gray Wolves arrived, **Coyotes** inhabited the park. They, too, were probably newcomers, because they were not known to be on the island in 1904. By 1957, not long after the arrival of wolves, Coyotes disappeared, perhaps as a direct result of competition and aggression from this closely related species.

Red Foxes are common on Isle Royale, and several color phases (including black) occur. Red Foxes are often seen by hikers, and there is usually one or more of these animals around the lodge and docks at Rock Harbor; visitors should not feed them.

The **Red Squirrel** is a common, abundant, diurnal species that is often seen foraging on the ground or heard chattering in the trees. **Snowshoe Hares** are also relatively common, but in summer they keep to brushy areas, where they are hard to see. Visitors do see these hares fairly often, however, especially at dawn or dusk.

Although once almost extirpated on Isle Royale, **Beavers** are now abundant, numbering in the hundreds. Their dams, houses, and cuttings are evident throughout the park, but Beavers are not usually seen except by quiet watchers at isolated ponds, where the animals may venture out while it is still light. Although it might seem that there is much **Muskrat** habitat on the island, Muskrats are not common and are seldom seen. It may be that the ponds and lakes are not productive of the kinds of vegetation that Muskrats need for food.

Several species of the weasel family are present in the park, but none is common. **Ermine** and **River Otters** are rare and are not usually seen by visitors. **Mink**, too, are present but seldom seen. Whether **Martens** are present is now questioned, although they were definitely present early in the twentieth century.

The only other carnivore that may be found on Isle Royale is the **Lynx**. It was present and trapped commercially around 1900, but afterward its presence was doubted until one was observed in 1970. Lynx may no longer be part of the park's resident fauna, and recent records may represent transients from the mainland.

CHECKLIST

___ Snowshoe Hare	___ Ermine
___ Red Squirrel	___ Mink
___ Beaver	___ River Otter
___ Muskrat	___ Lynx
___ Coyote (E)	___ White-tailed Deer (I, E)
___ Gray Wolf	___ Moose
___ Red Fox	___ Caribou (E)
___ Marten (?)	

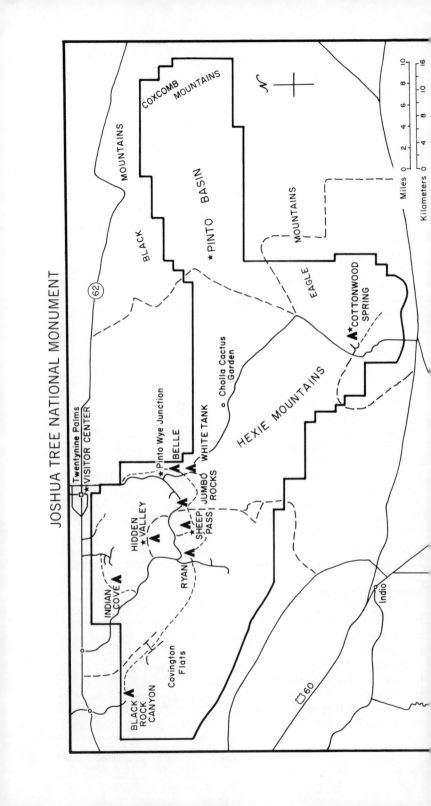

JOSHUA TREE NATIONAL MONUMENT

Joshua Tree National Monument

74485 National Monument Drive
Twenty-nine Palms, California 92277

Featuring biota of both the Colorado and Mojave deserts, Joshua Tree National Monument draws its name from the large (to 40 feet; 12 meters) member of the agave family, the branched yucca called "Joshua Tree" or "Praying Plant." It is found mainly in the higher elevation (above 3,000 feet; 914 meters) of the western part of the monument and blooms in March and April, but not each year. In these months, as well as May and June, there is also a magnificent display of desert wildflowers. Within Joshua Tree's 559,960 acres (226,610 hectares) of desert (the annual rainfall is less than 5 inches; 120 millimeters) are altitudinal ranges from 1,000 to nearby 6,000 feet (305 to 1,829 meters). Visitors must be prepared for flash floods during the rainy season (July to September), and for temperatures above 100° F. (38° C.) in summer and down to 0° F. (−18° C.) in winter.

There are roads through the central parts of Joshua Tree, and numerous campgrounds and picnic sites. Trails lead to lookout points and oases, as well as to historic sites. The mammals of the monument are typically desert forms, most of which are nocturnal, and include Mountain Sheep and Kit Foxes.

Although the hot days tend to encourage most of the mammals to be nocturnal, some of the members of the squirrel family are tolerant of heat and can be seen scurrying about even at midday. One such diurnal species is the **White-tailed Antelope Squirrel**, which generally can be seen throughout the lower elevations of the monument, especially around campsites and picnic areas. Another ground squirrel, the **Round-tailed Ground Squirrel**, is also active by day, but seems to do much less foraging, and it is more often seen sitting upright in sandy areas. It lacks the white side stripe and white underside of the tail of the Antelope Squirrel and is less often seen. Among the rock outcrops, the **California Ground Squirrel**, a larger, grayer species, can be found, and, in the higher pinyon-juniper areas, **Merriam's Chipmunks** are common. They can be distinguished from Antelope Squirrels by the stripe through the eye, as well as not having a white underside to the tail.

Although usually active at night, **Desert Cottontails** are often seen early in the morning or evening. The same is true of the **Black-tailed Jack Rabbit**, a lanky species that hikers sometimes startle from its daytime hiding places in the shade.

Coyotes are abundant and are often seen, even in the daytime, anywhere in this monument. Around some campgrounds, they are pests, and campers should neither feed them nor encourage their presence and should always remember that they are wild animals. **Gray Foxes** are present, although not as frequently seen as Coyotes, and are more nocturnal. Also present, but mainly in the lower, sandy areas, is the small, big-eared **Kit Fox**. It is sometimes seen at night in the southern parts of the monument.

There are several herds of **Mountain Sheep** in Joshua Tree, but their coloration blends so well with the rocks of the hills they inhabit that they are not often seen. The best places to look for them are between Twentynine Palms and Pinto Wye Junction, and in Hidden Valley and the Wonderland of Rocks. A herd also occurs at Stubbe Springs, 3 miles (5 kilometers) northwest of Keys View. **Mule Deer** are present, but not numerous or often seen. They live mainly in the higher, pinyon-juniper areas.

Bobcats live throughout Joshua Tree, but are not often observed, being both secretive and nocturnal. There are no definite records of **Mountain Lions** in the monument, but there have been reports of a large cat near Cottonwood Springs and also near Hidden Valley. Although both **Spotted** and **Striped Skunks** occur here, they are rare. The odor of skunk often detected in Joshua Tree usually emanates from the plant called Skunk Sage, which, when crushed, especially in springtime, gives off a skunklike odor.

Badgers occur throughout the monument, but are seldom seen. **Ringtails** are not common, but they do live in rocky areas and sometimes make their dens in abandoned buildings or mines. **Raccoons**, probably descended from released pets, have been recorded near the monument, but not actually in it. Although **Desert Woodrats** are found throughout Joshua Tree, and their large nests of sticks, stones, and pieces of cacti are often seen, the animals themselves are not often sighted. The Cholla Cactus Garden is a good place to see these nests, and there often are some in abandoned buildings and mines. **White-throated Woodrats** and **Dusky-footed Woodrats** also occur in the monument, but neither is widespread or common.

CHECKLIST

___ *Desert Cottontail*	___ *Kit Fox*
___ *Black-tailed Jack Rabbit*	___ *Gray Fox*
___ *Merriam's Chipmunk*	___ *Ringtail*
___ *White-tailed Antelope Squirrel*	___ *Raccoon (I?)*
___ *California Ground Squirrel*	___ *Badger*
___ *Round-tailed Ground Squirrel*	___ *Spotted Skunk*
___ *Desert Woodrat*	___ *Striped Skunk*
___ *White-throated Woodrat*	___ *Mountain Lion (?)*
___ *Dusky-footed Woodrat*	___ *Bobcat*
___ *Coyote*	___ *Mule Deer*
	___ *Mountain Sheep*

Katmai National Park

P.O. Box 7
King Salmon, Alaska 99613

At the base of the Alaska Peninsula in south-central Alaska, huge (3,829,151 acres; 1,549,619 hectares) Katmai National Park is one of the largest areas administered by the National Park Service. Originally proclaimed to protect, for purposes of study, the volcanic area that resulted from the eruption that blasted off the top of Mount Katmai in 1912, the park has been enlarged to include a diversity of environments of rivers, lakes, glaciers, forests, and coastline, as well as ash fields and volcanoes. The park is topped by 7,606 foot (2,318 meter) Mount Denison.

Virtually inaccessible by land, Katmai is best reached by plane or boat, usually from King Salmon. Summer is the prime time to visit, although insects are an annoyance; rainy, overcast, and cool weather is to be expected; and it can be windy. Katmai is, however, one of the best places to see Brown Bears, Moose, and Beavers, with fair opportunities of seeing Lynx, Porcupines, and Ermine.

The huge **Brown Bear** is common in Katmai and, although it may prefer brushy country and grasslands, is usually seen around Brooks River Camp, especially when the salmon swim upstream. Trails in the park are mainly bear trails, and hikers on them should be alert and try not to surprise bears.

Arctic Ground Squirrels are common residents around the buildings at Brooks River and also live in suitable places on the valley floor. **Red Squirrels** live in the forested parts of Katmai, and they are often heard and seen foraging on the ground at Brooks Camp. **Hoary Marmots** prefer rocky outcrops and are rarely seen and then only by hikers who venture into the higher, rugged, northern country of Katmai.

Beavers inhabit many of the lakes and streams of the park. Their characteristic domed lodges are readily seen in the ponds in the Margot Creek Valley, along the road to Valley of Ten Thousand Smokes. Toward the end of the long summer days, Beavers can often be seen swimming, as in Naknek Lake and directly in front of the lodge at Brooks River. **Mink** live in these same places, but are seldom seen. **River Otters** are also not often seen, except perhaps in lakes near the Shelikof Strait.

Moose are often seen in Katmai, usually feeding at lake sides. They are often present and visible from the road along Margot Creek, as well as at Naknek Lake, Savonoski River, and right around Brooks River. The only other species of hoofed mammal in Katmai, the **Caribou**, is absent from the park in summer, but migrates into the southwestern portion in winter.

Snowshoe Hares are active early and late in the day and are often seen along the Margot Creek road. The **Arctic Hare** is known to inhabit Katmai, but seems very rare. **Porcupines** are common in the forests and, although nocturnal, can often be seen during the day. There are usually some around or under the buildings at Brooks River.

KATMAI NATIONAL PARK

Mt. Douglas

Kaguyak

Hallo Bay

Cape Nukshak

Mt. Denison

Mt. Katmai

Crater Lake

Katmai

SHELIKOF STRAIT

Nonvianuk Lake

Lake Grosvenor

SAVONOSKI RIVER

BROOKS RIVER

VALLEY OF 10,000 SMOKES

Lake Coville

North Arm

Iliuk Arm

MARGOT CR.

KEJULIK MOUNTAINS

Naknek Lake

Lake Brooks

King Salmon

Becharof Lake

Miles
0   5   10   15   20

Kilometers
0   5   10   15   20   25   30

Gray Wolves are fairly common at Katmai, and they are sometimes encountered along the Margot Valley road or even near Brooks River. Backcountry hikers have the best chance of seeing wolves, however. Red Foxes are also common, mainly along the coast and western edge of the park, but they are seldom seen by visitors.

Ermine are common and they are frequently seen around Brooks River. The Wolverine occurs throughout the park, but is seldom seen. Observing one is quite a matter of luck. Wolverines have been seen along the Margot Valley road, but more often their presence is proclaimed by a somewhat skunklike odor. The only member of the cat family in Katmai is the Lynx. Although nocturnal, it is occasionally seen along the valley road late in long summer days, and even on the beach at Brooks River.

Various marine mammals are common residents of the shores and waters of the park at Shelikof Strait, an area not readily accessible to most park visitors. Sea Otters, Northern Sea Lions, and Harbor Seals are present, some in large numbers, and occasionally there are Whales present.

CHECKLIST

___ Snowshoe Hare	___ Brown Bear
___ Arctic Hare	___ Northern Sea Lion
___ Hoary Marmot	___ Ermine
___ Arctic Ground Squirrel	___ Mink
___ Red Squirrel	___ Wolverine
___ Beaver	___ River Otter
___ Porcupine	___ Sea Otter
___ Whales	___ Harbor Seal
___ Harbor Porpoise	___ Lynx
___ Gray Wolf	___ Moose
___ Red Fox	___ Caribou

Kings Canyon National Park

Three Rivers, California 93271

Contiguous with Sequoia National Park to the south, Kings Canyon's 460,136 acres (186,213 hectares) encompass mainly the Sierra Nevada wilderness peaks and the enormous canyons of the Kings River. A separate western section has a grove of giant sequoia trees. This grove includes the General Grant Sequoia, which, with a height of 267 feet (81 meters) and

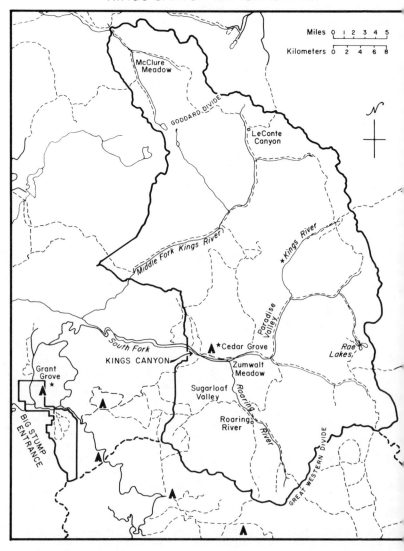

a circumference of 108 feet (33 meters), is second in volume only to Sequoia National Park's General Sherman tree. Most of Kings Canyon is accessible only by foot or horseback; Grant Grove and Cedar Grove are available to automobile traffic.

Although there is a diversity of wildlife in the park, the status of many species is not yet known, and comparatively few mammals are readily observable in Kings Canyon. The park is open throughout the year, although the road to Cedar Grove is closed in winter, and the road to Grant Grove is sometimes temporarily closed by snow. Snowshoe nature walks are held on winter weekends at Grant Grove.

Ground squirrels and chipmunks are the most frequently seen mammals at Kings Canyon. At Cedar Grove and Grant Grove, and in other suitable places up to about 8,000 feet (2,438 meters), the **California Ground Squirrel** is the common species. Higher, in the mountain meadows above 9,000 feet (2,743 meters), **Belding's Ground Squirrels** are usually seen standing conspicuously erect in grassy areas during the day. One or more species of chipmunk occurs in the park, from the lowest elevations to over 11,000 feet (3,353 meters). Below 5,000 feet (1,524 meters), on the western side of the Sierra Nevada, is the domain of **Merriam's Chipmunk**, and elevations of 5,000–11,000 feet (1,524–3,353 meters) are the range of the **Lodgepole Chipmunk**. The **Alpine Chipmunk** lives above timberline, and a fourth species, the **Colorado Chipmunk**, may occur on the eastern side of the Sierra. Although there are slight differences in size and color, as well as other distinctive characteristics between these species, they cannot readily be distinguished in the field. Around campgrounds and especially the buildings at Grant Grove, chipmunks are common. **Golden-mantled Ground Squirrels** also frequent woodlands, picnic areas, and campsites, at elevations up to about 11,000 feet (3,353 meters).

Of the two species of tree squirrel in Kings Canyon, only **Douglas' Squirrel**, which inhabits coniferous forests up to 10,500 feet (3,200 meters), can be regarded as common, and in some places it is readily seen. The **Western Gray Squirrel** inhabits low-elevation forests, but is not usually seen.

In the high mountain meadows and rocky areas, from timberline down to about 7,000 feet (2,134 meters), the **Yellow-bellied Marmot** is often seen, frequently sunning itself on rocks on bright summer days. Another inhabitant of high, rocky regions in the park is the **Pika**. Pikas live in talus slopes and around high basins up to 13,000 feet (3,962 meters), where patient observers, alerted by their shrill cries, can usually see them.

Mule Deer are widespread in the park, but, strangely, they are not as commonly seen as in other parks. The other hoofed animal here is the **Mountain Sheep**, found only at three localities in the extreme eastern and southern parts of the park. The usual visitor to Kings Canyon never sees these animals.

Of the two species of skunks in the park, the small **Spotted Skunk** is commoner, and at night around campgrounds and at Cedar and Grant groves, it is frequently seen. The **Striped Skunk**, which has been found only in the low-elevation areas on the western border of the park, is very rare; none has been seen in recent years. Of the many other members of the weasel family in the park, only the **Marten** is seen with some regularity. On summer days, back-country hikers sometimes see Martens in the high coniferous forest. Although **Fishers** probably also live in these forests, there seem to be no recent observations of them. **Long-tailed Weasels**, which live at elevations from 5,000 to over 10,500 feet (1,524 to over 3,200 meters), probably are not rare, but are seldom seen. The **Ermine**, however, is rare, and there is only a single record of it in the park near East Lake, at 9,000 feet (2,743 meters). Diggings indicate that **Badgers** range throughout much of the park, but the animals themselves are not usually seen. The **Wolverine**, however, is not at all abundant and, although primarily a high-country inhabitant, has been recorded as low as 3,100 feet (945 meters). Wolverines are seldom seen, and even their tracks are rare.

Coyotes roam throughout Kings Canyon, at elevations of 5,000–13,000 feet (1,524–3,962 meters). They are frequently seen, day or night. Of the two kinds of fox in the park, the **Gray Fox** is common. It inhabits the low chapparal country and, in winter, is frequently seen along the roads at night. The **Red Fox**, which is rare, generally inhabits the timberline zone, although it has also been seen within a few miles of Grant Grove. Although **Gray Wolves** might have once lived in the park, they are now extinct.

Although **Black Bears** are found throughout the park and are not scarce, they are not often seen. **Raccoons**, known only from lower elevations on the western side of Kings Canyon, are rare. **Ringtails**, which are also rare, live in the brush country at low elevations near Cedar Grove, and in some abandoned buildings near Lewis Creek.

Rabbits and hares are extremely scarce in this park. Although both the **Audubon Cottontail** and the **Brush Rabbit** may be present, there are no definite records of them, and any observations of cottontails should be reported to park rangers. Jack Rabbits are also seldom seen. At high elevations, the **White-tailed Jack Rabbit** is present, but scarce. Whether the **Black-tailed Jack Rabbit**, a low-elevation species, actually occurs in the park is uncertain.

Porcupines live mainly in the lodgepole pine areas of the park, but seem scarce and are not usually observed. The only woodrat in Kings Canyon is the **Dusky-footed Woodrat**, which is known to live around Cedar Grove and similar low-elevation sites. The populations of **Mountain Beavers** are low, but their sign—tunnels—appears at Grant Grove, Sequoia Creek, and Sheep Creek. **Opossums** are sometimes found at lower elevations in the park, but these are descendants of stock introduced from eastern states many years ago.

In the lower parts of the western sections of Kings Canyon, **Bobcats** are numerous, and it is not uncommon to see them along roads there at night. **Mountain Lions** are resident in the park, but are seldom seen. They live mainly in the yellow pine zone and have been reported around Cedar Grove.

CHECKLIST

___ Virginia Opossum (I)	___ Desert Cottontail
___ Pika	___ White-tailed Jack Rabbit
___ Brush Rabbit (?)	___ Black-tailed Jack Rabbit (?)
	(continued)

(continued)

- Mountain Beaver
- Alpine Chipmunk
- Merriam's Chipmunk
- Colorado Chipmunk (?)
- Lodgepole Chipmunk
- Yellow-bellied Marmot
- Belding's Ground Squirrel
- California Ground Squirrel
- Golden-mantled Ground Squirrel
- Western Gray Squirrel
- Douglas' Squirrel
- Dusky-footed Woodrat
- Porcupine
- Coyote
- Gray Wolf (E)
- Red Fox

- Gray Fox
- Black Bear
- Ringtail
- Raccoon
- Marten
- Fisher (?)
- Ermine
- Long-tailed Weasel
- Wolverine
- Badger
- Spotted Skunk
- Striped Skunk (E?)
- Mountain Lion
- Bobcat
- Mule Deer
- Mountain Sheep

Lassen Volcanic National Park

Mineral, California 96063

Until the eruption of Mount St. Helens in 1980, Lassen Peak was the only recently active volcano in the lower forty-eight states. Starting in 1914, Lassen Peak began erupting and continued volcanic activity for seven years. Even today, activity is still evident from hot springs at Bumpass Hell and the Sulphur Works, as well as steam vents, which sometimes emerge from the craters formed more than sixty years ago. The park's 106,373 acres (43,048 hectares) are traversed by more than 150 miles of hiking trails, and the Lassen Park Road circles three sides of Lassen Peak (10,457 feet; 3,187 meters). Although the park is open throughout the year and is the site of much winter visitation, the road across it is generally closed from late October until early June. The main winter center is near the southwest entrance, where downhill and cross-country skiing are popular.

Lassen Peak is the southern end of the Cascade Mountain chain, and, in addition to plant succession taking place on areas denuded by vulcanism, there are extensive stands of conifers, as well as aspens and willows, and many clear lakes and streams. There is much wildlife with, as usual, the chipmunks, ground squirrels, and deer being the most readily seen.

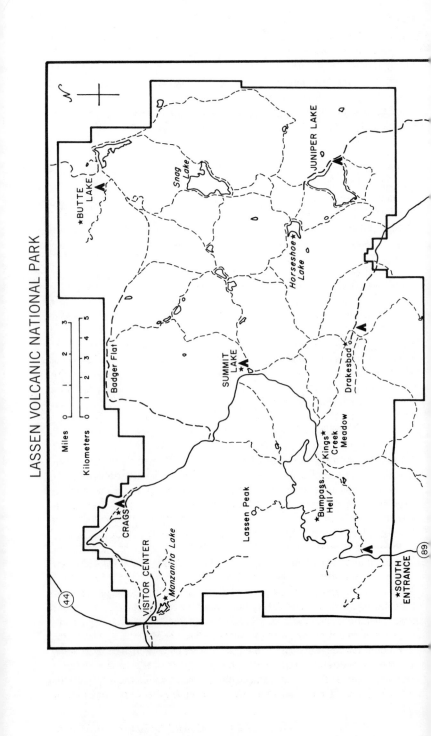

LASSEN VOLCANIC NATIONAL PARK

Yellow-bellied Marmots are common in Lassen, readily seen on rocky slopes near the south entrance road, along the road to Summit Lake, and on many of the trails. They generally can be found at the parking lot at the Summit Trail. Equally common and even more widespread are Golden-mantled Ground Squirrels. They are usually around campgrounds and picnic areas and are abundant by Manzanita Lake. In the park, there are three species of chipmunk, difficult to differentiate. The Lodgepole Chipmunk, generally the brightest chipmunk, with much deep red coloration, occurs on both eastern and western slopes of Lassen, up to about 8,500 feet (2,590 meters). The Yellow-pine Chipmunk may range even higher, but prefers open forest areas, whereas Townsend's Chipmunk inhabits pine forests and brushier country, also as high as 9,000 feet (2,743 meters).

Douglas' Squirrel, very common in Lassen, is readily seen throughout the forested areas, and is abundant around Manzanita Lake. The Western Gray Squirrel is also recorded in Lassen, but seems rare and is seldom sighted. It occurs only at the lower elevations on the western side of the park.

In summer, both varieties of Mule Deer are in the park. From Lassen Peak westward, the Black-tailed subspecies is the common deer and is frequently seen, except at the highest elevations. At dawn or dusk, they can almost always be found around the King's Creek Meadow, Manzanita Lake, and the Summit Lake region. Mule Deer, not the Black-tailed subspecies, inhabit the eastern part of the park and are often seen by hikers in such places as around Snag Lake. Some "stragglers" of the Black-tailed Deer are seen in the eastern section, but Mule Deer evidently do not wander to the west of Lassen Peak. Warner Creek is a good place to see Mule Deer. Although Pronghorns are listed as occurring in the park, there seem to be no recent records of them, and it seems unlikely that they occur, because they are best known to the east, at lower elevations.

In the high talus slopes, down to about 5,300 feet (1,615 meters), Pikas are common. They are often seen by hikers, and their ventriloqual whistle is the surest sign of their presence. Nuttall's Cottontails inhabit the lower elevations and seem not at all common in the park, although they have been observed around the Manzanita Lake Campground. Snowshoe Hares are not often seen, but inhabit the forests and thickets, and the likeliest places to find them are Warner Creek or Summit Creek.

Belding's Ground Squirrels are inhabitants of the high meadows and can be found near King's Creek Meadow and Summit Lake. They are conspicuous when above ground; however, they may begin to estivate as early as July, and thus may not be seen during the summer, or when hibernating in winter. The California Ground Squirrel is a resident of much lower elevations, and there is little habitat for it in the park, although some individuals might occur on the extreme western edge.

Porcupines seem not to be abundant in Lassen and perhaps are mainly at lower altitudes. Gnawed bark, sign of the Porcupine, has been seen at Warner Creek, and Porcupines probably can best be located in Forests in the 5,300 foot (1,615 meters) elevation range. Bushy-tailed Woodrats are common in the rock slides and around boulders but are seldom seen. Their nests of twigs and brush, and even stones, can be found at Warner Creek, Drakesbad, and at elevations to more than 8,000 feet (2,438 meters). **Muskrats** are not common in the park, although there may be some in Manzanita Lake. There seem to be no records of **Beavers**. **Mountain Beavers** which are also recorded from the east side of Lassen, in moist, deep-soiled valleys, are not ordinarily seen.

Carnivores are not notably abundant in Lassen. **Martens** are sometimes seen in mountain meadows, such as at King's Creek, and even at the Manzanita Lake Campground, but **Ermine** and **Long-tailed Weasels** are rarely seen. **Coyotes**, which are sometimes observed in the Warner Valley, are probably the most abundant members of the dog family in the park. **Gray Foxes** prefer lower elevations and are possibly not actually within the park's borders. **Red Foxes** are recorded in the park, ranging at elevations to about 8,000 feet (2,438 meters), including at Warner Creek, but are not thought to be common in Lassen. **Black Bears** are rare but have been seen, especially in the Warner Valley, Butte Lake, and near Hot Boiling Springs, although not often. **Grizzly Bears** are extinct here. **Mountain Lions** are present, living in the more remote areas of the park, and, as is true in other national parks, are a very rare sight. **Bobcats** probably prefer lower elevation chapparal and are rarely, if ever, seen in the park.

Both **Spotted** and **Striped Skunks** are rare in Lassen. The former prefers lower elevations and may not actually occur in the park, whereas the latter is probably not found above 6,000 feet, so there is little habitat for them in Lassen. **Badgers**, which are commoner, are sometimes seen hunting in ground squirrel colonies and where there are pocket gophers, especially in the Warner Valley. **Mink** may inhabit stream sides at lower elevations of the park, but are seldom seen. **River Otters** are known to inhabit Butte and Snag lakes, but are a rare sight. **Raccoons** are commoner, especially at lower elevations, and usually can be found in the Warner Valley. **Virginia Opossums**, from introduced stock, live near the park, but because they prefer low elevations, they are probably not within its borders.

CHECKLIST

___ Virginia Opossum (?)
___ Pika
___ Nuttall's Cottontail
___ Snowshoe Hare
___ Mountain Beaver
___ Yellow-pine Chipmunk
___ Townsend's Chipmunk
___ Lodgepole Chipmunk
___ Yellow-bellied Marmot

___ Belding's Ground Squirrel
___ California Ground Squirrel
___ Golden-mantled Ground Squirrel
___ Western Gray Squirrel
___ Douglas' Squirrel
___ Bushy-tailed Woodrat
___ Muskrat
___ Porcupine

(continued)

(continued)

- ___ Coyote
- ___ Red Fox
- ___ Gray Fox
- ___ Black Bear
- ___ Grizzly Bear (E)
- ___ Raccoon
- ___ Marten
- ___ Ermine
- ___ Long-tailed Weasel

- ___ Mink
- ___ Badger
- ___ Spotted Skunk
- ___ Striped Skunk
- ___ River Otter
- ___ Mountain Lion
- ___ Bobcat
- ___ Mule Deer
- ___ Pronghorn (E)

Mammoth Cave National Park

Mammoth Cave, Kentucky 42259

Part of the longest recorded cave system in the world, with more than 226 miles (364 kilometers) explored and mapped, Mammoth Cave National Park's 52,162 acres (21,109 hectares) also include rivers and forest-covered hills. Although the main visitor attractions are tours oriented toward the history and geology of the cool (54° F.; 12° C.) and humid (87 percent) limestone caves, there are ample trails for above-ground hiking, as well as boating opportunities, in the park.

The oak-hickory forested hills of Mammoth provide good opportunities for seeing a number of kinds of mammals both day and night. There are easily observed White-tailed Deer, Raccoons, and even Spotted Skunks in the park; the best localities for watching them are near the visitor center, the campground, and along Joppa Ridge; the best times are at dawn or dusk and at night.

Raccoons are abundant in the woods of Mammoth, and they boldly search for food around the lodge, campground, and visitor center. At dusk and after dark, they can readily be found by the cabins to the southwest of the visitor center. Another omnivore that is not often seen in other parks, but that may be common around the campground, is the **Spotted Skunk**. It searches for food there, near the garbage cans. The other skunk in the park, the **Striped Skunk**, is less common, but it is also sometimes encountered at night near the lodge or campground.

The **White-tailed Deer** is the commonest large mammal at Mammoth. These deer are often seen along the roads in morning and evening, as well as at night, and almost always can be seen near the campground and lodge. The original deer population at Mammoth was wiped out. After deer were restocked in the 1940s, their population grew excessively, and many were live-trapped and redistributed to other parts of Kentucky.

MAMMOTH CAVE NATIONAL PARK

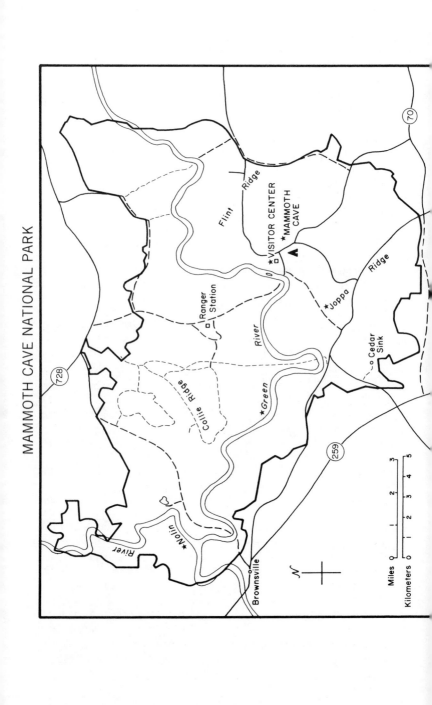

Gray Squirrels are common residents of the forests, and they are readily seen during the day near the lodge, foraging on the ground. Another species, the **Fox Squirrel**, may be a resident of the park, but few have been seen in recent years. **Eastern Chipmunks** reside in the forests, but do not seem abundant and are seldom seen in the cave area.

In the grassy clearings, **Woodchucks** usually can be seen grazing. There are generally some in the meadows around the lodge on summer mornings and afternoons. In winter they hibernate. **Eastern Cottontails** are present throughout the park, but do not seem to be especially common, and are usually observed only at night or at dawn along the dirt roads. The same is true of the **Virginia Opossum**.

Despite the extensive cave system, bats are not abundant at Mammoth and are not usually seen by visitors. More than a dozen species of bat have been recorded in the park, but during summer there are few bats present. In winter, the **Indiana Bat** uses the caves as a hibernaculum; some groups of males may remain through the summer and are sometimes seen by visitors.

Beavers were extirpated from the park's rivers in colonial times, but were restocked and are now regular inhabitants of the Green River. Visitors on the evening boat trips sometimes see them, as well as **Muskrats**. The Mammoth Park Beavers do not build lodges or dams, but live in burrows in the riverbanks. Another denizen of stream sides, the **Mink**, is rarely seen. **River Otters**, which once probably lived by the rivers, are no longer present.

Eastern Woodrats make their nests in the mouths of caves and are relatively common in the park. There is usually a nest just inside the new entrance, and visitors who wish to see it should ask a guide to point it out.

Both **Red** and **Gray Foxes** are fairly common at Mammoth. Red Foxes are more often seen, especially at night along the roads, and one often lives around the staff residence. Gray Foxes may be less abundant than Red Foxes. **Long-tailed Weasels** are present in the park, but observations of them are rare.

Black Bears, Bobcats, and **Mountain Lions** all were once a part of the fauna of this part of Kentucky. Black Bears have been gone from the region for a long time, and recent sightings of **Bobcats** are rare. Although there is no resident Mountain Lion population, recent sightings of this large cat near the park have raised hopes that it will once more become part of the fauna.

CHECKLIST

___ Virginia Opossum	___ Eastern Chipmunk
___ Bats	___ Woodchuck
___ Eastern Cottontail	___ Gray Squirrel

(continued)

(continued)

___ Fox Squirrel   ___ Long-tailed Weasel
___ Beaver (RI)   ___ Mink
___ Eastern Woodrat   ___ Spotted Skunk
___ Muskrat   ___ Striped Skunk
___ Red Fox   ___ River Otter (E)
___ Gray Fox   ___ Bobcat (E?)
___ Black Bear (E)   ___ Mountain Lion (?)
___ Raccoon   ___ White-tailed Deer (RI)

Mesa Verde National Park

Mesa Verde National Park, Colorado 81330

Pre-Columbian cliff dwellings and other archaeological artifacts lie on the tops of forested mesas and in the eroded walls of the 600–900 foot (183–274 meter) deep canyons carved by the Mancos River tributaries. Mesa Verde's 52,036 acres (21,059 hectares) are a semiarid environment with about 18.5 inches (470 millimeters) of annual precipitation. On the mesa tops, pinyon-juniper woodlands are the main plant community and Gambel oak and serviceberry form thickets. Only two of the mesas, Chapin and Wetherill, are accessible by automobile, and travel to these may be limited in order to protect the archaeological sites from vandalism. Similarly, camping and hiking are restricted, and back-country exploration is prohibited. The park is open in winter; there are guided tours to a cliff house, and, with permission, cross-country skiing and snowshoeing are allowed in some areas.

The mammals of Mesa Verde are typical of the semiarid southwest. The Ground squirrel and chipmunk are the most frequently seen species; Gray Foxes are often observed at night; and Mule Deer are common. Part of the Mancos River borders the park, but aquatic species such as Beaver, Muskrat, and Mink are sporadic or nonexistent within most of Mesa Verde.

Rock Squirrels are common throughout Mesa Verde and are often seen along the trails and around the cliff dwellings. **Golden-mantled Ground Squirrels** are only occasionally seen, usually at elevations higher than 7,000 feet (2,137 meters), and mainly in oak-scrub habitat. The two species of chipmunk in Mesa Verde are altitudinally distributed: the **Least Chipmunk**, which is the common species, in higher, brushy areas, and the

MESA VERDE NATIONAL PARK

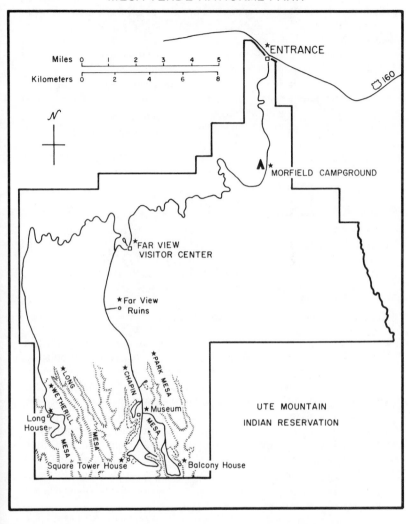

Colorado Chipmunk at the lower elevations, as well as among the pinyons and junipers and on the cliffs in higher zones. As usual with chipmunks, the two species are difficult to distinguish in the field. **Yellow-bellied Marmots** occur in the park, but they are rare and mainly confined to the higher elevations. The Prater Canyon area is one of the places where they have been seen.

Neither of the two species of tree squirrel in Mesa Verde is especially common. **Red Squirrels**, which are restricted mainly to high elevations, live in a side canyon of Prater Canyon, and the population is small. **Abert's Squirrels**, which generally inhabit ponderosa pine woods, are rare in this park, although they are sometimes seen near the park headquarters on Chapin Mesa.

Mule Deer are common throughout the park and are usually seen at dawn or dusk, often along roadsides. **White-tailed Deer** disappeared from the area in historic times, and so did **Pronghorns** from the surrounding valleys. **Mountain Sheep** had also been extirpated in the park, but were restocked in 1946 and, although not often seen, are not rare in Mesa Verde. **Elk**, however, are rare and not resident; occasionally a few transients may be seen, and some may winter in the park.

None of the park's carnivores is seen frequently, although some species are common. **Coyotes** are widespread and are heard howling more often than they are observed. **Gray Foxes** are abundant on the mesa and frequently seen at night along the roads. **Red Foxes** are scarce; however, some of the observations of this species have been of the black color phase. **Gray Wolves** were extirpated in the park, but there have been occasional observations that have led to the belief that they may still sometimes enter the park. Both **Spotted Skunks** and **Ringtails** are common in rock rubble, especially around ruins, but they are wholly nocturnal and thus seldom seen by visitors. **Striped Skunks** are fairly common at the mesa's base but are less abundant and only occasionally seen on its top. **Raccoons** and **Long-tailed Weasels** are not common. **Badgers** are sometimes common enough around the ruins to cause problems for archaeologists who are excavating. There are a few observations of **Black Bears** each year. **Mink** are known to occur only in the Mancos Canyon and are not common there; they are rare in the park.

Two species of cottontail are found in Mesa Verde, where they are often seen at dawn or dusk. The **Desert Cottontail** is the lower-elevation inhabitant, while **Nuttall's Cottontail** resides high on the plateau. Although rare on top of the mesa, **Black-tailed Jack Rabbits** are common at the base and are often seen at night or early and late in the day.

The **Porcupine** population seems to fluctuate. In some years, Porcupines are abundant and frequently encountered; in others, they are rare and seldom seen. **Beavers** are known to occur only in the Mancos Valley and in some years are absent even there. **Muskrats** are not rare in the Mancos River area, and there is a record of one from the top of a mesa. **Gunnison's Prairie Dogs** used to inhabit the broad canyon heads, but died out, and attempts to reintroduce them in the 1950s were unsuccessful. The **Mexican Woodrat** is the common species in the park, building its nests in crevices in the cliffs. The **Bushy-tailed Woodrat**, which is less common, utilizes vertical cracks in the rocks at high elevations. The **White-throated Woodrat** has been recorded from close to the park's southern border and may be the predominant species at low elevations.

Bobcats are common throughout Mesa Verde, but usually are not observed. **Mountain Lions**, although not abundant, are not rare here; there are occasional observations of these large cats.

CHECKLIST

___ Nuttall's Cottontail
___ Desert Cottontail
___ Black-tailed Jack Rabbit
___ Least Chipmunk
___ Colorado Chipmunk
___ Rock Squirrel
___ Golden-mantled Ground Squirrel
___ Gunnison's Prairie Dog (E)
___ Abert's Squirrel
___ Red Squirrel
___ Beaver
___ Bushy-tailed Woodrat
___ Mexican Woodrat
___ White-throated Woodrat (?)
___ Muskrat
___ Porcupine
___ Coyote
___ Gray Wolf (E?)
___ Red Fox
___ Gray Fox
___ Black Bear
___ Ringtail
___ Raccoon
___ Long-tailed Weasel
___ Mink
___ Badger
___ Spotted Skunk
___ Striped Skunk
___ Mountain Lion
___ Bobcat
___ Elk
___ Mule Deer
___ White-tailed Deer (E)
___ Pronghorn (E)
___ Mountain Sheep (RI)

Mount Rainier National Park

Tahoma Woods, Star Route
Ashford, Washington 98304

Dominated by the snow- and glacier-covered dormant volcano named Mount Rainier, this national park's 235,404 acres (95,266 hectares) are home to a variety of mammals and plants, inhabiting four ecological zones over a span of elevations from 2,000 feet (610 meters) to 14,410 feet (4,392 meters). Dense coniferous forests of fir, cedar, and hemlock are characteristic of the lower elevations, while white pine, fir, and hemlock predominate at the middle zones. Above the 6,500 foot (1,981 meters) timberline is the Arctic Alpine Zone, with snowfields and glaciers, and meadows in which some mammals live. Mount Rainier, though dormant, still has active warm springs flowing near Longmire and Ohanapecosh, and steam vents near the summit.

Mount Rainier is one of the better parks in which to see Mountain Goats, Hoary Marmots, Beavers, as well as the black-tailed subspecies of the Mule

MOUNT RAINIER NATIONAL PARK

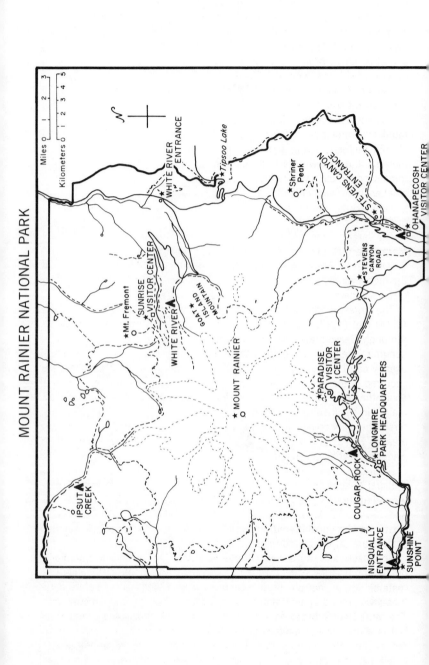

IPSUT CREEK

NISQUALLY ENTRANCE

SUNSHINE POINT

COUGAR ROCK

LONGMIRE PARK HEADQUARTERS

PARADISE VISITOR CENTER

MOUNT RAINIER

Mt. Fremont

SUNRISE VISITOR CENTER

WHITE RIVER

GOAT ISLAND MOUNTAIN

WHITE RIVER ENTRANCE

Tipsoo Lake

Shriner Peak

STEVENS CANYON ROAD

STEVENS CANYON ENTRANCE

OHANAPECOSH VISITOR CENTER

Miles 0 1 2 3

Kilometers 0 1 2 3 4 5

Deer. The park is open throughout the year, although many roads, especially on the eastern side, are closed from late November until June.

The most commonly seen mammals of Mount Rainier Park are diurnal members of the squirrel family. The only tree squirrel is **Douglas' Squirrel**, which is found throughout the forested parts of the park and is common and often seen. Around campgrounds and picnic areas, especially in the eastern part of the park, the **Cascade Golden-mantled Ground Squirrel** is abundant; it can usually be found around Sunrise and Box Canyon, until it begins hibernating in October.

Two species of chipmunk inhabit the park. The larger **Townsend's Chipmunk** is common at the lower forested elevations—up to about 5,000 feet (1,524 meters). Living at elevations of 3,000–8,000 feet (914–2,438 meters), the **Yellow-pine Chipmunk** prefers more open situations in dry forests and meadows, including alpine meadows. In addition to being smaller than Townsend's Chipmunk, the Yellow-pine species is also paler in color. Both species hibernate in winter, but occasionally venture out briefly in snow.

The largest member of the squirrel family in the park is the **Hoary Marmot**, a rock-pile and meadow dweller at elevations mainly of 4,000–8,000 feet (1,219–2,438 meters). This large rodent is usually seen along the higher roads, and it makes a characteristic clear whistle that has given it the nickname "whistler." It hibernates throughout the winter.

The largest rodent in the park is the **Beaver**, found along streams below 4,500 feet (1,372 meters), but mainly below 3,000 feet (914 meters). Although Beavers are not abundant and not usually seen, the colony at Longmire Meadow across from the inn is readily accessible, and Beavers usually can be seen by a quiet, patient watcher at dusk. **Muskrats** once occurred in the park, but are now extinct. The **Mountain Beaver**, which is about the size of a Muskrat, is a burrowing mammal that lives in the park mainly in thick-soiled, low-elevation areas below 4,000 feet (1,219 meters). There are colonies along the Wonderland Trail in the Stevens Canyon area, as well as in Box Canyon, but the animals are seldom seen; the main evidence of their presence is extensive tunnel systems.

Mountain Goats are a special feature of Mount Rainier, and it is not unusual to see some on Tum-Tum Peak from the Sunshine Point Campground at the Nisqually entrance. Goat Island Mountain, Mount Fremont, and Skyscraper Mountain are other promising locations, but Mountain Goats may be encountered almost anywhere in the higher alpine reaches of the park during the summer.

The black-tailed subspecies of the **Mule Deer** is commonly seen thoughout Mount Rainier Park, at elevations up to about 6,000 feet (1,829 meters). Mule Deer are usually along roadsides at dawn and dusk. The meadows around Longmire are a good place to find them in the evening. The Mule Deer subspecies—which differs in that the upper surface of its tail is white, except for the tip—is less common and less often seen in Mount Rainier, occurring only on the eastern border at high elevations.

Elk had probably been extirpated from Mount Rainier when the park was founded, but now there are several herds present in the eastern part, especially near Sunrise, Owyhigh, Shriner Peak, the Sourdough Mountains, and Mount Fremont lookout. These Elk are thought to be descendants of animals that were introduced elsewhere in the Cascades at the beginning of the twentieth century and that have thrived and expanded their range.

Although **Black Bears** may roam anywhere in the park, they live mainly below 6,500 feet (1,981 meters). They are not numerous and are not often seen, the best opportunities being in high meadows and places where there are ripe berries. The Black Bears of Mount Rainier remain in their dens from December to March. Another large carnivore is the **Mountain Lion**, even less frequently seen than the Black Bear. The park has a small resident population that inhabits mainly the forested sections, but because of better visibility, most of the few sightings of Mountain Lions have been in the high, open areas.

Pikas are abundant in Mount Rainier Park, living in rock slides from 2,500 feet (762 meters) to above timberline. The Pikas' characteristic summer hay piles and ventriloqual whistles are often the first indications of their presence, and the patient observer can usually see them. The **Snowshoe Hare** is the only member of the rabbit family in the park, and it undergoes cyclical population fluctuations. Snowshoe Hares live from the lowest elevations, mainly in the forests, to alpine meadows above 6,000 feet (1,829 meters) and, when abundant, can usually be seen along roadsides at dawn or dusk, or around campgrounds, such as the one at White River. Interestingly, the Snowshoe Hares living below 3,000 feet (914 meters) in the park do not turn white in winter.

Mount Rainier has many species of small- and medium-sized carnivore, but none are commonly seen. **Raccoons** usually live in the lower valleys, mainly near water, and they are often seen around Longmire and campgrounds. The only skunk inhabiting the park, the **Spotted Skunk**, is mainly a low-elevation inhabitant of the western side, and is neither common nor often seen. Perhaps the most abundant and frequently seen carnivore is the **Marten**, which lives mainly in forests at elevations approaching 6,500 feet (1,981 meters) and which is often active during the day. The **Ermine** is also common in the park below 6,000 feet (1,829 meters), especially near forest streams, and has been seen with some frequency around Sunrise. The **Long-tailed Weasel** is much less common and perhaps is found only at elevations above 6,000 feet (1,829 meters). **Mink** live along the streams and lakes at low elevations, but are seldom encountered. The **Fisher, River Otter**, and **Wolverine**, all of which once inhabited this part of the Cascade Mountains, are now considered extinct in the park.

Coyotes, Red Foxes, and **Bobcats** occur throughout the park below timberline, but none of these predators is seen very often, although Coyotes may be heard howling at night and have been seen sometimes along the road to Sunrise and to Paradise. **Gray Wolves** and **Lynx** have been gone from the park for many years.

Bushy-tailed Woodrats are not abundant, and although they sometimes make their stick nests around camps or in cabins, they are rarely seen because of their nocturnal habits. **Porcupines** are relative newcomers to Mount Rainier, having been unknown here at the beginning of the twentieth century. They are not especially abundant, and although most sightings have been on the western side of the park, below 4,500 feet (1,372 meters), they probably now occur throughout the forested areas.

Virginia Opossums, Eastern Cottontails, California Ground Squirrels, and **Striped Skunks**, while not known within the park, occur near enough to its borders that they may eventually be inhabitants.

CHECKLIST

___ Pika
___ Snowshoe Hare
___ Mountain Beaver
___ Hoary Marmot
___ Cascade Golden-mantled
    Ground Squirrel
___ Yellow-pine Chipmunk
___ Townsend's Chipmunk
___ Douglas' Squirrel
___ Beaver
___ Bushy-tailed Woodrat
___ Muskrat (E)
___ Porcupine
___ Coyote
___ Gray Wolf (E)
___ Red Fox

___ Black Bear
___ Raccoon
___ Marten
___ Fisher (E)
___ Ermine
___ Long-tailed Weasel
___ Mink
___ Wolverine (E)
___ Spotted Skunk
___ River Otter (E)
___ Mountain Lion
___ Lynx (E)
___ Bobcat
___ Elk (RI?)
___ Mule Deer
___ Mountain Goat

North Cascades National Park

800 State Street
Sedro Woolley, Washington 98264

North Cascades National Park consists of two units, North and South, which are separated by the 117,574 acre (47,581 hectare) Ross Lake National Recreation Area. On the southern border is another national recreation area, Lake Chelan (61,890 acres; 25,046 hectares). The national park itself, which has virtually no roads, comprises 504,785 acres (204,282 hectares) of the Cascade Mountains. The park is replete with more than three hundred glaciers, peaks of over 9,000 feet (2,743 meters), dense vegetation in the moist lower elevations, and alpine meadows and lakes above.

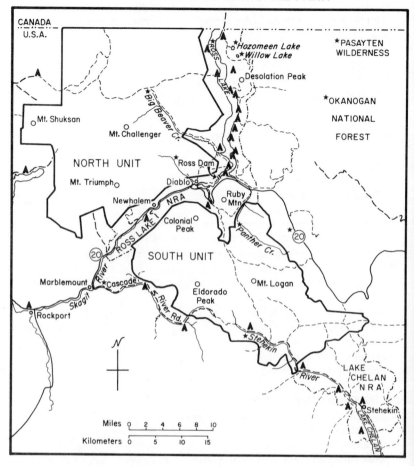

Because North Cascades is a relatively new park, and without the usual auto routes for visitors, its mammals have not been well surveyed. Hikers who are quiet and patient have opportunities to see a number of shy species in the back country, including such relative rarities as the Grizzly Bear, Gray Wolf, Lynx, and Mountain Lion, all of which are believed still to occur in the park. Summer is the best time to view these animals, but visitors should expect rain, and even snow at high elevations, and insects can be a major annoyance at times.

Black Bears are found throughout the park, especially in summer, when meadow bushes are ripe with berries. These bears are sometimes pests around campsites, and back-country campers must suspend food out of their reach. **Grizzly Bears** once inhabited the park; it is hoped that some may still persist in the eastern parts and in South Unit. A summer high-country inhabitant is the **Mountain Goat**, which is occasionally seen on rocky ledges or in high meadows by back-country hikers.

Mule Deer, as well as the Black-tailed subspecies, are present in the park. The former are commoner in the eastern sections of North Cascades, and the latter predominate in the western and central areas. Deer are often seen along State Highway 20 at Ross Lake and near trail heads. These deer also hybridize in the Ross Lake area. **Moose** are also North Cascade inhabitants, but are not at all common. They prefer moist areas; the best chances to see them are in the northern parts of the park, especially near Hozomeen, Ross, and Willow lakes. **Elk** are in the park, but seem to be very rare. Some have moved into the park from outside areas, so that their numbers are expected to increase. Although **Mountain Sheep** have not actually been recorded in the park, they occur in the Pasayten Wilderness at the northeastern edge of North Cascades.

Douglas' Squirrels are common residents of the park's forests, where they are frequently seen and heard. At the Ross Dam overlook on State Highway 20, Douglas' Squirrels can be seen foraging for scraps, along with chipmunks. The closely related **Red Squirrel** also lives in North Cascades, but in the pine belt on the eastern side. The **Western Gray Squirrel** is thought to inhabit the Stehekin area in the South Unit, but it has not yet actually been documented in the park itself and is not common. **Fox Squirrels**, which have been introduced in the Okanagan National Forest to the ·east of the park, have not been definitely seen in the park.

Hoary Marmots are common residents of the talus slopes and meadows of the higher parts of the park and are often seen by hikers who are patient. Another common inhabitant of the talus slopes is the **Pika**, which is found throughout the park. Pikas can readily be seen from the Ross Dam Overlook on State Highway 20.

The **Cascade Golden-mantled Ground Squirrel** is widespread in North Cascades. It commonly forages at campgrounds and trail heads, where it is sometimes mistaken for a chipmunk. Chipmunks are smaller, however, and there are several species in the park. In the eastern portions is the **Yellow-pine Chipmunk**, while the **Townsend's Chipmunk** lives in the western parts of the park. It can be seen at the Ross Dam overlook on State Highway 20. The **Least Chipmunk** may occur in the southeast part of the park. Identification of these chipmunks is mainly a matter for the scientists; they look very much alike. **Columbian Ground Squirrels** may occur in the eastern part of the park, although there are no actual records of them there.

Beavers are present locally in ponds and streams, such as the aptly named Big Beaver Creek. They are shy and nocturnal, and their dams and lodges are commoner sights than the animals themselves. But on swift streams they den in banks burrows rather than lodges. **Muskrats** are present in the park, but are uncommon and seldom seen. **River Otters** are also present at the larger rivers and lakes, and they are sometimes seen by fishermen. The **Mink** is another species that lives near streams and lakes, but there are few records of Mink, and those are mainly from the lower elevations. They are rarely seen. The same is true of **Raccoons**, also residents of low-elevation watercourses. Even around campgrounds, they are not abundant.

Coyotes are not rare in North Cascades and are sometimes seen in mountain meadows by hikers, or heard howling at night. **Gray Wolves** are extremely rare, but some are believed still to inhabit the park. The only other member of the dog family in the park is the **Red Fox**, which lives in the forests, but is infrequently seen in the high open meadows. The Red Fox is considered rare in North Cascades.

There are two species of skunk in the park, both living mainly at the lower elevations. The small **Spotted Skunk** is fairly common in the western sections, especially around campsites. **Striped Skunks** are considered rarer in North Cascades than the spotted species and are not often seen.

Bushy-tailed Woodrats are widespread in the park, and their stick nests are often seen in abandoned buildings or cabins. The animals themselves are rarely seen. **Mountain Beavers** have a very local distribution in the park, but their diggings can be found in low, wet, wooded areas and along the Cascade River Road.

Snowshoe Hares are seldom encountered in the park, but may be fairly common. **White-tailed Jack Rabbits, Eastern Cottontails** (introduced), **Brush Rabbits**, and **Nuttall's Cottontails** all are known to occur near the park boundaries, but have not yet been recorded within the park itself.

Mountain Lions have been reported from North Cascades and are probably park residents. For back-country hikers, seeing one is mainly a matter of luck. Mountain Lions are sometimes seen in the Ross Lake–Diablo area. Another cat, the **Lynx**, is present but rare; its usual sign is tracks in the snow. Widespread, but also seldom seen, are **Bobcats**.

Several members of the weasel family which are known to live in the park are rarely seen, even though they may be abundant in places. These are the **Marten, Ermine**, and **Long-tailed Weasel**. Of these, the Marten is the most frequently encountered—by hikers in low, forested regions. **Wolverines** are also known to live in North Cascades, by their tracks in the snow, but they are extremely rare. Two other members of the weasel family may live in the park, but there are few actual records of them. One is the **Fisher**, which lives in forests, and the other is the **Badger**, which is known to occur near the eastern boundary. **Porcupines** are generally rare west of the Cascades, and there are no valid records of them from the park. **Virginia Opossums**, descendants of introduced stock, have been seen near the park's western border, but not in it.

CHECKLIST

___ Virginia Opossum (I?)	___ Eastern Cottontail (I?)
___ Pika	___ Nuttall's Cottontail (?)
___ Brush Rabbit (?)	___ Snowshoe Hare

(continued)

___ White-tailed Jack Rabbit (?)
___ Mountain Beaver
___ Least Chipmunk (?)
___ Yellow-pine Chipmunk
___ Townsend's Chipmunk
___ Hoary Marmot
___ Columbian Ground
   Squirrel (?)
___ Cascade Golden-mantled
   Ground Squirrel
___ Western Gray Squirrel (?)
___ Fox Squirrel (I?)
___ Red Squirrel
___ Douglas' Squirrel
___ Beaver
___ Bushy-tailed Woodrat
___ Muskrat
___ Porcupine (?)
___ Coyote
___ Gray Wolf
___ Red Fox

___ Black Bear
___ Grizzly Bear
___ Raccoon
___ Marten
___ Fisher
___ Ermine
___ Long-tailed Weasel
___ Mink
___ Wolverine
___ Badger (?)
___ Striped Skunk
___ Spotted Skunk
___ River Otter
___ Mountain Lion
___ Lynx
___ Bobcat
___ Elk
___ Mule Deer
___ Moose
___ Mountain Goat
___ Mountain Sheep (?)

Olympic National Park

600 East Park Avenue
Port Angeles, Washington 98360

Occupying much of the Olympic Peninsula of western Washington, this huge (901,216 acres; 364,713 hectares) park includes habitats as diverse as glaciers, rain forests, and ocean beaches. The highest peak, Mount Olympus (7,965 feet; 2,428 meters) intercepts moisture-laden clouds blowing eastward and, with 200 inches (5 meters) of annual precipitation—more than 40 feet (12 meters) of which is snowfall—is the wettest spot in the continental United States. There are two parts to Olympic National Park: the central mountainous section and the long, narrow strip along the Pacific Coast.

The abundant moisture and relatively mild climate provide an environment for diverse and luxuriant vegetation. Huge trees—Sitka spruce, west-

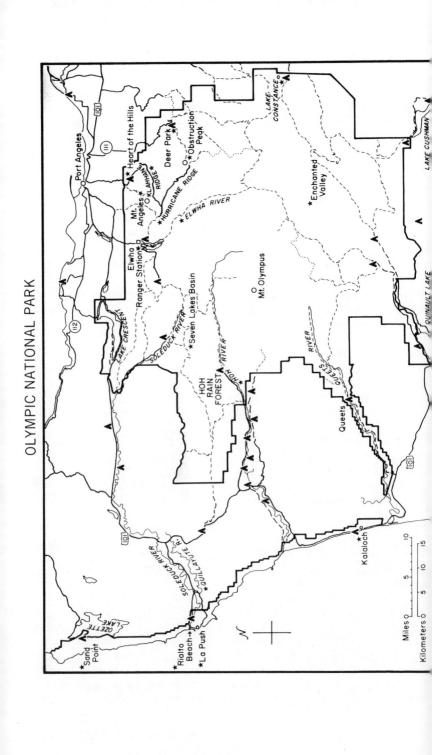

OLYMPIC NATIONAL PARK

ern hemlock, Douglas fir, and western red cedar—reaching heights of more than 200 feet (61 meters) and diameters of 8–21 feet (2.4–6.4 meters), are characteristic of the fern- and moss-laden Olympic rain forest. Hemlock, fir, cedar, and pine are typical at higher elevations. One mammal is unique to the park, the Olympic Marmot, and it is readily seen. There are good opportunities for visitors to see Snowshoe Hares, Harbor Seals, Gray Whales, Elk, River Otters, and Spotted Skunks. Summer is the best time to visit; Hurricane Ridge, the Hoh Rain Forest, and Rialto Beach are among the good places to see mammals.

The black-tailed subspecies of the **Mule Deer** is abundant in Olympic Park and may be encountered almost anywhere. Mule Deer can almost always be seen along the road to Hurricane Ridge, at the ridge itself (where they are quite unafraid of human beings), on the Hoh Nature Trails, and in meadows and forests throughout the park. In winter, the Elwha Ranger Station and the coast at Sand Point are good places to see them. The famed large **Elk** of the Olympics are not frequently seen. In summer, these animals are usually in the open meadows of the high mountains, especially in the Seven Lakes Basin, the High Divide area, and the western parts of the park. There is usually, however, a small population that is resident around the Hoh Rain Forest Campground. Elk can generally be seen here early in the morning. In winter, Elk descend to low elevations and are most likely to be encountered near the Elwha Ranger Station and along the Hoh River Road. **Mountain Goats**, which are not native to the Olympics, were reintroduced in the park in the 1920s and increased to more than five hundred animals by 1980. Because these animals are causing damage to the habitats of Olympic Park, and because it is now park service policy to reduce the effects of exotic animals in the parks, plans are presently being made to control or remove Mountain Goats from Olympic.

The **Olympic Marmot**, a species found only on the Olympic Peninsula, and in no other national park, lives either in the high country, where there is suitable soil for burrowing, or in talus slopes, as on Hurricane Ridge, Hurricane Hill, and Obstruction Peak. Chipmunks, probably both the **Yellow-pine** and **Townsend's Chipmunks**, occur throughout the park. They are usually seen at auto lookouts along the roads, and around campgrounds. Hurricane Ridge is always a good place to see them. Both the Olympic Marmot and the chipmunks hibernate during the winter, but the **Douglas' Squirrel**, an inhabitant of the lower forests, is active throughout the year. It can often be seen along trails, such as the Hoh Nature Trail, and in the coastal sector near Rialto Beach.

Although **Black Bears** are not rare in the park and occur throughout it, they are not often seen. In July and August, they are found in the high subalpine zone, especially in the Hurricane Ridge, Destruction Point, High Divide, Seven Lakes Basin, and Enchanted Valley areas. They are also found along the coast, especially when berries are ripe, and Sand Point is a good place to see them.

Raccoons are common in most low-elevation habitats of the park and are often seen. The area around the Elwha Ranger Station is an especially good place, and they are usually around low-elevation campgrounds and beaches at night. Another abundant species in the same habitats is the **Spotted Skunk**, often seen around campgrounds at night. The larger **Striped Skunk** is also common in the lowlands, but seems not as abundant as the spotted species. It, too, is a nighttime invader of campgrounds.

The coastal sector provides a home for a number of aquatic species. The **Harbor Seal**, a small species, is common along the coast and is often seen at such places as Rialto Beach, frequently swimming quite close to the shore. **Northern Sea Lions** are resident on a few of the coastal islets, such as those off the mouth of the Quileute River, where there is probably a nearby breeding colony. Male **Northern Elephant Seals**, from the increasing populations to the south, are now sometimes found on the beaches in the coastal sector of the park. The **California Sea Lion** is occasionally present here, but, like the Elephant Seal, it is considered an occasional visitor.

Gray Whales swim by the coastal section of Olympic Park each year during spring and fall migrations, and some may be in the area at any time. The best places to watch for migrating Gray Whales is between La Push and Kalaloch. **Humpback Whales, Sperm Whales, Killer Whales**, and **Common Pilot Whales** are sometimes seen, as well as **Pacific White-sided Dolphins** and **Harbor Porpoises**.

Sea Otters were extirpated on the Washington coast by the beginning of the twentieth century. A few have been seen lately, probably descendants of animals reintroduced at Point Grenville. Much more frequently seen, and often mistaken for Sea Otters, are **River Otters**. They are abundant in the coastal section of the park, on rocks offshore north of Sand Point, as well as in large rivers throughout Olympic Park. River Otters have been seen at elevations as high as 5,200 feet (1,585 meters) on Hurricane Ridge, and they are also sometimes seen in the Hoh River and its tributaries. **Mink** are common in the park, but are not often seen. They live along watercourses, especially the larger ones, from salt water to the higher elevations. In local situations, **Beavers** may be fairly common, although the animals themselves are seldom seen. Signs of their presence are dams, but Olympic Beavers seem not to build lodges; they den in burrows in the riverbanks. **Muskrats** are spotty in their distribution, although common locally, especially in lowland rivers. They are not often seen.

Snowshoe Hares may be found anywhere below timberline and are seen fairly often, especially along Hurricane Ridge and the Hoh Rain Forest Nature Trails, and in the brush along the beaches. They are usually seen at dawn or dusk. Snowshoe Hares in the park do not turn white in winter. **Bushy-tailed Woodrats** live in the middle elevations and subalpine areas, occasionally in abandoned cabins and sometimes in occupied ones. They are not often seen, although their stick nests can be found. **Mountain Beavers** have a very local distribution in the park. Although they are seldom seen, signs of their presence, burrows, can be found in moist areas in forest and thickets, such as near Heart-of-Hills Campground, and on the Switchback Trail on Mount Angeles.

Mountain Lions are resident in the park, but the size of the population is not known. Because Mountain Lions often range outside the park, they are vulnerable to hunters. They are sometimes seen at night along roads, and there have been sightings of them along Hurricane Ridge and in the western valleys, as well as in the Elwha Valley. **Bobcats**, although common, are so secretive that they are rarely seen. **Coyotes** are common throughout the park, although they are not regularly seen. More often, they are heard howling at night. The **Gray Wolf** formerly lived in the park, but the last authentic record of one was in 1922.

Martens occur throughout the coniferous forest, from salt water to timberline. They are not usually seen, and their numbers may be low. Most records are from their tracks in the snow, and the best areas in which to see them seem to be the high country around Lake Constance and Royal Basin. **Long-tailed Weasels** are fairly common throughout the park, although they are not regularly encountered. **Fishers** are rare, but they have been seen at the Lower Elwha River. Ermine are also rare in the park, and, although these weasels molt in the fall, they do not turn white in winter.

Virginia Opossums have been seen just outside the park, but none has been recorded from within its boundaries. Virginia Opossums are not native here, but are descended from ones introduced from eastern states early in the twentieth century. **Porcupines**, which were unknown in the area in the early days of the park, are now resident, but uncommon. They have been seen at Deer Park and may be more abundant in the southwestern and northern parts of the park.

CHECKLIST

___ Virginia Opossum (I)	___ Black Bear
___ Snowshoe Hare	___ Northern Sea Lion
___ Mountain Beaver	___ California Sea Lion
___ Yellow-pine Chipmunk	___ Raccoon
___ Townsend's Chipmunk	___ Marten
___ Olympic Marmot	___ Fisher
___ Douglas' Squirrel	___ Ermine
___ Beaver	___ Long-tailed Weasel
___ Bushy-tailed Woodrat	___ Spotted Skunk
___ Muskrat	___ River Otter
___ Porcupine	___ Sea Otter (RI)
___ Sperm Whale	___ Harbor Seal
___ Pacific White-sided Dolphin	___ Northern Elephant Seal
___ Killer Whale	___ Mountain Lion
___ Common Pilot Whale	___ Bobcat
___ Harbor Porpoise	___ Elk
___ Gray Whale	___ Mule Deer
___ Coyote	___ Mountain Goat (I)
___ Gray Wolf (E)	

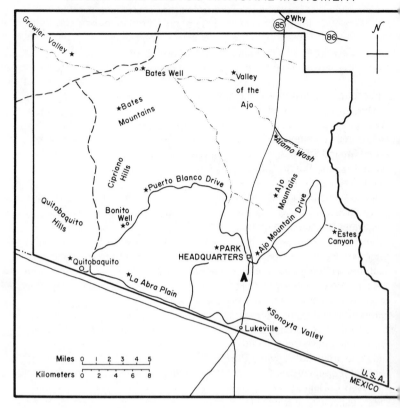

Organ Pipe Cactus National Monument

P.O. Box 38
Ajo, Arizona 85321

With Mexico forming its southern border, Organ Pipe Cactus National Monument's 330,689 acres (133,826 hectares) preserve characteristic Sonoran Desert. Diverse kinds of cacti and other desert plants—saguaro, prickly pear, cholla, agave, elephant tree, senita cactus, and creosote bush—as well as the Organ Pipe Cactus (for which the monument is named) are to be found here, along with many typically arid-land species of mammals.

Daytime temperatures in summer may be higher than 104° F. (40° C.); this is the season of thunderstorms, which account for half of the annual 9 inches (229 millimeters) of rainfall. In winter, temperatures may drop below freezing at night, and many species of rodent hibernate at this time. Organ Pipe is one of the few places where Antelope Jack Rabbits may be seen, and is a good park in which to see Coyotes, Gray Foxes, and Kit Foxes.

Despite the daytime heat, ground squirrels are active, and the commonest of them, **Harris' Antelope Squirrel**, is generally to be seen along roads and drives, especially near rocky sites such as along the Ajo Mountain Drive. Less common, and preferring sandy or silty areas, is the **Round-tailed Ground Squirrel**, which can be found near Bates Well and appropriate habitat on the Puerto Blanco Drive. The large **Rock Squirrel** lives, as its name implies, in rocky places throughout the monument and has been seen at Alamo Canyon above Growler Wash, as well as on the Puerto Blanco Drive about 0.5 miles (0.8 kilometers) north of the visitor center.

Coyotes are common and are found throughout the monument. They are often seen early in the morning along roadsides; Alamo Wash and the Valley of the Ajo are particularly good places to find them. **Gray Foxes**, too, are abundant near the mountainous portions of Organ Pipe, and they are sometimes seen by campers at night. This is one of the few park areas where **Kit Foxes** are fairly common, and they are often seen at night around Bates Well and even at the visitor center. Visitors should be wary of overly friendly Kit Foxes or animals otherwise behaving oddly, because Kit Foxes with rabies have been found in the monument. **Gray Wolves** have been recorded from Bates Well, Growler Wash, and Bonito Well, but are not thought to reside in the monument, but rather to be transients from Mexico.

Desert Cottontails are readily seen at dawn, dusk, or night along roadsides, especially at Bates Well, La Abra Plain, and along the Ajo Mountain Drive. The commonest hare in Organ Pipe is the **Black-tailed Jack Rabbit**, which is abundant on the valley floors and plains. It is readily seen at night or early in the morning along roadsides. The other species of hare in the monument inhabits low, grassy areas near Quitobaquito and the Sonoyta Valley. It is the large **Antelope Jack Rabbit**, which, now that cattle grazing has ceased in the monument, may be increasing in numbers. The best place to find Antelope Jack Rabbits is probably near Armenta Well, in the northern part of the monument.

Four kinds of skunk may inhabit Organ Pipe, but all are rare except the **Spotted Skunk**, which inhabits rocky areas and is commonly seen at night around the campground. **Striped Skunks** and **Hog-nosed Skunks** are extremely rare. Although the **Hooded Skunk** should be present in the monument, on the basis of habitat and general distribution, there are no valid records of it.

Collared Peccaries live mainly in the Ajo Mountains, which are in the eastern part of the monument, and although not often seen, they have been observed just west of the visitor center, in the Alamo Canyon area, and near the Victoria Mine and Dripping Spring. About fifty **Mountain Sheep**

are known to live in Organ Pipe. They inhabit the Ajo Mountains and the Bates Mountains, but are rarely seen. **Mule Deer** occur throughout Organ Pipe, but except for some near the visitor center and at Quitobaquito, they are not often seen. A small, isolated population of **White-tailed Deer** inhabits the Ajo Mountains. They are seldom seen and should not be disturbed by searchers. There seem to be no more than ten or twelve animals in the herd.

Although formerly well-established, **Pronghorns** are now scarce in the monument and exist mainly along the western side. Pronghorns are generally uncommon, and few visitors see them.

Bobcats are widespread, but there are few sightings or records of them in the monument. **Mountain Lions** are rare here, but there are usually several observations of them each year. Quitobaquito Pond is one of the better places to find them, although they evidently roam throughout the monument. **Jaguars** have not been recorded from Organ Pipe, although it is possible that transient animals from Mexico might rarely range into the monument.

White-throated Woodrats are widely distributed in the monument, but are seldom seen. Their characteristic stick-and-cactus houses are seen in many places, such as near State Road 85 just south of the visitor center. The **Desert Woodrat** may occur near Growler Mine. **Porcupines** occur near the monument but, if within its borders, have not been reported.

Badgers are not abundant, although their diggings have been seen in Ajo Valley, Abra Valley, Sonoyta Valley, and north of Bates Well. **Ringtails** are probably not rare in the monument, but are seldom seen, except around the visitor center at night. **Raccoons** are rare and are known only from a few records near the Sonoyta River and its washes, and Quitobaquito and Dripping Springs. **Coatis** are probably not resident in Organ Pipe, and observations in the Sonoyta Basin and Ajo Mountains are believed to be of transient individuals.

CHECKLIST

___ Desert Cottontail	___ Gray Wolf
___ Black-tailed Jack Rabbit	___ Kit Fox
___ Antelope Jack Rabbit	___ Gray Fox
___ Harris' Antelope Squirrel	___ Ringtail
___ Rock Squirrel	___ Raccoon
___ Round-tailed Ground Squirrel	___ Coati
___ White-throated Woodrat	___ Badger
___ Desert Woodrat	___ Spotted Skunk
___ Porcupine (?)	___ Striped Skunk
___ Coyote	___ Hooded Skunk (?)
	___ Hog-nosed Skunk

(continued)

(continued)

____ Jaguar (?)
____ Mountain Lion
____ Bobcat
____ Collared Peccary

____ Mule Deer
____ White-tailed Deer
____ Pronghorn
____ Mountain Sheep

Petrified Forest National Park

Petrified Forest National Park, Arizona 96028

In sharp contrast to the humid and well-watered forest that existed 200 million years ago, present day Petrified Forest National Park is arid and has few trees (cottonwood, pinyon pine, and juniper). The park's 93,493 acres (37,836 hectares) are essentially high desert grassland at 5,350–6,235 feet (1,631–1,900 meters) elevation, which is dissected by wind- and water-eroded valleys. These eroded areas have exposed the fossilized remains—the petrified tree trunks—of the ancient forest. A 27-mile (43-kilometer) road bisects most of the park and gives access to many petrified wood sites and ruins of prehistoric Indian villages and petroglyphs. The northern Painted Desert sector is a roadless wilderness area available for overnight camping by backpackers with permits.

The mammals of Petrified Forest are of limited diversity, but visitors often see Pronghorns and White-tailed Antelope Squirrels, as well as Desert Cottontails and Black-tailed Jack Rabbits. There are also several colonies of Gunnison's Prairie Dogs, but they are not easily accessible to visitors. The park is open throughout the year.

Pronghorns are conspicuous residents of Petrified Forest's open grasslands and are usually seen by visitors. The middle of the park is a good place to find them, especially around the road to Blue Mesa. The only other hoofed animal here is the **Mule Deer**, which is not numerous and is rarely seen.

White-tailed Antelope Squirrels are abundant and are usually seen by visitors throughout the park in summer. **Spotted Ground Squirrels**, which prefer sandy soil for their homes, are much rarer, and the same is true of the **Rock Squirrel**, which, as its name implies, inhabits rocky places. There are no tree squirrels in the park.

Desert Cottontails are common throughout the park and are often seen, especially at dawn and dusk along the roads. The large **Black-tailed Jack**

PETRIFIED FOREST NATIONAL PARK

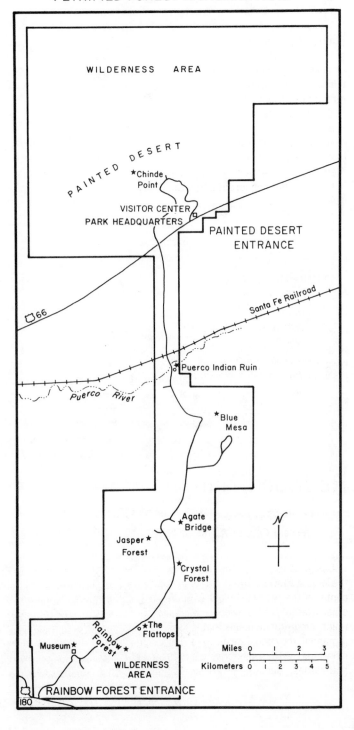

Rabbit is also abundant and frequently observed, although it is somewhat more nocturnal than the cottontail.

The park contains several colonies of the rare, white-tailed **Gunnison's Prairie Dog**. Because of the precarious status of this species, the locations of the prairie dog towns are not publicized and they are rather inaccessible to visitors. **Porcupines** occur in Petrified Forest, but are seldom seen. **White-throated Woodrats** are common, but because of their nocturnal habits are not usually observed.

Coyotes are fairly common here and are sometimes seen by visitors or heard howling at night. Much less common is the **Gray Fox**, which is rarely seen; the same is true of the **Kit Fox** and **Bobcat**. **Striped Skunks** are not rare, especially in the northern part of Petrified Forest, but because of their nocturnal activities, are not often seen. **Spotted Skunks** also occur in the park, but are considered rare. **Badgers**, likewise, are rare and seldom seen.

CHECKLIST

___ *Desert Cottontail*	___ *Coyote*
___ *Black-tailed Jack Rabbit*	___ *Kit Fox*
___ *White-tailed Antelope Squirrel*	___ *Gray Fox*
___ *Spotted Ground Squirrel*	___ *Badger*
___ *Rock Squirrel*	___ *Spotted Skunk*
___ *Gunnison's Prairie Dog*	___ *Striped Skunk*
___ *White-throated Woodrat*	___ *Bobcat*
___ *Porcupine*	___ *Mule Deer*
	___ *Pronghorn*

Redwood National Park

1111 Second Street
Crescent City, California 95531

Designed to protect the huge redwood trees and the adjacent coastline, the long, narrow Redwood National Park (106,000 acres; 32,309 hectares) contains the world's tallest (367.8 feet; 112.1 meters) tree and many other examples of redwoods taller than 300 feet (91 meters), as well as much clear-cut acreage that shows too well the fate from which these remanent redwood forests were saved. The park is a melange of administrations, with some private holdings and three state parks within its boundaries. Because of an annual rainfall of 80–100 inches (2,032–

REDWOOD NATIONAL PARK

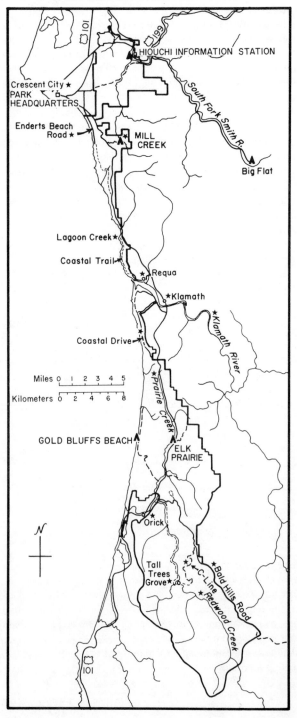

2,540 millimeters), summers are foggy and cool, and winters are rainy with high winds. Spring and fall are the clearer seasons, cool and pleasant.

With its seacoast, Redwood Park offers both marine and terrestrial mammal viewing. It is one of the better parks in which to see Elk and Brush Rabbits, and there are fair opportunities to see Gray Foxes, Bobcats, and Porcupines.

Elk are common in the central part of the park and can almost always be seen along Prairie Creek, just to the south of Elk Prairie Campground along U.S. Highway 101, as well as near Redwood Creek in the southern part of the park. The black-tailed subspecies of the **Mule Deer**, which is common in the park, is often seen, especially near Mill Creek Campground, near the C-line Road, and on the Enderts Beach Road.

Although both **Douglas' Squirrels** and **California Gray Squirrels** occur in the park, the former is not at all common and is seldom seen. There is a better chance of seeing the latter, although it is also not abundant; the most likely place in which to see it seems to be on the bluff road near Enderts Beach. In this same area, as well as other drier sites along the coast, the California Ground Squirrel is a common species. **Townsend's Chipmunk** is abundant and often seen in the wooded areas of the park, especially near campgrounds and picnic areas. If the **Golden-mantled Ground Squirrel** actually occurs in Redwood Park, its presence is undocumented; observations of this species should be reported to park rangers.

Bobcats are common in Redwood Park and are often seen early in the morning on the Coastal Drive, or on the Enderts Beach Road. **Mountain Lions** are resident in the park, though rare, and recent sightings have been considered valid evidence of their presence. **Gray Foxes** are among the commoner medium-sized carnivores and can be encountered almost anywhere in the park, but are most often seen at night along the roads. **Coyotes**, too, are far from rare here and may be met with almost anywhere in the park.

Black Bears are common, and there is a fair chance of seeing one almost anywhere from the beaches to the forests. When the berries are ripe along the edge of the beach, bears come to feed and are commonly seen. The **Raccoon** is another abundant species that is found throughout the park. As usual, Raccoons are often found around campgrounds, such as at Prairie Creek, as well as at Redwood Creek and Mill Creek.

The small **Brush Rabbit** is abundant and usually can be seen early in the morning or at dusk along the side of the Bald Hills Road, in the Redwood Creek area. There are no other rabbits or hares known from the park. **Porcupines** abound in this park and are often seen. The best areas in which to find them are along the Bald Hills Road and the Geneva Road. Another rodent is the **Mountain Beaver**, but it is rarely seen. Colonies are known to be in the moist ground just to the south of Vista Point in the Crescent City area, as well as near Enderts Beach and Nichols Creek. **Dusky-footed Woodrats** are fairly common in cutover areas, mainly inland, but are seldom seen. Although the **Bushy-tailed Woodrat** could occur here, there seem to be no valid records of it.

Beavers are known to inhabit Marshall Pond, Redwood Creek, and Lagoon Creek, but they are not common and are seldom seen. Another aquatic rodent, the Muskrat, is uncommon, although some Muskrats are known to live around the Crecent Beach lagoon. There is some question whether the Muskrats of Redwood are from native or (re)introduced stock, and Beavers might also have been reintroduced. Mink do live along watercourses, but are uncommon and not likely to be seen. The Redwood Creek area seems the most likely place to find them. River Otters are frequently seen in the same area, as well as along Lagoon Creek.

Both Spotted and Striped Skunks live in the park and are considered common; however, neither is often seen. The same is true of the Long-tailed Weasel, but the Ermine, Marten, and Ringtail are all considered most uncommon in this park and are rarely seen. Whether the Fisher is part of the park's fauna is not clear, and any observations of this species should be reported to park rangers.

The only marine mammal that is resident in the park's waters is the Harbor Seal. These seals may be sighted swimming almost anywhere along the coastline; Enderts Beach is one of the more promising locales. Other seals and sea lions also visit the Redwood coastline, and Northern Elephant Seals, Northern Sea Lions, and California Sea Lions are now regularly seen, although none yet has a resident population. The California Sea Lion is usually observed near the mouth of the Klamath River and from the Coastal Drive. The Sea Otter, exterminated from this coast many years ago, now is occasionally seen. These sightings are probably of populations expanding from the south. Northern Fur Seals have sometimes been observed in coves north of Requa Road.

Gray Whales are regular visitors of the coastal waters during their fall and spring migrations, and sometimes can be observed from the overlook above Enderts Beach. Pacific White-sided Dolphins, Harbor Porpoises, and Dall's Porpoises are other marine mammals that sometimes are seen along this coast. Sperm Whales, Hump-backed Whales, Fin Whales, Grampuses, Goose-beaked Whales and False Killer Whales are sometimes seen from the park's coast, but less frequently.

CHECKLIST

___ Brush Rabbit
___ Mountain Beaver
___ Townsend's Chipmunk
___ California Ground Squirrel
___ Golden-mantled Ground
    Squirrel (?)
___ Western Gray Squirrel
___ Douglas' Squirrel

___ Beaver
___ Dusky-footed Woodrat
___ Bushy-tailed Woodrat (?)
___ Muskrat (RI?)
___ Porcupine
___ Goose-beaked Whale
___ Sperm Whale
___ Pacific White-sided Dolphin

(continued)

(continued)

___ Grampus
___ False Killer Whale
___ Harbor Porpoise
___ Dall's Porpoise
___ Fin Whale
___ Hump-backed Whale
___ Coyote
___ Gray Fox
___ Black Bear
___ Northern Fur Seal
___ Northern Sea Lion
___ California Sea Lion
___ Ringtail
___ Raccoon
___ Marten

___ Fisher (?)
___ Ermine
___ Long-tailed Weasel
___ Mink
___ Spotted Skunk
___ Striped Skunk
___ River Otter
___ Sea Otter
___ Harbor Seal
___ Northern Elephant Seal
___ Mountain Lion
___ Bobcat
___ Elk
___ Mule Deer

Rocky Mountain National Park

Estes Park, Colorado 80517

The Rocky Mountain National Park, which is astride the Continental Divide in central Colorado, is composed of 265,679 acres (107,518 hectares). The park has a range of habitats, which extend from the relatively low-elevation foothill brushlands, through ponderosa pine and Douglas fir forests, lodgepole pine, Englemann spruce, subalpine fir, and limber pine, to alpine meadows above 11,300 feet (3,444 meters). More than 107 of the park's peaks exceed 11,000 feet (3,353 meters) in altitude, the highest being Long's Peak at 14,256 feet (4,345 meters). Access to the heights is available via Trail Ridge Road (U.S. Highway 34), which crosses the park and reaches 12,183 feet (3,713 meters) above sea level.

With so many habitats, Rocky Mountain has a varied mammal fauna, mainly of animals characteristic of coniferous forest zones. There are many species of the squirrel family, and Mountain Sheep, Elk, and Mule Deer, as well as Black Bears and many smaller carnivores. The park is open throughout the year, although Trail Ridge and other high roads are closed from October until June. Many of the rodents hibernate in winter. There are more than 300 miles (483 kilometers) of hiking trails in Rocky Mountain, and, in winter, Hidden Valley is accessible for downhill skiing.

ROCKY MOUNTAIN NATIONAL PARK

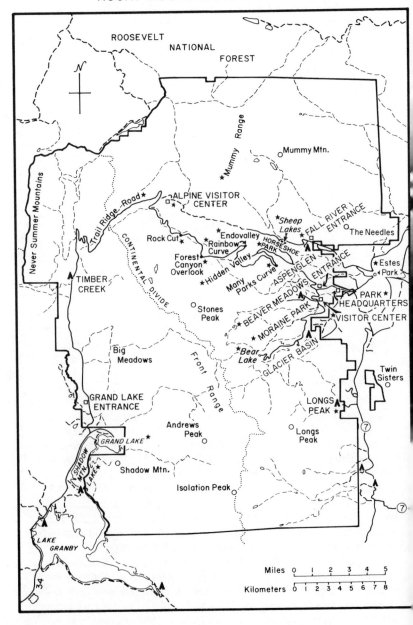

Two species of chipmunk, largely indistinguishable in the field, are abundant in the park. The **Colorado Chipmunk** is not definitely recorded from within the park's boundaries, preferring rocky country generally below 7,500 feet (2,286 meters), but may eventually be found there. Above 8,000 feet (2,438 meters) to timber line, both the **Uinta** and **Least Chipmunks** occur, and it is not clear what habitat preferences, if any, distinguish them. As its name implies, the Least Chipmunk is slightly smaller than the Uinta species, and the latter sometimes has slightly duller color and less distinct stripes on its back. Either or both species are commonly seen at automobile turnouts on Trail Ridge Road, such as at Rainbow Circle.

Sometimes confused with chipmunks, but larger and without a conspicuous stripe through the eye, is the **Golden-mantled Ground Squirrel**. It ranges throughout the park from the lowest elevations to above timberline, and, like Chipmunks, is often seen around campsites and picnic sites, as well as viewing places along the roads. **Richardson's Ground Squirrel** is locally abundant in areas that are open and have well-drained soils. Although they are known to occur at 6,000–12,000 feet (1,829–3,658 meters), they are mainly found in low-elevation valleys and meadows, such as around the Lawn Trailhead, Horseshoe Meadow, and the Moraine Park Visitor Center. The **Rock Squirrel** lives below 7,500 feet (2,286 meters) and therefore may not actually occur within the park's borders.

The only other ground squirrel in the park is the **Yellow-bellied Marmot**, a large species that prefers rocky areas near lush herbage and that ranges from the foothills to above timberline. It is most abundant in subalpine rockpiles and meadows, where it usually can be seen foraging early and late in the day and sunbathing at midday. The turnout at Rock Cut on Trail Ridge is a good place to see these marmots.

In the forests of middle elevations in the park, **Red Squirrels** are common, and they are often heard chattering by hikers in the woods. They also make conspicuous caches of conifer cones (called middens), some up to 30 feet (9 meters) in diameter. In ponderosa pine woodlands on the eastern side of the park, up to 8,500 feet (2,591 meters), another species of squirrel lives. It is the **Abert's Squirrel**, nearly twice as long and heavy as the Red Squirrel. Its large size and tufted ears generally distinguish it from the Red Squirrel, but it is seldom seen in the park.

Beavers are abundant in Rocky Mountain, and, on most low-elevation streams of moderate grade, there are numerous dams and lodges. Although the animals themselves are mainly nocturnal and not often seen, this park is an excellent place to examine Beaver dams and lodges and to see the effect that Beavers have on the environment. Among the better sites for observing Beaver works are Trail Ridge Road near Lower Hidden Valley, Bear Lake Road, Endovalley Picnic area, and the Kawuneeche Valley on the western side of the park. **Muskrats** also inhabit most of the low-elevation Beaver ponds, but are also nocturnal and therefore not usually seen. The other large rodent in the park is the **Porcupine**, which lives in open pine woodlands at middle elevations and in willow thickets, and which may be encountered, although not frequently, anywhere in Rocky Mountain Park.

Mule Deer are abundant in the park in most places except dense forest or open parkland. In summer they are mainly in high meadows near treeline, and in winter they live at lower elevations and usually can be found on moraines at Mill Creek, Moraine Park, and Beaver Meadows. **Elk**, too, are numerous, and they spend the summer days in high-elevation meadows and at the forest edge. In fall they come to the low meadows to mate, and they winter in Horseshoe Park, Moraine Park, and Hallowell Park. **Mountain Sheep** are another attraction here, and they range from alpine tundra to the lower meadows. They inhabit the Never Summer Range and Mount Craig and can sometimes be seen from the Alpine Visitor Center, as well as at Sheep Lakes at the Fall River entrance road above Horseshoe Park. **Bison** and **Pronghorn**, both of which probably inhabited the park, were wiped out about a century ago. A few remnant **Moose** may occasionally have been in the park, but in 1979 Moose were reintroduced into the Arapaho National Forest, and some of these animals have recently been sighted in the western part of the park.

In the talus slopes, from above timberline down to about 8,500 feet (2,591 meters) patient observers should be able to hear and see the **Pikas**, which are abundant in these areas. Some usually can be seen at Rock Cut on the Trail Ridge Road. The only cottontail present in the park is **Nuttall's Cottontail**, which may be found from the low elevations to timberline, but it is not especially abundant. The **Snowshoe Hare**, one of two hares in the park, prefers dense forest with a shrubby undergrowth, as well as shrub lands in burned-over areas, mainly from 8,000–11,000 feet (2,438–3,353 meters). **White-tailed Jack Rabbits** prefer open country in summer and shrubby areas in winter, and they are known from the lowest to the highest altitudes, but are probably more abundant in the lower zones.

Coyotes inhabit most areas of the park and sometimes can be heard howling at night. They seem less abundant in the dense, forested areas and are sometimes seen during the day hunting in open meadows. **Gray Wolves** are believed to be extinct in Rocky Mountain, but there have been some reports indicating that they are still present. There are some fifty **Black Bears** in the park, mainly in the middle, forested elevations, but they are seldom seen. **Grizzly Bears** are extinct here.

Although **Raccoons** have been seen at elevations as high as 11,000 feet (3,353 meters), they live mainly at the lower elevations, both east and west, where locally they are abundant and readily seen. Whether the **Ringtail**, which appears near the park's eastern borders, actually occurs within its boundaries is not known. **Gray Foxes**, too, have been seen near the park's eastern side, but not actually in it. **Red Foxes** are locally present at the forest edge and clearings near water, but are seldom seen.

There are many species of the weasel family in Rocky Mountain, but few are often seen. **Martens** inhabit mainly subalpine fir areas and are sometimes observed hunting in meadows during the summer days. **Long-tailed Weasels, Ermine**, and **Mink** are all known to occur in the park, but are rarely seen. **Striped Skunks**, which live mainly at lower elevations, are not numerous and are not usually seen. **Spotted Skunks** occur near the eastern border of the park, but have never been reported in it. **Badgers** inhabit open lands, even above timberline, but are rarely observed. **Wolverines** are very rare, and there have been no recent sightings of them. **River Otters** are not surely known to be native to the park, but some were introduced in the Kawuneeche Valley on the west side of the park and are now established there.

Bobcats probably occur in shrubby areas throughout the park, but are seldom seen because they are so adept at avoiding human beings. **Lynx** may also occur in the park, but they are rare and their present status is unknown. **Mountain Lions** are resident in Rocky Mountain, and there are usually several observations of this large cat annually, mainly from the lower elevations on both eastern and western sides.

Two species of woodrat inhabit the park. The commoner is probably the **Bushy-tailed Woodrat** which ranges from timberline down to the lowest elevations. The **Mexican Woodrat** lives at lower elevations—up to 8,000 feet (2,438 meters)—and usually makes its nest where there are horizontally overhanging rocks, whereas the Bushy-tailed species prefers vertical clefts. Neither woodrat is often seen.

CHECKLIST

___ Pika
___ Nuttall's Cottontail
___ Snowshoe Hare
___ White-tailed Jack Rabbit
___ Least Chipmunk
___ Uinta Chipmunk
___ Colorado Chipmunk (?)
___ Yellow-bellied Marmot
___ Richardson's Ground Squirrel
___ Rock Squirrel (?)
___ Golden-mantled Ground Squirrel
___ Abert's Squirrel
___ Red Squirrel
___ Beaver
___ Mexican Woodrat
___ Bushy-tailed Woodrat
___ Muskrat
___ Porcupine
___ Coyote
___ Gray Wolf (E?)
___ Red Fox

___ Gray Fox (?)
___ Black Bear
___ Grizzly Bear (E)
___ Ringtail (?)
___ Raccoon
___ Marten
___ Long-tailed Weasel
___ Ermine
___ Mink
___ Wolverine (?)
___ Badger
___ Spotted Skunk (?)
___ Striped Skunk
___ River Otter (I)
___ Mountain Lion
___ Lynx (?)
___ Bobcat
___ Elk
___ Mule Deer
___ Moose (RI)
___ Pronghorn (E)
___ Mountain Sheep
___ Bison (E)

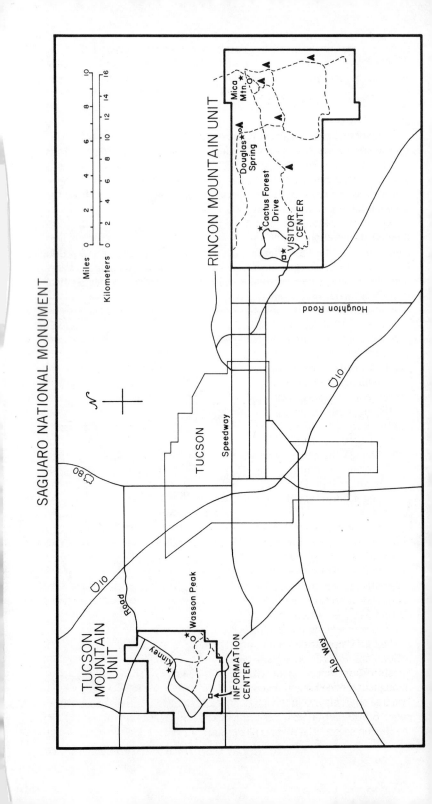

SAGUARO NATIONAL MONUMENT

Saguaro National Monument

Route 8, P.O. Box 695
Tucson, Arizona 85730

The two sections of Saguaro National Monument lie about 17 miles (27 kilometers) to the east and to the west of Tucson. The western sector, called the Tucson Mountain Unit, is about one-third the size of the eastern sector, Rincon Mountain Unit, and the total size of the monument is 83,576 acres (33,822 hectares). Both parts of the monument feature saguaro cacti. Characteristic representatives of this part of the Sonoran Desert, these giants grow in hills that are up to about 4,500 feet (1,372 meters) in elevation. In the higher parts of the Rincon Mountain Unit, which reaches its highest point on Mica Mountain (8,666 feet; 2,641 meters), there are Douglas and white firs and, below them, ponderosa pine and oak which grade into the desert scrub zone at elevations of about 4,000 feet (about 1,200 meters).

Relatively little of the large Rincon Mountain Unit is accessible by automobile, but the Cactus Forest Drive (8 miles; 13 kilometers) usually provides some glimpses of mammals, even at midday. There are more roads in the less diverse Tucson Mountain Unit. Because of the varied habitats in the monument, mammals characteristic of both cooler and eastern habitats (Black Bears, White-tailed Deer, and Eastern Cottontails) are present, as well as desert and southern species (Antelope Jack Rabbits, Coatis, and Collared Peccaries). This is one of the few park areas in which four species of skunk occur and through which Jaguars and Jaguarundis have, at times, wandered. The monument is open throughout the year, and although the cooler winter and spring are the main seasons, many visitors come in summer. The peak of the wildflower bloom is in April. Because of the high daytime temperatures, the best times to see mammals are at night, dusk, and dawn, but only the Tucson Mountain Unit is open at night.

Ground Squirrels endure heat well and are active during summer days and are often seen. The **Harris' Antelope Squirrel**, characterized by its white side stripe, is abundant in the desert and foothills and can usually be seen near the visitor center and the Cactus Forest Drive, which are on the western side of the Rincon Mountain Unit. Equally common, but limited to the desert sections below 3,200 feet (975 meters) is the **Round-tailed Ground Squirrel**, which does not have a side stripe. It, too, can be seen in the daytime along the Cactus Forest Drive.

The **Cliff Chipmunk** is common in the foothills (above 4,500 feet; 1,372 meters) and is a customary visitor at campgrounds in the higher-elevation forests of the Rincon Mountain Unit. The large **Rock Squirrel** is found throughout the monument, mainly in rocky habitats, and is commonly seen. Two tree squirrels live in the monument, both in the higher, forested areas above 6,500 feet (1,981 meters), which are in the eastern mountains of the Rincon sector. The **Arizona Gray Squirrel**, a large species, exists in low numbers in the mountain and is seldom seen. In the same area are **Abert's Squirrels**, which seem equally rare and which may represent a population that was introduced in the Rincon Mountains from elsewhere.

The **Black-tailed Jack Rabbit** is the more frequently seen hare in the monument and is commonly observed in the foothills areas at night and early in the morning. The **Antelope Jack Rabbit** inhabits the lower (below 2,900 feet; 884 meters) desert realms and is sometimes seen on Cactus Forest Drive. This large, pale hare has pale sides and lacks the black-tipped ears of the Black-tailed Jack Rabbit. The commoner of the two cottontails in Saguaro is the **Desert Cottontail**, readily seen at night and early in the morning in the desert and foothill areas. High (above 6,500 feet; 1,981 meters) in the eastern Rincons are **Eastern Cottontails**, but they are not often seen and are probably few in number.

Mule Deer are common in Saguaro. They are often seen, especially at dawn and dusk, throughout the desert and foothills habitats. Much rarer is the secretive **White-tailed Deer**, which is found only in the oak and pine woodlands of the mountains of the Rincon Unit. Their numbers are low and they are seldom seen. Another hoofed animal in Saguaro is the **Collared Peccary**. It is common in the desert and foothills habitats and, during the winter, usually can be seen along the Cactus Forest Drive and even around the visitor center in the Rincon Mountain Unit. Peccaries are less visible in summer, because they usually spend their days in shaded gullies and beneath vegetation. **Mountain Sheep** once lived in the mountains here, but they have been extinct in the monument since the 1940s.

Coyotes roam throughout Saguaro and are common. They are often seen by visitors anywhere in the monument, and Cactus Forest Drive is where many visitors encounter them. The **Gray Fox** is also relatively common, ranging from the desert to the mountains, and may be encountered anywhere. Hikers see them fairly often. The **Kit Fox** lives in the monument, but is rare and inhabitants of the desert zone of the Tucson Mountain Unit. **Gray Wolves** probably once inhabited this area, but are no longer present.

In the higher portions of the Rincon Mountain Unit there are **Black Bears**. They are rarely encountered by campers and are not numerous. **Grizzly Bears**, which once roamed through the Rincon Mountains, have been extinct in Arizona for more than a half-century. Four species of skunk are known to live in Saguaro. Although none is particularly abundant, the most frequently seen is the **Striped Skunk**, which may be found anywhere at night. **Spotted Skunks** also live in any of the monument's habitats, but prefer gullies and rocky areas. They are less often seen than Striped Skunks. From the Rincon Mountains down to the upper foothills is also the range of the **Hog-nosed Skunk**, a large, white-backed species with no white on its nose. It is seldom seen. In the desert is the rarest of the four Saguaro skunks, the **Hooded Skunk**. In common with the Striped Skunk, this species has a narrow white strip down the middle of its forehead, but its back is either gray mantled (mixed black and white hairs) or black, with a thin white stripe on each side. The Spotted Skunk is the smallest of the four species and has many broken stripes and spots on the body.

The **Coati**, which is a relative of the **Raccoon**, is only occasionally seen, but occurs in both the eastern and western units. **Ringtails** are fairly common in the desert and foothills of the monument, but are seldom seen. Raccoons, however, are not particularly abundant, but may be encountered anywhere. **Badgers** are not rare in the desert and foothills, but are not usually seen.

Two kinds of woodrat live in Saguaro. Below 6,000 feet (1,829 meters) in the foothills and desert zones, the **White-throated Woodrat** is fairly common. Although it is seldom seen, its houses are often visible under cacti or spiny shrubs, and there are some woodrat houses near the visitor center, where the animals are occasionally seen at night. High in the mountains of the Rincon sector is the **Mexican Woodrat's** range, but it is also rarely seen. **Porcupines** once occurred throughout the monument, from desert to mountains, but are now extremely rare.

The most abundant member of the cat family is the **Bobcat**, which ranges through all the habitats, but, as elsewhere, is rarely seen. **Mountain Lions** also range throughout the monument and, although not common, are occasionally seen, even along as popular a route as the Cactus Forest Drive. In the past, **Jaguars** have been seen in Saguaro. These animals undoubtedly were transients that had wandered in from Mexico, and as those populations diminish and the human population of the area increases, the likelihood of Jaguar sightings in the monument becomes most unlikely. Similarly, the small, long-bodied, unspotted cat, the **Jaguarundi**, was recorded in the monument many years ago, probably representing a transient from Mexico, and no longer is believed to be present in the area.

CHECKLIST

___ Eastern Cottontail	___ Grizzly Bear (E)
___ Desert Cottontail	___ Black Bear
___ Black-tailed Jack Rabbit	___ Ringtail
___ Antelope Jack Rabbit	___ Raccoon
___ Cliff Chipmunk	___ Coati
___ Harris' Antelope Squirrel	___ Badger
___ Rock Squirrel	___ Spotted Skunk
___ Round-tailed Ground Squirrel	___ Striped Skunk
___ Abert's Squirrel (I)	___ Hooded Skunk
___ Arizona Gray Squirrel	___ Hog-nosed Skunk
___ White-throated Woodrat	___ Jaguar (E)
___ Mexican Woodrat	___ Mountain Lion
___ Porcupine	___ Jaguarundi (E)
___ Coyote	___ Bobcat
___ Gray Wolf (E)	___ Collared Peccary
___ Kit Fox	___ Mule Deer
___ Gray Fox	___ White-tailed Deer
	___ Mountain Sheep (E)

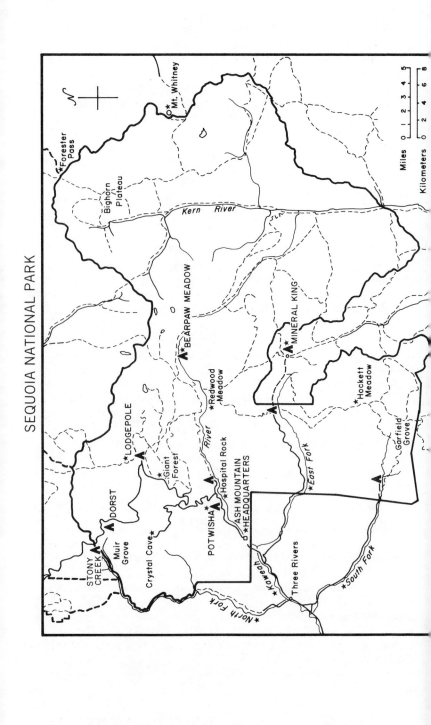

SEQUOIA NATIONAL PARK

Sequoia National Park

Three Rivers, California 93271

Featuring the General Sherman Tree, the world's largest living thing, and Mount Whitney, the highest peak in the contiguous United States, Sequoia National Park's 386,823 acres (156,543 hectares) include the glacier-scarred Sierra Nevada, roaring streams and waterfalls, vast forests (including groves of huge sequoias), and low-elevation chaparral-oak habitats. Only the western part of the park is accessible by automobile, but roads will take the visitor to the featured attraction, the giant sequoia trees. The General Sherman Tree is 275 feet (84 meters) tall, 103 feet (31 meters) in circumference, 2,500–3,000 years old. It has a trunk volume and weight of 52,500 cubic feet (1,487 cubic meters) and 1,385 tons (1,256 metric tons). There are taller, wider, and older trees, but none more voluminous.

Although the many environments of this park provide habitats for a great diversity of mammals, most of the mammals are nocturnal and secretive and therefore are not usually seen. Sequoia is open throughout the year, and many of the mammals signify their presence by their tracks in snow. Most of the mammals are characteristic of the Sierra Nevada, and none is unique to this park. The northern border of Sequoia is contiguous with Kings Canyon National Park, which is even less developed than Sequoia, and together the two comprise a marvelous Sierra Nevada wilderness.

Diurnal members of the squirrel family are among the most frequently seen mammals in Sequoia. **California Ground Squirrels** are common at lower elevations around Ash Mountain Headquarters, Potwisha, and Hospital Rock, and also in the Giant Forest area near Sunset and Beetle Rocks, and Lodgepole. **Western Gray Squirrels** are found in the same localities, mostly in black oak and yellow pine, to elevations of about 7,700 feet (2,347 meters) at Paradise, Three Rivers, and Beechey Flat. The populations of this species fluctuate, however, and at times the animals may be quite scarce. More common, usually, is **Douglas' Squirrel**, which ranges in the conifers up to elevations of 10,000 feet (3,048 meters) and usually can be seen foraging on the ground at Giant Forest.

Four species of chipmunk inhabit the park, one or more being found from elevations as low as 1,500 feet (457 meters) to those as high as 12,500 feet (3,810 meters). **Merriam's Chipmunk** occupies the lowest zones, in chaparral, and is not especially common. In elevations from 5,000 feet (1,524 meters) to about 11,000 feet (3,353 meters), the **Lodgepole Chipmunk** is abundant and often seen, whereas the common species at still higher elevations is the **Alpine Chipmunk**. Also along the crest of the Sierra, especially on the eastern side, is the domain of the **Colorado Chipmunk**, which lives in the limber and white-bark pine zones. It is virtually impossible to distinguish these species from one another in the field although they have distinguishing differences in dimensions, color, and voice.

Also commonly seen, but generally less abundant than the chipmunks, are **Golden-mantled Ground Squirrels**. They range from the middle elevations up to timberline and usually can be seen foraging at Giant Forest and around campsites.

The largest member of the squirrel family in Sequoia is the **Yellow-bellied Marmot**. It frequents rock outcrops from timberline down to about 7,000 feet (2,134 meters) and usually can be observed at Hockett Meadow, Alta Peak, or Hamilton Lakes. Another rock-pile denizen is the **Pika**, which is common in the talus slopes up to 13,000 feet (3,962 meters) and usually can be observed by the patient watcher.

There are few kinds of hoofed animals in the park, and only the **Mule Deer** is common, especially at middle elevations in forests above 5,000 feet (1,524 meters). Some can usually be seen at Giant Forest and Crescent Meadows early or late in the day. **Mountain Sheep** also occur in the park, but are rarely seen. They live in the high country and are known from Mount Langley, Baxter Pass, Mount Gould, and Bighorn Plateau. East of the Sierra crest, they move in and out of the park and are very seldom encountered.

Coyotes are common in Sequoia at elevations from 1,700 feet (518 meters) to 14,000 feet (4,267 meters) and are often heard at night and sometimes seen, even in Giant Forest meadow. **Gray Wolves** are extinct in the park, but **Gray Foxes** are common at the low elevations, below 5,000 feet (1,524 meters), especially around Ash Mountain Headquarters, Potwisha, and Hospital Rock. **Red Foxes** are rare in Sequoia and are reported only from high elevations, such as at Wallace Lake, at 11,200 feet (3,414 meters).

Both **Spotted** and **Striped Skunks** live in the park at lower elevations. They usually can be seen at night around Ash Mountain Headquarters and Potwisha and at campgrounds in these areas. The smaller Spotted Skunk ranges at elevations as high as 4,500 feet (1,372 meters), whereas the Striped Skunk is common mainly lower than 3,000 feet (914 meters). The **Virginia Opossum** is also fairly common in the lower, southwestern part of the park, but is not seen regularly. These opossums are not native to California, but are descendants of animals introduced from the eastern states early in the twentieth century.

The largest carnivore in Sequoia is the **Black Bear**, which is not uncommon and is seen in forested areas; Giant Forest meadow is a likely place to find these bears. The **Mountain Lion** is resident in Sequoia, mainly in the western ponderosa pine areas, but is not numerous and sightings of this large cat are few. Much more abundant, but also seldom seen, is the **Bobcat**, which is known to occur as low as 1,600 feet (488 meters) on the Kaweah River, up to 11,000 feet (3,353 meters) at Cottonwood Lakes.

Three species of woodrat live in the park. At middle elevations, the **Bushy-tailed Woodrat** is common, and, at low elevations, the **Dusky-footed Woodrat** is abundant. Much rarer is the **Desert Woodrat**, which has been recorded from Giant Forest only. None of these nocturnal rodents is regularly seen, although their nests often are. **Porcupines** range throughout the park, down to about 3,000 feet (914 meters), but are seldom seen and are regarded as uncommon. Another rarity is the **Mountain Beaver**, which inhabits moist meadows in the middle elevations at Giant Forest, Clover Creek, and Wolverton Creek, but is hardly ever seen. The true **Beavers**, which live in some of the park's rivers, are not native to the area, but are descendants of transported stock.

Two species of cottontail, the **Desert Cottontail** and the **Brush Rabbit**, live in Sequoia, but neither is often seen. They are restricted to lower elevations on the western side, where they are considered rare. Also living at low elevations, mainly in valleys with good grass, are **Black-tailed Jack Rabbits**, but they are not numerous. **White-tailed Jack Rabbits** are found at higher elevations, which range from 13,000 feet (3,962 meters) down to about 8,000 feet (2,438 meters), and they have been seen at Mineral King, Hackett Meadow, and even at the General Sherman Tree.

A number of small carnivores inhabit Sequoia, but most of them are rare, or, if fairly abundant, are not often seen. **Raccoons**, however, may be abundant locally at low elevations below 6,000 feet (1,829 meters) and are sometimes seen around Ash Mountain Headquarters and even at Giant Forest. Ranging over the same area and those a bit higher is the Raccoon's relative, the **Ringtail**. Although common around rocks, caves, and even old buildings, Ringtails are wholly nocturnal and are seldom seen. Among the other seldom seen members of the weasel family in the park are **Ermine** (high country), **Long-tailed Weasels** (middle elevations), **Fisher** (middle elevations), and **Marten** (high country in summer, middle elevations in winter). Both **Wolverines** and **Badgers** have been recorded from low to high altitudes in the park, but they are considered rare, and seeing either of these members of the weasel family is a matter of luck.

CHECKLIST

___ Virginia Opossum (I)
___ Pika
___ Brush Rabbit
___ Desert Cottontail
___ White-tailed Jack Rabbit
___ Black-tailed Jack Rabbit
___ Mountain Beaver
___ Alpine Chipmunk
___ Merriam's Chipmunk
___ Colorado Chipmunk
___ Lodgepole Chipmunk
___ Yellow-bellied Marmot
___ Belding's Ground Squirrel

___ California Ground Squirrel
___ Golden-mantled Ground Squirrel
___ Western Gray Squirrel
___ Douglas' Squirrel
___ Beaver (I)
___ Desert Woodrat
___ Dusky-footed Woodrat
___ Bushy-tailed Woodrat
___ Porcupine
___ Coyote
___ Gray Wolf (E)
___ Red Fox

(continued)

(continued)

___ Gray Fox   ___ Wolverine

___ Black Bear   ___ Badger

___ Grizzly Bear (E)   ___ Spotted Skunk

___ Ringtail   ___ Striped Skunk

___ Raccoon   ___ Mountain Lion

___ Marten   ___ Bobcat

___ Fisher   ___ Mule Deer

___ Ermine   ___ Mountain Sheep

___ Long-tailed Weasel

Shenandoah National Park

Route 4, P.O. Box 292
Luray, Virginia 22835

Long, narrow Shenandoah National Park, some 75 miles (121 kilometers) in length and about 4 miles (6 kilometers) in average width, is comprised of 194,078 acres (78,541 hectares), which encompass elevations of 600–4,049 feet (183–1,234 meters) in the northern Blue Ridge Mountains. The park is bisected by 105-mile (169-kilometer) Skyline Drive along the crest of the mountains, which gives access to more than 500 miles (805 kilometers) of hiking trails, including the Appalachian Trail.

The wildlife and vegetation of Shenandoah are fine examples of nature's resiliency following overexploitation by human beings. At the beginning of the twentieth century, the future park's area was biologically depauperate. Forests had been destroyed, wildlife eradicated, soil depleted and eroded. Today, deer and bears are numerous, as are wild turkeys, and perhaps the Mountain Lion once more roams the lush, regrowing forests. Within a day's drive of major eastern population centers, Shenandoah attacts visitors throughout the year, especially when spring flowers or autumn foliage are prominent.

The **White-tailed Deer** is one of the species that had been wiped out of the park's area before its establishment in 1935. From several reintroduced populations, starting in 1934, deer have thrived; they number in the thousands. Visitors can readily see deer at night along the Skyline Drive, and at dawn and dusk there, as well as at such places as Big Meadows and Loft Mountain, Elk Wallow, and Piney River. Another species with a similar history is the **Black Bear**. Thought not to be present in the park in 1935, bears reinvaded the area from surrounding country by 1937, when two

SHENANDOAH NATIONAL PARK

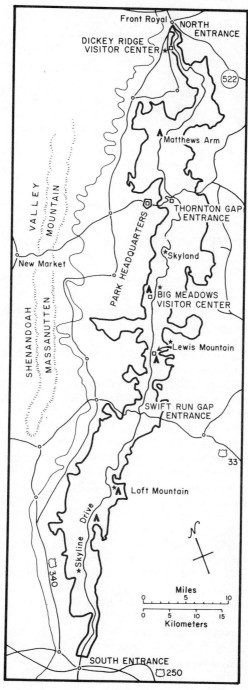

were in the park. By 1951, the bear population was up to thirty, and at present is at least one hundred, and perhaps as high as three hundred. Despite their relative abundance, bears are not often seen by visitors. They do sometimes raid campgrounds, and most observations have been around Loft Mountain, Elk Wallow, Matthews Arm, and Skyland.

Eastern Cottontails occur at all elevations in Shenandoah and are mainly seen along roadsides early and late in the day. Also thought to occur in the park is the **New England Cottontail**, a species that prefers woods at elevations above 3,000 feet (914 meters), but it is not distinguishable in the field from the Eastern Cottontail.

Gray Squirrels inhabit the park's hardwood forests at all elevations, but they are not commonly seen. Another hardwood-inhabiting squirrel is the **Fox Squirrel**, which is considered very rare in the park. The third species of squirrel here is the **Red Squirrel**, an inhabitant of coniferous tree areas. Red Squirrels are not common, but usually some live in the conifers near Byrd Visitor Center at Big Meadows. Red Squirrels usually advertise their presence with noisy chattering.

Eastern Chipmunks are more commonly seen than the tree squirrels in this park, mainly in woods and clearings above 2,000 feet (610 meters) elevation, and are often seen around buildings, camps, and picnic grounds. They may even be active on mild, snowless days in winter. **Woodchucks** occur at elevations up to 3,700 feet (1,128 meters) and usually can be seen in grassy areas along Skyline Drive and in such places as Big Meadows near the Rapidan Road. Woodchucks are generally in hibernation from the end of December to the end of February in Shenandoah.

Striped Skunks are common and are often seen along Skyline Drive at night. They can usually be seen around campgrounds and buildings, and in the Big Meadows area at night. The **Spotted Skunk** is less common in the park, but at times has been regularly observed at the Big Meadows Campground in the evening. **Raccoons** live mainly along streams at lower elevations, but may be found at night at any elevation. Their populations seem to fluctuate, and at times they are quite uncommon.

Gray Foxes and **Bobcats** are numerous in the park, living at all elevations. Despite their abundance, Bobcats are seldom seen, most observations having been on Skyline Drive at night, in the Lewis Mountain area, and north of Skyland. Gray Foxes are seen (more frequently than Bobcats) at night along Skyline Drive and at forest clearings such as Big Meadows, where some years one has wandered around the amphitheater before the nightly ranger talks. **Red Foxes** are rare in Shenandoah and, if present, are mainly at the lower elevations.

Along the streams at low elevations, **Beavers** and **Muskrats** may occur. Beavers were extirpated from the park area many years ago, and attempts to reintroduce them have not been markedly successful. There may be some Beavers on the streams in the northeastern sector of the park, and the best place to look for them, as well as for the equally uncommon Muskrat, is along the Thornton River, at the park boundary east of Thornton Gap. **Mink** are also seldom seen semiaquatic inhabitants of the lower stream sides; **River Otters** are thought no longer to be a part of the Park's fauna.

Virginia Opossums live only in the lower forested zones, below 2,000 feet (610 meters) in elevation. They are fairly common in wooded areas and around streamsides. The **Long-tailed Weasel** and perhaps also the **Least Weasel** are both rare in the park. The **Eastern Woodrat** is another uncommon species, which usually lives along cliffs at higher elevations, but sometimes makes its nests in buildings.

Many other species are known to have lived in the park area in the past but no longer occur here. The **Fisher, Elk, Gray Wolf**, and **Bison** all were eradicated by human activity. **Mountain Lions** also once occurred in Shenandoah, and although there have been some reports of them, valid evidence of their presence is still lacking.

CHECKLIST

___ *Virginia Opossum*

___ *Black Bear*

___ *Raccoon*

___ *Fisher (E)*

___ *Least Weasel (?)*

___ *Long-tailed Weasel*

___ *River Otter (?)*

___ *Spotted Skunk*

___ *Striped Skunk*

___ *Red Fox*

___ *Gray Fox*

___ *Gray Wolf (E)*

___ *Mountain Lion (E?)*

___ *Bobcat*

___ *Woodchuck*

___ *Eastern Chipmunk*

___ *Red Squirrel*

___ *Gray Squirrel*

___ *Fox Squirrel*

___ *Beaver (RI)*

___ *Eastern Woodrat*

___ *Muskrat*

___ *Eastern Cottontail*

___ *New England Cottontail*

___ *Elk (E)*

___ *White-tailed Deer (RI)*

___ *Bison (E)*

Theodore Roosevelt National Park

Medora, North Dakota 58645

Comprised of three separated units that total 69,528 acres (28,137 hectares), Theodore Roosevelt National Park features the badlands of the Little Missouri River, Theodore Roosevelt's ranch site, and a northern plains

THEODORE ROOSEVELT NATIONAL PARK

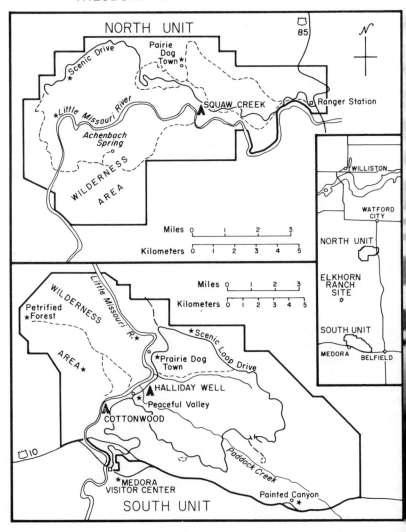

flora and fauna. The small Elkhorn Ranch site is undeveloped and relatively inaccessible, but the North Unit and South Unit, each astride the Little Missouri River, have paved roads, hiking trails, campgrounds, and picnic areas. The predominant vegetation is grassland, with sagebrush in the drier areas, willow and cottonwood along the river, and juniper on the higher slopes, Although the park is open throughout the year, winter visitation is reduced by extreme cold (to −30° and −40° F.; −34° to −40° C.) and occasional snow-blocked roads; the park is utilized mainly in summer, when the days are warm and the nights are cool.

In keeping with the park's role in historical education, a small non-breeding herd of Longhorn cattle is maintained in the North Unit and usually can be found between the visitor center and the Squaw Creek Campground. In the South Unit, there is a herd of feral horses, usually east of Painted Canyon. Bison, Black-tailed Prairie Dogs, Mule and White-tailed Deer, and reintroduced Mountain Sheep are among the more frequently and readily seen mammals in this park.

Both species of North American deer are present in the park and are often seen. **White-tailed Deer** live mainly in the bottom lands along the streams and rivers, where there are deciduous woods. **Mule Deer** are more commonly seen on the drier, higher sites and in broken country. **Elk**, which once roamed the grasslands, are extinct in the park, but **Pronghorns**, another open-country species, persist, although they are not common. On the high plateaus and in the Petrified Forest in the South Unit are places where they are sometimes seen.

Several hundred **Bison** are present in both the North and South units. Although the last native Bison in North Dakota was supposedly killed in 1884 near Dickinson, it is not clear whether the Roosevelt Park herd is wholly of reintroduced animals, or if it is derived from a few badlands survivors. Bison are now seen regularly in both park units. **Mountain Sheep** also once occurred here, but were killed off by 1905. In 1956, Mountain Sheep were reintroduced in the badlands and have thrived. They are sometimes seen on the cliffs opposite Peaceful Valley, and also near the Beef Corral and prairie dog town in the South Unit.

The most frequently seen rodent in the park is the **Black-tailed Prairie Dog**; there are several colonies in both the North and South units, some right at the sides of paved roads. It is hoped, although there is no evidence, that the rare **Black-footed Ferret**, a weasel whose life is tied to prairie dogs, may persist in this park, but it has not been seen since the 1960s. Prairie dog colonies also attract other predators, one of which, the **Coyote**, is common in the park. Although it is not often seen, is heard howling at night. **Gray Wolves** roamed the area when Bison were abundant, but were exterminated later by cattlemen. Another predator that frequents the colonies is the **Badger**, which has also been seen around the Beef Corral, but more often its presence is revealed by its extensive diggings.

Both **Beavers** and **Muskrats** live along the river, but they are not often seen. Also present in this area, but also scarce, are **River Otters** and **Mink**, neither of which is customarily observed by visitors. **Raccoons** are common in these river bottoms but are seldom seen, but **Porcupines**, which are found here and also in the juniper uplands, are relatively abundant and occasionally observed.

Least Chipmunks are abundant and are often seen in the park, especially in broken, rocky areas. The **Thirteen-lined Ground Squirrel** is a short-grass inhabitant that is abundant and is often seen. It is questionable whether **Franklin's Ground Squirrel** occurs here at all. Another questionable inhabitant of the park is the **Red Squirrel**, and any sighting of this species should be reported to park rangers.

Cottontail are scarce, even though three species might be present. The **Eastern Cottontail** is the most likely resident of the park; the **Desert Cottontail** as well as **Nuttall's Cottontail**, may also occur. Also present but common is the large **White-tailed Jack Rabbit**, which inhabits the open country and can be seen along roadsides in the evening. Among the rocks, the **Bushy-tailed Woodrat** dwells, but the usual evidence of its presence is the pile of debris that it accumulates for its nest.

Two species of fox live in the park, but neither is seen often. The **Red Fox** is the larger of the two and is characterized by its white-tipped tail. The diminutive **Swift Fox** is a rare species and can be distinguished by its black-tipped tail. **Bobcats** seem more abundant, although rarely seen, except for some sightings of them around the campgrounds. Tracks in the snow indicate that the **Lynx** may also inhabit this park, but it, like its large relative the **Mountain Lion**, must be considered a very rare species here.

There are frequent nighttime observations of **Striped Skunk** in the park, and these animals are abundant. **Least Weasels** and **Long-tailed Weasels** also live here, but are not often seen. **Black Bears**, as well as **Grizzly Bears**, once inhabited this region, but have been extirpated.

CHECKLIST

___ Eastern Cottontail (?)	___ Grizzly Bear (E)
___ Nuttall's Cottontail (?)	___ Raccoon
___ Desert Cottontail (?)	___ Least Weasel
___ White-tailed Jack Rabbit	___ Long-tailed Weasel
___ Least Chipmunk	___ Black-footed Ferret (E?)
___ Thirteen-lined Ground Squirrel	___ Mink
___ Franklin's Ground Squirrel (?)	___ Badger
___ Black-tailed Prairie Dog	___ Striped Skunk
___ Red Squirrel (?)	___ River Otter
___ Beaver	___ Mountain Lion
___ Bushy-tailed Woodrat	___ Lynx (E?)
___ Muskrat	___ Bobcat
___ Porcupine	___ Feral Horses (I)
___ Coyote	___ Elk
___ Gray Wolf (E)	___ Mule Deer
___ Red Fox	___ White-tailed Deer
___ Swift Fox	___ Pronghorn
___ Black Bear (E)	___ Longhorn Cattle (I)
	___ Bison (RI?)
	___ Mountain Sheep (RI)

Virgin Islands National Park

P.O. Box 7789, Charlotte Amalie
St. Thomas, Virgin Islands 00801

Covering almost two-thirds of St. John Island, Virgin Islands National Park's 14,488 acres (5,863 hectares) include palm-rimmed white sand beaches, blue-water coves, and forested hills. Although the park headquarters are on St. Thomas, there is a visitor center at Cruz Bay on St. John. Daily ferry service is available. There are hiking trails as well as an underwater trail in the park, which is open year-round.

The park site was once a plantation and much of its vegetation is second growth rather than native. Exotics such as breadfruit, frangipani, bougainvillea, and hibiscus are evident. Similarly, the only predominant mammal in the park is the **Asian Mongoose**, a species that was introduced in the nineteenth century to control rats (also introduced) by plantation owners. The only native mammals are six species of bat. Bird life is rich, with both aquatic and terrestrial species evident, and snorkelers see a varied array of tropical reef fish and invertebrates.

Voyageurs National Park

P.O. Box 50
International Falls, Minnesota 56649

Voyageurs National Park was historically important as a canoe route of the French-Canadian fur trappers. Its 219,128 acres (88,679 hectares) are about one-third water. The land areas are heavily forested with fir, spruce, pine, aspen, and birch, and very little of the park is accessible by automobile. Much of the land within this relatively new park is still privately owned, although the National Park Service continues to acquire property, and visitors should respect the rights of the owners. All of the park campsites are primitive and can be reached only by water.

As might be expected, Voyageurs Park is rich in aquatic species; Beavers, River Otters, Mink, and Muskrats are common in the lowlands, while White-tailed Deer, Red Squirrels, and Black Bears are numerous in the forested uplands. Much remains to be learned about the status of the park's mammals. Voyageurs is open throughout the year and is the site of much winter recreation.

White-tailed Deer, which are abundant in both the forested lowlands and uplands, are often seen, especially around Sullivan Bay, Black Bay, State Point, and Neil Point. **Moose** are not common in the park, however, and are seldom seen, although the Lost Bay area seems a good place to find them. **Caribou** once inhabited the park area, but are no longer present.

VOYAGEURS NATIONAL PARK

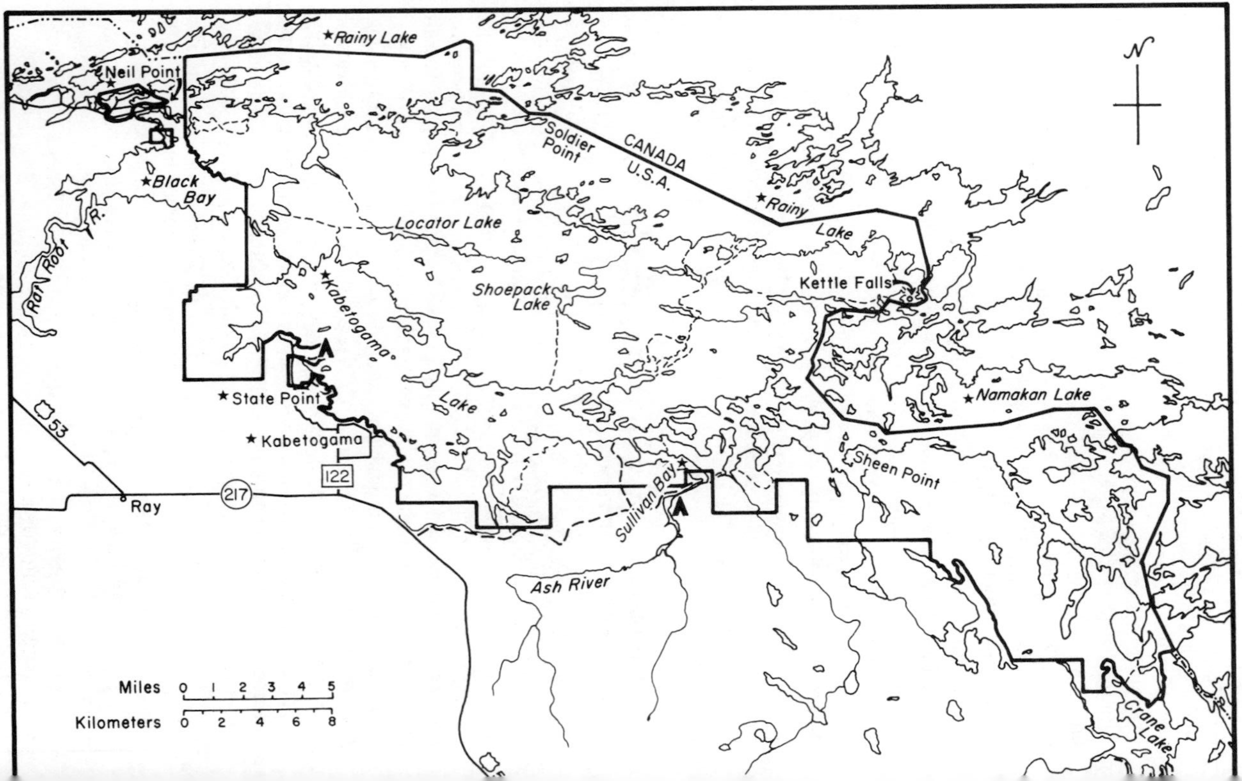

Crane Lake

Namakan Lake

Sheen Point

Kettle Falls

★Rainy Lake

Rainy Lake

CANADA
U.S.A.

Soldier Point

Sullivan Bay

Ash River

Locator Lake

Shoepack Lake

Lake

★Rainy Lake

★Kabetogama

State Point ★

122

★Kabetogama

217

Ray

★Black Bay

★Neil Point

Rat Road

53

Miles
0 1 2 3 4 5

Kilometers
0 2 4 6 8

Beavers are numerous in the wetlands, and their dams and lodges can readily be seen throughout the park. In remote, quiet places, Beavers are often active during the day and are easily seen by quiet and patient observers. **Muskrats** also inhabit the park's waters and are common. In the same aquatic habitats are **River Otters**, which are common in Voyageurs and are occasionally seen by canoeists. The **Mink** is also abundant here, although less of a swimmer than the River Otter, and has been observed in such places as Sullivan Bay.

The commonest tree squirrel in Voyageurs is the **Red Squirrel**, which is mainly an inhabitant of the coniferous forests; it can readily be seen at Neil Point, Black Bay, State Point, Sullivan Bay, and Whispering Pines. The commonest chipmunk is the **Least Chipmunk**, often seen at Neil, Black Bay, State Point, and Sullivan Bay. Also present is the **Eastern Chipmunk**, a slightly larger species, which seems to prefer hardwood and brushy areas.

The **Snowshoe Hare** is common in most places in the park and is often seen early in the morning. In winter, its tracks in the snow proclaim its presence. If **Eastern Cottontails** are present in Voyageurs, they are seldom seen and probably live at the edge of clearings. Such sites are also the habitat of the **Woodchuck**, another species considered uncommon elsewhere within the park's boundaries.

Black Bears are common mainly in the uplands and are regularly seen at such places as Neil Point, Black Bay, State Point, and Sullivan Bay. Most of the other carnivores in the park are seldom seen. **Bobcats** are common, but not often observed. **Lynx** are rare. **Mountain Lions** are extinct. **Gray Wolves** are probably present, as are **Coyotes**, but neither is regularly observed. Both **Red** and **Gray Foxes** are present; the former is seen more often than the latter, but neither is common enough to be seen regularly.

Small members of the weasel family such as **Ermine, Least Weasels**, and **Long-tailed Weasels** are seldom seen, even if common. **Striped Skunks** are common in places such as Black Bay, but are more often smelled than seen. **Marten** are so rare that they are believed to be extinct in the park; **Fishers**, though rare, are present. **Badgers** are reported to be present, but their status is unknown; **Wolverines** are extinct here.

Porcupines are common and are sometimes observed; the same is true of **Raccoons**, especially around Neil Point and State Point, and both of these species can sometimes be pests around cabins and camps.

CHECKLIST

___ Eastern Cottontail (?)	___ Woodchuck
___ Snowshoe Hare	___ Red Squirrel
___ Eastern Chipmunk	___ Beaver
___ Least Chipmunk	___ Muskrat

(continued)

(continued)

___ Porcupine
___ Coyote
___ Gray Wolf
___ Red Fox
___ Gray Fox
___ Black Bear
___ Raccoon
___ Marten (E?)
___ Fisher
___ Ermine
___ Least Weasel
___ Long-tailed Weasel

___ Mink
___ Wolverine (E)
___ Badger
___ Striped Skunk
___ River Otter
___ Mountain Lion (E)
___ Lynx
___ Bobcat
___ White-tailed Deer
___ Moose
___ Caribou (E)

Wind Cave National Park

Hot Springs, South Dakota 57747

A small park of 28,052 acres (11,352 hectares), at the southern edge of the Black Hills, Wind Cave is famous for its extensive limestone caverns, of which 35 miles (56 kilometers) have been explored. Three-fourths of the park is grassland, while the forested area, mainly in the west and northwest parts, is largely ponderosa pine and Rocky Mountain juniper. There are no high mountains, and the park's elevation ranges from 3,650 feet (1,113 meters) to 5,000 feet (1,524 meters).

For its size, Wind Cave is one of the best mammal-viewing places in the United States. It is excellent for seeing Bison, Black-tailed Prairie Dogs, and Pronghorns, as well as Coyotes. This is one of the few national parks where the extremely rare Black-footed Ferret may still exist. The best places to observe mammals are along the main road (U.S. Highway 385), at the Prairie Dog Exhibit, and near Elk Mountain Campground; the best time is from late spring to early autumn.

Bison are readily seen at Wind Cave along and to the east of State Road 87 and U.S. Highway 385, especially between the visitor center and Rankin Ridge. There are several hundred of these large mammals in fewer than 41 square miles (66 square kilometers); thus they cannot be missed. There are often cows with calves along the northern part of Park Road 5 in summer. The original Bison of Wind Cave were killed off in the nineteenth century. The Bison now in the park are descendants of stock reintroduced in 1913. **Pronghorns** are also readily seen at Wind Cave, almost anywhere

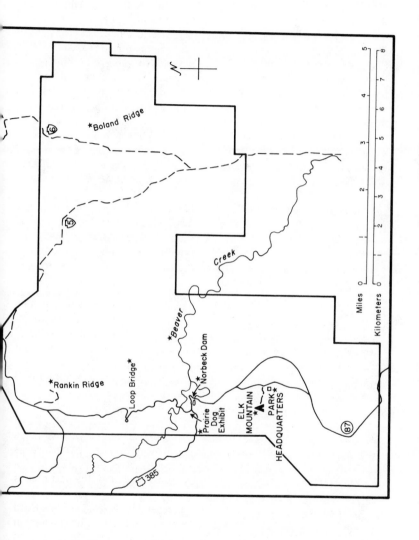

in the grasslands, especially in the eastern sector between Park Roads 5 and 6. These, too, are descended from animals reintroduced in 1914, after the original Pronghorns had been killed off.

The other hoofed animals of the park as easily seen as Bison and Pronghorn. **Mule Deer** are common, however, and usually can be seen at dawn or dusk near the visitor center and Elk Mountain Campground. **White-tailed Deer** are rare and possibly are no longer resident in the park. The northern border and brushy ravines are the best places to look for them. The native **Elk** were extirpated, and in 1914 some were reintroduced from Wyoming; Elk now number 400–500. At Wind Cave, Elk are present in three main herds: the largest lives around Beaver Creek, in the west-central part of the park; the second is near Broken Ridge, in the eastern sector; and the third resides around Shirttail Canyon, in the southeastern part. Elk are not readily seen, except along roads before sunrise, but the bulls can be heard bugling in late August and September. **Mountain Sheep** may not have been a part of the park's original fauna, but some from nearby stock that was reintroduced were seen in the park, but they seem not to have survived. There are Mountain Sheep just to the north of Wind Cave, in Custer State Park, which are also descendants of introduced stock.

At least ten major colonies of **Black-tailed Prairie Dog** are present in the park. They inhabit about 6 percent of the park's unforested land, a total of nearly 3 square miles (5 square kilometers). A good viewing point is at the Prairie Dog Exhibit at the junction of U.S. Highway 385 and State Road 87. Prairie dogs in Wind Cave are active all winter. Another rodent often seen in prairie dog towns, as well as throughout the grassland, is the **Thirteen-lined Ground Squirrel**. Prairie dog towns also attract predators, and **Badgers**, although seldom seen, are found there. Sign of Badger activity—large excavations—are seen in the prairie dog towns, especially in Bison Flats and Shirttail Canyon.

Wind Cave may still have a few **Black-footed Ferrets**, one of the rarest North American weasels. The last valid sightings were in the late 1960s and early 1970s, near Norbeck Dam. If still present, these Black-footed Ferrets may be the descendants of some reintroduced in the park (two males, one female) in 1953.

Two species of rabbit are present in the park. The **Desert Cottontail**, a pale-colored species, is more commonly seen in prairie dog towns and open grasslands. The darker **Eastern Cottontail** is less often seen because it prefers brushy ravines. Dawn and dusk are the best times for seeing either of these rabbits, and both have been seen on the lawns at the visitor center. The large **White-tailed Jack Rabbit** also lives in the high grassland and prairie of Wind Cave, but its numbers seem to be low, and there are few sightings, mainly along roads at night.

Coyotes are common in Wind Cave and are often seen, especially at dawn, near prairie dog colonies. The open country along Park Road 6 in the eastern part of the park is another good place to find Coyotes early in the morning, and they are often heard howling at night. A predator that is seldom seen in the park is the **Bobcat**, which is nocturnal and secretive.

The **Bobcats** that have been seen were mainly in the northwest sector, north of U.S. Highway 385 on State Road 87. **Mountain Lions** are not known to be resident in the park, but they have been observed near the western and northern boundaries. Tracks indicate that they sometimes roam into the park from the north in winter.

The **Least Chipmunk** is present in forested and brushy parts of Wind Cave and can be seen near the visitor center, Elk Mountain Campground, Beaver Creek Canyon, and elsewhere, up to 4,200 feet (1,280 meters) elevation. In the conifer forests are **Red Squirrels**, which are often seen or heard chattering in the woods, especially around Loop Bridge. The **Fox Squirrel** may occur in the park, although neither sightings nor actual specimens have been recorded from Wind Cave, one was "heard" at Beaver Creek. **Porcupines** are abundant in the woods, but they are not often seen. Debarked trees, which are a sign of their activity, are evident from the roads. The **Bushy-tailed Woodrat** is another common but seldom seen rodent of Wind Cave. Woodrats ranges down to 4,000 feet (1,219 meters) elevation and also inhabit the cave itself, near the entrance. Elsewhere, evidence of their presence is their brush and twig nests at rock outcrops and in old buildings.

Raccoons were formerly common around campgrounds and moist areas, but are now rare and seldom seen. The **Long-tailed Weasel** has been seen near the visitor center, but there are no other records of this weasel in the park. **Striped Skunks** also have been recorded in the park, but seem extremely rare. The **Spotted Skunk** may also occur here, but there are no valid records. Similarly, there are no records of two foxes, the **Red Fox** and the **Swift Fox**, either of which might be part of the park's fauna.

About ten species of bat have been recorded from Wind Cave, but these are present, if at all, only in winter and thus are seldom seen by visitors. Some are usually seen flying around streetlights in the campground on summer nights.

Gray Wolves must have been part of the original fauna of Wind Cave, and the **Grizzly Bear**, the **Black Bear, Mink, Beaver**, and **Muskrat** may have also occurred here; none lives here today.

CHECKLIST

___ Bats	___ Fox Squirrel (?)
___ Eastern Cottontail	___ Red Squirrel
___ Desert Cottontail	___ Beaver (E)
___ White-tailed Jack Rabbit	___ Bushy-tailed Woodrat
___ Least Chipmunk	___ Muskrat (E)
___ Thirteen-lined Ground Squirrel	___ Porcupine
___ Black-tailed Prairie Dog	___ Coyote
	___ Gray Wolf (E)

(continued)

(continued)

___ Red Fox (?)	___ Striped Skunk
___ Swift Fox (?)	___ Mountain Lion
___ Black Bear (E)	___ Bobcat
___ Grizzly Bear (E)	___ Elk (RI)
___ Raccoon	___ Mule Deer
___ Long-tailed Weasel	___ White-tailed Deer
___ Black-footed Ferret (E?)	___ Pronghorn (RI)
___ Mink (E)	___ Bison (RI)
___ Badger	___ Mountain Sheep (E)
___ Spotted Skunk (?)	

Yellowstone National Park

P.O. Box 168
Yellowstone National Park, Wyoming 82190

The world's first national park, dating from 1872, Yellowstone is also one of the largest national parks, with 2,219,823 acres (898,340 hectares). It boasts of more than 10,000 geysers and hot springs; a colorful and dramatic canyon with 300-foot (94-meter) waterfalls; mountains more than 10,000 feet (3,048 meters) high; vast, forested plateaus; petrified trees; lakes; and spectacular wildlife. The park is mainly in northwestern Wyoming, but its northern and eastern edges are partially in Montana and Idaho. Summer days are cool and nights may be freezing. July and August are the sunniest and warmest months, but winter visitation has been increasing. Winter temperatures are usually below freezing, and temperatures as low as $-66°$ F. ($-54°$ C.) have been recorded. Most of the annual precipitation, which ranges from 14 to 38 inches (356 to 965 millimeters), is in the form of snow—12.5 feet (3.8 meters) is average on the plateaus.

Yellowstone's famed Black and Grizzly Bears are now seldom seen, but there are many other readily observable species, including Bison, Elk, Mule Deer, Moose, and many smaller mammals. The park's main highway circles its central portions and gives access to its major features, including Old Faithful, the Grand Canyon and Falls of the Yellowstone River, Yellowstone Lake, hot springs, geysers, and many of the better places to view mammals.

Large mammals are common sights in Yellowstone. **Elk** are abundant and readily seen, especially in summer in the meadows (Gibbon Meadow, Elk Park) near Madison Junction, in the Hayden Valley south of Canyon Junction, and at Fountain Flats north of Old Faithful. In winter, they are regularly seen on the hills in the north-central part of the park, from the north entrance at Gardiner to the Lamar Valley. The **Moose** is another large

YELLOWSTONE NATIONAL PARK

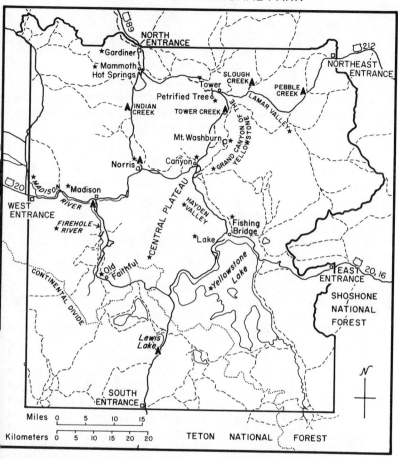

mammal that is readily observed in the park. Although they are not as social as Elk and are not seen in herds, Moose are large and conspicuous. They feed, in summer, on vegetation in and near streams and ponds, and they can usually be found among the willows along Pelican Creek near Fishing Bridge; on the shores of Yellowstone Lake; near the Indian Creek Campground, south of Mammoth; along the Lewis River in the south; and along the Madison River in the west.

The **Bison** is another hoofed animal that is readily seen in the park. Herds of these shaggy beasts usually can be found in the meadows near the fountain geysers between Old Faithful and Madison Junction, and in the Lamar and Hayden valleys throughout the year. **Mountain Sheep** are less readily observed, but in summer they sometimes can be seen on Mount Washburn, from the road south of Tower Junction, and hikers may encounter them along the east side of the Yellowstone River, southeast of Tower Junction. In winter, these sheep are often seen on the east side of the road, from Mammoth to Gardiner. They are found only in the northern parts of Yellowstone.

Mule Deer occur throughout Yellowstone and may be seen in summer almost anywhere at dawn or dusk. There are usually some around Yellowstone Lake and its campsites, in meadows near Canyon Village, and near Mammoth Hot Springs. In winter, they usually can be found around Old Faithful and the Gardiner River Valley, north of Mammoth. **White-tailed Deer** disappeared from Yellowstone in the 1920s, when their habitat—the willows along the Gardiner and Lamar rivers—was destroyed. **Pronghorns** are not common in Yellowstone and occur only in its northern part. There are usually some in the Gardiner Valley in the sagebrush areas between Mammoth Hot Springs and Gardiner, and, especially in winter, there may be some in the Lamar Valley.

Bears were once the major attraction of Yellowstone. **Black Bears** stood by the roads soliciting—and receiving—food from tourists, and, for many years, **Grizzly Bears** were encouraged by being fed nightly by the park service. However, increasing problems resulting from interactions between bears and people, including the loss of human life, led to a controversial reevaluation of the bears' role in the park and the discontinuance of sources of food provided by people. As a result, the bear population has diminished and the animals no longer congregate around human settlements, but remain in the back country, where natural sources of food are available. Visitors to Yellowstone see bears only occasionally now, and hikers are discouraged or prohibited from going into areas where Grizzly Bears are known to be. Travelers in the back country should be extremely cautious when in the vicinity of any bears. Although most Black Bears in the park are black, other color variations, including cinnamon and pale brown, occur; color is not a certain guide to identifying bears. Grizzlies in the park tend to be brownish in color, often with gray tinges, but also vary considerably. The large size, hump on the shoulders, and dished profile of the Grizzly will usually distinguish it from the Black Bear.

Chipmunks and ground squirrels are found throughout the park. The **Least Chipmunk, Yellow-pine Chipmunk**, and **Uinta Chipmunk**, difficult to distinguish from one another, are found from the low sagebrush flats to above timberline, and one or more species is often seen around campgrounds, parking lots, and lookouts. **Golden-mantled Ground Squirrels** inhabit much of the same area and are often seen in the same places as chipmunks. **Uinta Ground Squirrels** generally are found, and often seen, at the lower elevations of the park, and especially good places to find them are along the path to the Petrified Tree, near Tower Falls, and around the cabins at Mammoth Hot Springs.

Another common member of the squirrel family is the **Yellow-bellied Marmot**, a dweller in the rocky areas up to and above timberline. It is often seen sunning itself on a rock outcrop, and some can usually be seen along the road to Dunraven Pass. Another rock dweller, mainly in the high talus slopes, is the **Pika**, which is common. The only tree squirrel in Yellowstone is the **Red Squirrel**, abundant in the coniferous forests, where it often identifies itself by noisily chattering at hikers. Red Squirrels are readily seen near Yellowstone Falls. Coniferous forests are also the home of **Porcupines**, which are abundant in the park; they are occasionally encountered by hikers and sometimes wander through campsites at night. **Bushy-tailed Woodrats** are not common in Yellowstone and are rarely seen, although their large nests are sometimes found in rock crevices and abandoned cabins.

Snowshoe Hares are the most numerous of the park's representatives of the rabbit family. They live mainly in the coniferous forest and are not often seen, although their tracks in snow are indicative of their abundance. Sometimes they are seen along roads at night. Much less common, and mainly in the lower elevations near Mammoth, where there is sagebrush, are **Nuttall's Cottontails**; they are not often seen. Some also have been reported from around Old Faithful. **White-tailed Jack Rabbits** are a bit more abundant and widespread than the cottontail, but are an uncommon sight, nevertheless, preferring sagebrush and level grasslands. White-tailed Jack Rabbits are found in the same places as the cottontail, including around Old Faithful.

Coyotes are probably the most frequently seen carnivores in the park. Most observations of them have been in open situations, such as the Lamar and Hayden valleys, almost any meadow, and around Mammoth. They are often heard howling at night. **Gray Wolves** were thought to be extinct in the park, but it now seems that there may be a few in the northern part; sightings of them are extremely rare. **Red Foxes** probably occur in the coniferous forest, but are scarce and are rarely seen. Another rare inhabitant of the dense forest is the **Lynx**, and observations of it are few. **Bobcats** are probably more numerous and more widely distributed, but they are seldom seen. There is a small **Mountain Lion** population in the park, and most of the infrequent observations of them have been in the Hayden and Lamar valleys and south of Mammoth Hot Springs.

Of the smaller carnivores, the **Marten** is the only one that is occasionally encountered, usually in meadows in the coniferous forest, and **Badger** diggings are common in ground squirrel habitat, although few of the animals are seen. **Striped Skunks, Ermine**, and **Long-tailed Weasels** are seldom seen, and the **Wolverine** is rare. It is not certain if **Fishers** occur in the park.

Yellowstone National Park is connected with Grand Teton National Park, 7 miles (11 kilometers) to the south by the John D. Rockefeller, Jr., Memorial Parkway.

CHECKLIST

___ Pika	___ White-tailed Jack Rabbit
___ Nuttall's Cottontail	___ Least Chipmunk
___ Snowshoe Hare	___ Yellow-pine Chipmunk

(continued)

(continued)

___ Uinta Chipmunk
___ Golden-mantled Ground
    Squirrel
___ Yellow-bellied Marmot
___ Uinta Ground Squirrel
___ Red Squirrel
___ Beaver
___ Bushy-tailed Woodrat
___ Muskrat
___ Porcupine
___ Coyote
___ Gray Wolf (?)
___ Red Fox
___ Black Bear
___ Grizzly Bear
___ Marten
___ Fisher (?)

___ Ermine
___ Long-tailed Weasel
___ Mink
___ Wolverine
___ Badger
___ Striped Skunk
___ River Otter
___ Mountain Lion
___ Lynx
___ Bobcat
___ Elk
___ Mule Deer
___ White-tailed Deer (E?)
___ Moose
___ Pronghorn
___ Bison
___ Mountain Sheep

Yosemite National Park

Yosemite National Park, California 95389

Ranging in elevation from 2,000 feet (610 meters) to more than 13,000 feet (3,962 meters), Yosemite National Park's 760,917 acres (307,935 hectares) encompass five major plant communities in California's Sierra Nevada. Yosemite's attractions include major features of glacial geology, with domes, moraines, waterfalls, lakes, and rivers in spectacular settings, as well as vast coniferous forests, meadows, and giant Sequoia trees. The magnificence of these features attracts many visitors, and human needs and activities have now become an impediment to full enjoyment of this lovely park. Plans have been made to rectify this situation.

The varied habitats of Yosemite provide environments for many kinds of mammal, some readily viewed, others rarely seen. Even people who visit only Yosemite Valley cannot avoid seeing several species, including tree and ground squirrels, and deer. Yosemite is open throughout the year, although some roads and sections are closed in winter, and reservations are usually required, even for many of the campgrounds.

Diurnal members of the squirrel family are widespread throughout the

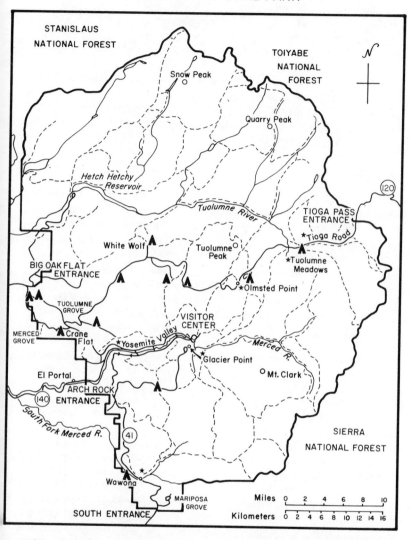

park, and, during summer, encounters with the nonhibernating species are frequent. At the lower elevations, from Wawona to Yosemite Valley, the **California Ground Squirrel** is common and can be seen in meadows and on rock piles, as well as foraging for scraps around buildings in the valley and at turnouts on the road to Glacier Point. Higher up, two other species of ground squirrel are abundant. **Belding's Ground Squirrel** is a high-country species, resident above 8,000 feet (2,438 meters) in meadows, where it frequently stands erect, looking like a stake in the ground. Tuolumne Meadows is a good place to see these squirrels. From 6,000 feet

(1,829 meters) to 9,000 feet (2,743 meters) the **Golden-mantled Ground Squirrel** is common and readily seen. This squirrel is often mistaken for a chipmunk, from which it can be distinguished by its larger size and lack of a stripe through the eye. It forages for scraps around campgrounds and at turnouts, for example, Olmsted Point on Tioga Road, west of Tuolumne Meadows.

Chipmunks are found at all elevations in Yosemite. Six species, indistinguishable to most viewers, live in the park, and, except for **Merriam's Chipmunk**, which lives below 4,000 feet (1,219 meters) elevation, all are common and are often seen around campgrounds and turnouts, and along trails. The **Alpine Chipmunk** lives at elevations from 11,000 feet (3,353 meters) up, wherever there is enough vegetation to provide food. The **Lodgepole Chipmunk** occurs at slightly lower altitudes from 7,000 feet (2,134 meters) to 9,000 feet (2,743 meters). At elevations of 4,000–7,000 feet (1,219–2,134 meters), **Long-eared, Yellow-pine,** and **Townsend's Chipmunks** make their homes.

Another diurnal member of the squirrel family which is easily seen in Yosemite is the **Yellow-bellied Marmot**. It lives mainly in rocky areas of about 7,000–10,000 feet (2,134–3,048 meters). It is usually seen at turnouts on the Tioga Road, especially just west of Tuolumne Meadows, where some individuals beg for food from travelers.

Two species of diurnal tree squirrel live in the park. The noisy **Douglas' Squirrel**, a dweller of the coniferous forest, is common at elevations of about 4,000–9,000 feet (1,219–2,743 meters). It is usually seen around buildings and campsites in Yosemite Valley, as well as at the turnouts on the Tioga Road, near Tuolumne Meadows, generally foraging on the ground. Less often seen is the larger **Western Gray Squirrel**, which is a forest dweller at elevations of 4,000–7,000 feet (1,219–2,134 meters). Periodically, its numbers seem to decrease markedly, perhaps from disease; it is then rarely seen. When the population is high, however, these squirrels can be seen in Yosemite Valley, foraging on the ground.

Mule Deer inhabit the lower-elevation scrub to the higher mountain meadows (in summer) and are common. They can easily be observed in Yosemite Valley early or late in the day, or at night, and are often seen along roadsides after dark almost anywhere in the park. The only other native hoofed animal in Yosemite is the **Mountain Sheep**. These sheep are found mainly on the eastern side of the Sierra Nevada, and it is questionable even if there is resident population in the park. Sightings of these Bighorns have been rare, and all sightings should be reported to park rangers.

Raccoons are common in Yosemite; they are found in elevations up to about 7,000 feet (2,134 meters). They are readily observed around cabins and tent cabins in the valley where, at night, they boldly forage for leftovers. The Raccoon's relative, the **Ringtail**, is also nocturnal, but is relatively shy and is less often seen, although it may also be common at the same elevations as the Raccoon. Ringtails prefer rocky areas, as well as buildings, and at times they have lived in hotels in the valley.

Striped and Spotted Skunks are both common at lower elevations in the park, up to about 7,000 feet (2,134 meters). Spotted Skunks prefer to live in rocky areas and sometimes in old buildings. Both kinds of skunk are frequently seen foraging at night around campgrounds and buildings. Other members of the weasel family are less readily seen in the park; however, the Marten, found in forests of 7,000–9,000 feet (2,134–2,743 meters), is sometimes active during the day, even foraging for chipmunks and other ground squirrels in campgrounds. In winter, Marten tracks are frequently seen in snow. Both Long-tailed Weasels and Ermine are common in Yosemite, but neither is often seen. The former is found at elevations of about 5,000–11,000 feet (1,524–3,353 meters); the latter occurs from 7,000 feet (2,134 meters) upward. The Long-tailed Weasel sometimes forages for chipmunks and mice in campgrounds during the day.

Much rarer members of the weasel family are the Fisher, a seldom seen resident of dense forests of 5,000–11,000 feet (1,524–3,353 meters), and the Wolverine, the largest of the weasel family. The main signs of the Wolverine's presence are usually tracks in snow in the high country near timberline, and fortunate indeed is the visitor who gets a glimpse of this rare animal. Neither Mink nor Badgers are often seen in the park. The former are found near water, at elevations from 4,000 feet (1,219 meters) upward, but are secretive. Badgers range throughout the park to the higher mountain meadows, where they seek the ground squirrels that make up most of their diet. They are rarely seen in Yosemite; their excavations are usually the only indication of their presence. River Otters may not have been a part of the original Yosemite fauna before the introduction of trout in the high mountain lakes. River Otters now inhabit lakes at elevations up to 9,000 feet (2,743 meters) in the park, though whether these are transient or resident animals is not yet known.

Coyotes are common from the lowlands to the high mountain meadows of the park, and are sometimes seen alongside roads or hunting in meadows, as well as heard howling at night. Gray Foxes are common below 7,000 feet (2,134 meters) elevation, but are not regularly seen. They are nocturnal but sometimes are seen alongside roads at night or at dawn, especially in brushy areas. The Red Fox is believed to inhabit high meadows at forest edges in the park, but, if so, is rare. No Gray Wolves occur in the park now.

Black Bears are widespread in Yosemite, mainly at elevations of 4,000–9,000 feet (1,219–2,743 meters), but they are not often encountered. There are usually some in Yosemite Valley, however, where they beg for food, despite park service discouragement. Black Bears are common enough in the high country so that campers should take precautions to protect their food and supplies by suspending them out of reach of bears. Grizzly Bears ("Yosemite" is derived from an Indian word for Grizzly Bear) were wiped out many years ago.

Pikas are common in talus slopes at elevations above 7,000 feet (2,134 meters), where the careful watcher can usually see them. Other members of the rabbit and hare group may be less common. Although rabbits are usually one of the most frequently seen mammals in many national parks, in Yosemite, the only rabbit—the **Brush Rabbit**—is among the rarest of mammals and is found only in the dense brush country at the lowest park elevations, below 4,000 feet (1,219 meters). Three kinds of hare occur in the park and are also not often seen. The **Black-tailed Jack Rabbit** occurs in brushy country below 6,000 feet (1,829 meters), but is seldom seen; most observations are at night or early in the morning along roadsides. The **White-tailed Jack Rabbit** is sometimes common above 7,000 feet (2,134 meters) elevation, but undergoes periodic population fluctuations, during the "lows" of which the rabbits are very scarce. The **Snowshoe Hare** may occur in the extreme northern part of Yosemite.

Porcupines are not rare in Yosemite at elevations from 6,000 feet (1,829 meters) to the timberline, but are not usually seen. Sometimes they wander into campsites in search of salty materials, but the usual indication of their presence is scarred trees, which they have stripped the bark to feed.

Two species of woodrat live in Yosemite. The **Dusky-footed Woodrat** lives below 4,000 feet (1,219 meters), where its presence is marked by large stick-and-debris nests, and occasional occupancy of old buildings. From 6,000 feet (1,829 meters) up is the domain of the **Bushy-tailed Woodrat**. Both species are nocturnal and are not usually seen.

Mountain Lions, although perhaps present anywhere in the park from 8,000 feet (2,438 meters) upward, are rarely seen and certainly are not common. **Bobcats** range throughout Yosemite to all but the highest elevations. Nocturnal and secretive, although probably abundant, **Bobcats** are, relatively, one of the least seen mammals of the park; most observations are of individuals crossing or walking along a road at nighttime.

The **Mountain Beaver** lives only in moist areas of 4,000–7,000 feet (1,219–2,134 meters) and is not common and is rarely seen. Extensive burrows in soggy ground are signs of its presence. True **Beavers** are not common in the park, the only population being derived from animals introduced on low-elevation rivers near Wawona. Another nonnative species is the **Virginia Opossum**, recorded from low elevations on the western side of the park and descended from stock introduced from the eastern states early in the twentieth century.

CHECKLIST

___ Virginia Opossum (I)	___ Yellow-pine Chipmunk
___ Pika	___ Townsend's Chipmunk
___ Brush Rabbit	___ Merriam's Chipmunk
___ Snowshoe Hare (?)	___ Long-eared Chipmunk
___ Black-tailed Jack Rabbit	___ Lodgepole Chipmunk
___ Mountain Beaver	___ Yellow-bellied Marmot
___ Alpine Chipmunk	___ Belding's Ground Squirrel

(continued)

___ California Ground Squirrel
___ Golden-mantled Ground
    Squirrel
___ Western Gray Squirrel
___ Douglas' Squirrel
___ Beaver (I)
___ Dusky-footed Woodrat
___ Bushy-tailed Woodrat
___ Porcupine
___ Coyote
___ Gray Wolf (E)
___ Red Fox (?)
___ Gray Fox
___ Black Bear
___ Grizzly Bear (E)
___ Ringtail

___ Raccoon
___ Marten
___ Fisher
___ Ermine
___ Long-tailed Weasel
___ Mink
___ Wolverine
___ Badger
___ Spotted Skunk
___ Striped Skunk
___ River Otter
___ Mountain Lion
___ Bobcat
___ Mule Deer
___ Mountain Sheep (?)

Zion National Park

Springdale, Utah 84767

Featuring a dramatic, high-walled canyon carved by the Virgin River, Zion National Park exhibits thousands of feet of layered strata, which represent deposits as old as 200 million years. Within the park's 146,547 acres (59,306 hectares) are four major plant zones, which extend over altitudes of 3,666–8,740 feet (1,117–2,664 meters). The steep walls of canyons and the shade and moisture provided by them compress these life zones and, in places, reverse the usual altitudinal sequence of vegetation. The main road in Zion, which traverses the southeastern section of the park, includes a twelve-mile (19-kilometer) round trip through Zion Canyon. Other roads from the west cross the center of the park and enter the northwestern part. There are hundreds of miles of hiking and horse trails. The park is open throughout the year. In winter, snow is rarely a major problem in the canyon. In summer, daytime temperatures in the canyon may exceed 100° F. (38° C.).

Although the park has diverse wildlife, mammal-viewing is not especially good here. Mule Deer are the most frequently seen large mammals and Rock Squirrels, chipmunks, and Red Squirrels are often observed during the day, and Gray Foxes at night.

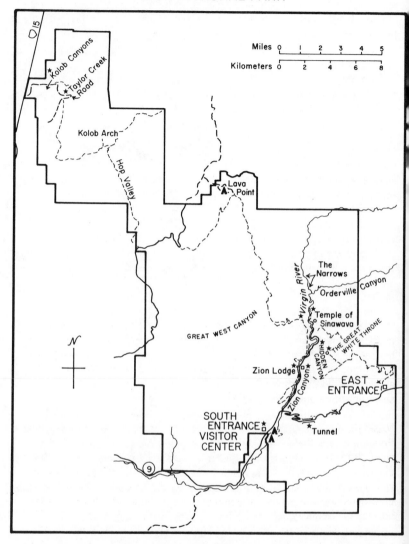

The squirrel family is well represented in Zion; there are four species of chipmunk. The **Least** and **Uinta Chipmunks** are denizens of the high plateau, the former in brushy areas and the latter in conifer forest. At lower altitudes, the main species is the **Cliff Chipmunk**, and, less abundant, the **Colorado Chipmunk**. Distinguishing these species in the field is nearly impossible. Probably the most often seen squirrel is the **Rock Squirrel**, which occurs throughout the park and is sometimes a pest in campgrounds. In the valleys to the west of the Kolob Canyon area is the **Townsend's**

Ground Squirrel, which is considered a rare species in Zion. In the hotter, lower parts of the park, **White-tailed Antelope Squirrels** occur but are not abundant. The largest of the ground squirrels in Zion is the **Yellow-bellied Marmot**, which lives in the middle and higher elevations, but is not common. The only tree squirrel in Zion is the **Red Squirrel**, which lives in the pines and firs on the plateau and is regularly seen.

Mule Deer are abundant throughout Zion and are frequently seen, especially early and late in the day. There are generally some to be seen at dusk around the lodge and on roadsides in the canyon. **Elk** are rare in the park and are not always present in the higher elevations, where they have been reported. **Mountain Sheep** were extirpated in the park, but, in 1973, twelve sheep were reintroduced and kept in an enclosure across from the visitor center. By 1976, when the population had reached twenty, sheep were released near the East Fork of the Virgin River and have survived and reproduced. They have been observed, especially near the switchback area in the southeastern part of the park.

In Zion Canyon at night, **Gray Foxes** are often seen along the roadside north of the lodge. **Kit Foxes** are known to inhabit the lowest parts of the park, and **Red Foxes** the higher elevations, but observation of either of these species is very rare. **Coyotes** are common throughout the park, but are not as often seen as heard howling at night. **Gray Wolves** are extinct in Zion.

The only member of the rabbit family that is regularly seen in Zion is the **Black-tailed Jack Rabbit**, which occurs throughout the park but ordinarily is seen in level, unforested areas. Both the **Desert Cottontail**, which lives at the lower elevations below 5,000 feet (1,524 meters), and **Nuttall's Cottontail**, which ranges above this altitude, especially in conifer forest, are considered uncommon in Zion. **Pikas** live in the high talus slopes, but seem not to be abundant or often seen.

Beavers live in the park's streams and rivers and are fairly numerous; **Muskrats** are not actually reported to be in the park, but live nearby. **Porcupines** are fairly abundant throughout the park, but sign of them—scarred trees—is more often seen than the animals themselves. Two species of woodrat inhabit Zion, the more common being the **Desert Woodrat**, which builds its conspicuous nests under rock overhangs at low and middle elevations; the **Bushy-tailed Woodrat** lives only at higher elevations. Neither species is often seen.

Neither **Black Bears** nor **Raccoons** are resident in Zion, although individuals of both species sometimes range into the park. Black Bears have been sighted, although rarely, on the high plateau. All observations of Raccoons are from along the Virgin River in the canyon. **Grizzly Bears** were wiped out here many years ago.

The commonest of the small carnivores in Zion is the **Ringtail**, which inhabits the low-elevation cliffs, but which, despite its abundance, is seldom seen because it is wholly nocturnal. **Long-tailed Weasels, Badgers,** and **Striped Skunks** have been reported throughout the park but are seldom seen, and the same is true of the **Spotted Skunk**, a low-elevation resident.

There are believed to be about twenty to thirty **Mountain Lions** resident in Zion, but there are only a few observations of them each summer. **Bobcats**, which also roam throughout the park, are likewise rarely seen.

CHECKLIST

___ Pika
___ Nuttall's Cottontail
___ Desert Cottontail
___ Black-tailed Jack Rabbit
___ Least Chipmunk
___ Cliff Chipmunk
___ Colorado Chipmunk
___ Uinta Chipmunk
___ Yellow-bellied Marmot
___ White-tailed Antelope
    Squirrel
___ Townsend's Ground
    Squirrel
___ Rock Squirrel
___ Red Squirrel
___ Beaver
___ Desert Woodrat
___ Bushy-tailed Woodrat
___ Muskrat (?)

___ Porcupine
___ Coyote
___ Gray Wolf (E)
___ Red Fox
___ Kit Fox
___ Gray Fox
___ Black Bear
___ Grizzly Bear (E)
___ Ringtail
___ Raccoon
___ Long-tailed Weasel
___ Badger
___ Spotted Skunk
___ Striped Skunk
___ Mountain Lion
___ Bobcat
___ Elk
___ Mule Deer
___ Mountain Sheep (RI)

Part 2

MAMMALS

Marsupials

Commonly called "pouched mammals," marsupials have only a single representative in America's national Parks—the Virginia Opossum. Although most people think of marsupials as being animals primarily of the Australian region, about 30 percent of the approximately 250 species of marsupial live in South and Central America. Compared with the placental mammals, marsupials generally have a smaller and less convoluted brain, relatively few chromosomes, and lower metabolism. Their most notable difference is their method of reproduction; the male has a forked penis, which is located posterior to the testes, and the female has a divided vagina and uterus. Despite the common name "pouched mammals," not all marsupials have a pouch (although they all have the supporting "marsupial bones"). Their young are born after a relatively short gestation period (8–40 days). With the exception of one family, the bandicoots, marsupials lack the kind of placenta that, in most other mammals, provides nutrition to the fetus. They have only a yolk-sac placenta, and the babies are born in a relatively undeveloped condition and complete their development attached to the mother's nipples, in the pouch for those species that have one. Marsupials are believed to represent a group that diverged from the other branch of mammals—the placentals—many millions of years ago (in the mid-Cretaceous) and, partly through isolation in South America and Australia, have adapted to most environments. The Virginia Opossum is a relatively recent inhabitant of North America, invading after the land bridge from Central America to South America was completed a few millions of years ago, and it is known to have increased its range northward in the past hundred years.

VIRGINIA OPOSSUM

There is no other mammal in the national parks quite like an opossum. With its pointed snout, naked ears, scaly prehensile tail, and hind feet with opposable big toes, the Virginia Opossum represents America's only native pouched mammal, the marsupial. The opossum, which is about the size of a domestic cat, has a body that is roughly one foot (305 millimeters) long and a naked, scaly, prehensile tail of the same length. Extremely fat individuals may weigh as much as 14 pounds (6.4 kilograms); most opossums weigh less than half this amount, males averaging 6 pounds (2.7 kilograms), and females 4 pounds (1.8 kilograms).

Virginia Opossums are as much at home in trees as on the ground, and, because they are nocturnal, the usual view of them is in the glare of a headlight or flashlight, when their pointed white faces are their most distinguishing characteristic. The body coloration is generally grayish, composed of black and white hairs, but black-phase coloration is common in some places, especially the south. Both phases, as well as all-white animals, may occur in the same litter. The ears are naked and black, with pink tips, and the tail is black for its basal half and white terminally.

Virginia Opossum with Young

Compared with most other mammals of equivalent size, opossums are slow-moving, running at about 4.5 miles (7.2 kilometers) an hour. When cornered, they may threaten—displaying many of their fifty teeth (they have more teeth than any other species of North American land mammal)—and hiss, but rarely bite. If further frightened, the animal may engage in its famed "playing possum" behavior, that is, feigning death. The animal falls over on its side, opens its mouth and lolls its tongue, and lies limp. If picked up and shaken, it does not react. However, if the animal, despite its seeming unconsciousness, is left for a few minutes, it may quickly resume its normal state and then scurry away. This behavior evidently serves the animal well because, as experienced observers have noted, opossums are seldom eaten by predators, probably because the odor from their glands makes them distasteful.

Opossums are omnivores. In summer, much of their diet is composed of insects, as well as fruits, berries, and vegetables, but they also eat mice and birds, eggs, and carrion. These habits often lead them to campsites in the parks, where they forage for leftovers. It is usually in these situations that tourists encounter opossums, as late-night visitors to a food or garbage supply.

The opossum is not especially social. An individual opossum ranges over an area of about 58 acres (23 hectares) or more, depending upon the habitat, and rarely roams more than half a mile (0.8 kilometers) from its den. It may share its home range with several other opossums. Its den is usually on the ground, under a log, in a culvert, under an old building, or in a hollow log or tree, and it is lined with dead leaves. Silvery opossum hairs at the mouth of a hole are often an indication of the animal's inhabitation. Although they do not hibernate, opossums put on huge layers of fat in the fall and, during cold weather, may remain in their dens for several days, but eventually venture out. Their tracks in the snow are often evident. They are not well adapted to northern climates and may freeze their ears and tail in these parts of their range.

Opossums mate in April or May in the north, and as early as January in the south. The male's forked penis has led to a common belief that breeding is through the female's nostrils, with her sneezing the sperm into her vagina. Actually, copulation is quite normal, the forked penis serving to introduce sperm into each of the female's separated vaginas. After a 12½ day gestation period, the female gives birth to up to twenty-five tiny young. The babies are naked, blind, earless, and only their mouth and forelegs are reasonably well developed. They are half an inch (12 millimeters) long and weigh one-fifteenth of an ounce (1.9 grams), and a whole litter can fit into a teaspoon. Using the claws on their front feet, they crawl through the mother's fur to the hair-lined pouch on her abdomen. There they find teats and attach themselves to them so firmly that blood is drawn if they are pulled off. Because there are usually only thirteen teats, excess babies die. The young remain attached to the nipple and continue their development for about two months, which is about the same length of time that a comparably sized placental mammal would gestate. The babies continue to suckle for another month or month-and-a-half, at which time they are weaned. They are, as is true of most mammals, capable of eating solid food long before they are weaned.

When they are three months old, the babies sometimes ride on the mother's back, but as soon as they are on their own, there is little contact between the mother and her offspring. Because opossums have such a high reproductive potential—those in the southern part of the range customarily have two litters a year—in any given area as many as 75 percent of the opossums may have been born that year. Females breed during the winter or spring following their birth, when less than one year old.

Most opossums do not live more than two years, and the maximum life span in captivity is about seven years. A high reproductive rate (with about six young surviving to weaning) and broad and diverse food habits have enabled the opossum to persist and to increase its range in the past few hundred years. It is now found throughout the eastern United States as far as the Canadian border, usually in wooded areas not far from water. In the western states, where it was introduced early in the twentieth century, it has thrived and spread, so that it is now found from Canada to Mexico in the Pacific states.

Visitors to the national parks do not usually see opossums, partly because the peak of their activity is from 11 P.M. until 2 A.M. and partly because they do not aggregate even in small numbers. They are usually seen in a headlight beam from a car and, all too often, are found dead on highways in the morning. Mammoth Cave, Everglades, Great Smoky Mountains, and Shenandoah are the better parks in which to find opossums.

Shrews and Moles

Shrews and moles are placed in the Order Insectivora, which, as the name implies, eat mainly insects and invertebrates. Some also prey on small vertebrates. There are about two dozen species of shrew found throughout North America. They are all small animals (the largest is about 10 inches—254 millimeters—in total length), with small eyes, velvety fur, a pointed snout, and five clawed toes on each foot. The smallest North American mammal is the Pygmy Shrew, which has a body less than 4 inches (102 millimeters) long and weighs about one-tenth of an ounce (2.8 grams).

Because of their small size, shrews have a proportionately large body-surface area and lose heat rapidly. To maintain their high body temperature, they must feed as often as every three hours, and thus are active day and night. In twenty-four hours a shrew may consume more than twice its own weight in insects, earthworms, centipedes, and other forms of protein. Some shrews have venomous saliva, which enables them to paralyze their prey, and some also make high-pitched shrieks that they use to find their way, just as bats use their own echo-location system.

One or more species of shrew occurs in most of the national parks, and shrews may be numerous. However, because of their small size and secretive habits, they are seldom seen by park visitors. When they are seen, they are often mistaken for mice. For these reasons they are not discussed in the accounts of the parks in this book. Identification of species of shrew in the wild is extremely difficult.

Moles are insectivores that, like shrews, have velvety fur, tiny eyes, and a pointed snout, and enlarged forefeet with strong claws. Moles have adapted for living underground and spend most of their time in burrows that they dig themselves. Although their eyes are tiny, they are sensitive to light, but moles do not rely on them much. Their best senses are touch and smell, and one mole, the Star-nosed Mole, has a ring of twenty-two tentacles on its nose that it uses to locate food by touch. There are seven species of mole in North America, mainly in the eastern half of the United States and the humid Pacific Coast.

Moles feed on invertebrates, especially earthworms; they eat little vegetation. Although active day and night, virtually all their time is spent underground. The usual evidence of their presence is either the surface ridges caused by their shallow burrows, or the conical mounds of dirt, 6–

Coast Mole

8 inches (152–203 millimeters) high, which are thrown up from their deep excavations.

Moles occur in some of the northeastern and Pacific Coast parks, but because they are so rarely seen, they are not discussed in this book.

Bats

It probably is a surprise to learn that bats are second only to rodents in numbers of species, and perhaps in numbers of individuals in the wild. The approximately 900 species of bat represent nearly one-quarter of the 3,800 species of living mammal, and, in some places, bat colonies have millions or even tens of millions of individuals. Bats are the only mammals capable of true, sustained flight. A bat's wings are formed by elongate fingers joined by a web of skin, and bats often have additional membranes enclosing the legs and tail. The hind legs are rotated 180 degrees at the knee, which enables bats to hang head down. Although there are species of bat elsewhere in the world that feed on fruit, small mammals and reptiles, and fish and blood, most North American bats are strictly insectivorous. There are a few in the southwestern states that feed on pollen and nectar. All bats are nocturnal and spend the day hanging in caves, on branches, under rocks, or under the bark of trees. Many species migrate long or short distances before winter, either to an area where food may be available or where there is a site for hibernation. Most North American bats have a complex and effective orientation system, which involves emitting high-pitched sounds and listening for the echoes that bounce back from objects in front of them. This sonarlike mechanism is called *echo location.*

In America north of Mexico, there are about forty species of bat, many of which inhabit national parks. All of them are nocturnal and most are insectivorous. Visitors to parks may often see bats at dusk, especially when they are silhouetted against the sky or as they flit over a lakeshore, but identification of flying bats at night defies even the experts. During the day, bats may be encountered hanging in caves, from trees in forests, or sometimes in buildings. The forest dwellers are usually Red Bats, Hoary Bats, or Silver-haired Bats, all characterized by the "frosted" appearance of their hair. Cave bats are usually members of the Myotis group, Big Brown Bats, or Brazilian Free-tailed Bats. In the northern parts of the ranges, insectivorous bats either spend the winter in hibernation or migrate to areas where insects may still be flying. The hibernaculum is usually a cave or other place where the temperature is constant, but well above freezing, and some species may assemble and cluster in the thousands or tens of thousands. The three forest dwellers mentioned above are migratory, flying south for the winter and returning to the northern forests in spring. Because there is only one species of bat that is a predominant feature of any national park, only the Brazilian Free-tailed Bat of Carlsbad Caverns National Park is discussed in detail in this book.

BRAZILIAN FREE-TAILED BAT

A dramatic and unique sight in the national parks is the evening flight of Brazilian Free-tailed Bats from Carlsbad Caverns National Park. During the summer, at dusk, bats spew forth from the mouth of the main cavern in huge numbers—up to 10,000 per minute—with a roar of flapping wings. Sometimes they emerge in long streams, wafting away like vapor toward the Pecos Valley to the east. At other times, they come out in separate groups, appearing like successive puffs of black smoke. The flight of bats led to the discovery of the caverns. These are small bats, with a total length of about 4 inches (102 millimeters), of which at least one-fourth is the tail. In flight their wingspread, although less than one foot (305 millimeters), makes them appear larger. They weigh about a half ounce (14 grams) and are strong fliers, cruising at speeds of 12–15 miles (19–24 kilometers) an hour. Their common name, "free-tailed bat," refers to the fact that—unlike many of the other North American bats, which have the tail entirely enclosed within the membrane that extends from ankle to ankle—only the basal half of their tail is within the membrane and the tip half is "free."

As the weather grows cooler in the fall, the Carlsbad bats migrate south to Mexico, where they spend the winter. Some bats that have been marked with metal bands are known to have migrated about 1,000 miles (1,609 kilometers). A very few of these bats may remain in Carlsbad throughout the year. It is in Mexico that the bats breed, and, when they arrive back at Carlsbad, starting in April, the females are pregnant. The single baby, naked and blind, is born in a colony that consists only of females (a "maternity colony"), starting in May. The males are generally in their own colony at this time, elsewhere in the main cavern. The babies are hung from the cavern roof, where the bats are so densely packed that there may be about 250 per square foot (0.09 square meter), and, in the jostling that occurs, many babies fall to the cave floor, where they die because the mothers make no attempt to retrieve them. The mothers do not carry them on their nightly flights to feed, and when they return, it is believed, they do not necessarily nurse their own offspring.

Brazilian Free-tailed Bats feed on insects, mainly moths, beetles, and leafhoppers, but relatively few mosquitoes. Each bat may consume from one-third to one-half its own weight each night; Carlsbad bats thus eat more than a ton of insects each night. The manure from these bats, guano, had accumulated in Carlsbad Caverns to such depths that, for many years, the deposits were mined for their rich minerals, which were used both as fertilizer and for the manufacture of gunpowder.

Fifty years ago, the number of bats in Carlsbad was estimated to be between 8 and 9 million, but the numbers have declined, and, at present, there are probably no more than a quarter of a million bats during the peak of the season. One reason for the decline in numbers may have been the use of insecticide, which diminished the bats' food supplies and also affected the survival of the bats themselves.

The observation of the flight of bats from the main cavern is facilitated by the park service. A small amphitheater above the cavern's mouth permits ready observation, and bulletin boards post the expected time of the emergence of the bats. However, it is a good idea to arrive earlier than scheduled, because the bats sometimes emerge unexpectedly, and someone who is fifteen minutes late may miss the entire flight. Equally worth seeing is the morning arrival of the bats back at the cavern. Just before dawn, they fly back high over it and then drop as much as 1,000 feet (305 meters) with their wings closed and open them with a pop just above the cavern mouth.

Brazilian Free-tailed Bat: Carlsbad Caverns

Nine-banded Armadillo

The Nine-banded Armadillo, a strange, armored mammal, is the only North American representative of a group that includes sloths, anteaters, and about twenty other species of armadillo. They are all, except for the Nine-banded Armadillo, South or Central American animals. However, giant ground sloths roamed the United States, especially the southwest, less than a million years ago. This group of mammals, called the edentates, is an ancient one, an early branch off the basic mammalian stock. The various families have peculiarities not found in any other kinds of mammal. Armadillos are characterized by a "shell" of bony plates covered by horny skin. Their teeth are peglike and absent from the front of the mouth. Their short legs are armed with strong claws, which aid them in their digging.

The Nine-banded Armadillo is about the size of a domestic cat, some

Nine-banded Armadillo

2½ feet (762 millimeters) in total length, of which about 15 inches (381 millimeters) is the tail. Yellowish-brown to almost black in color, this armadillo may weigh up to 17 pounds (7.7 kilograms) and is not likely to be mistaken for any other North American mammal. It is distributed around the Gulf Coast, and it ranges northward to southeastern Kansas, but in much of this area it is a relatively recent invader. It was introduced in Florida about 1917 and, from the vicinity of Tampa, spread both north and south. The western populations, originally in Texas and Louisiana, also spread, crossed the Mississippi River and moved eastward, where they have now met up with the ones that spread from stock introduced in Florida.

These animals are usually insectivorous but sometimes eat berries and fruit. They are active mainly at night and keep to brush and chaparral country. Armadillos have a good sense of hearing and probably also detect ground vibrations readily. They are excellent diggers, capable of quickly digging themselves out of sight in soft soil. Although some species of armadillo characteristically roll up like a ball when threatened, the Nine-banded species relies on running to a thicket or burrow. Throughout its home range of about 50 acres (20 hectares), it has many burrows—some of which are 25 feet (7.8 meters) long—into which it can escape. Armadillos do not hibernate and are sensitive to cold, so that winter temperatures are probably the main factor limiting their northward distribution.

One peculiarity of this species is that it always gives birth to quadruplets. Nine-banded Armadillos mate in midsummer, but implantation of the embryos is delayed. During this period of delay, however, the single fertilized egg divides and then splits until four embryos—identical quadruplets—are formed. After implantation, gestation takes 4 months, and the well-formed but soft-skinned babies, rather piglike in appearance, are born in an underground nest. Although they are nursed for two months, they start following their mother in search of insects when only a few weeks old. Because the babies are genetically identical, having come from a single fertilized egg, Nine-banded Armadillos are sometimes used as animals for laboratory research. The recent discovery that they can harbor the bacteria that cause leprosy has made them an important resource in the study of that disease.

Because of their bony covering, armadillos do not ordinarily float. They cross small streams by walking across the bottom. However, they are able to swallow air and thus inflate their intestines, and this gives them enough buoyancy so that they can swim across large expanses of water. Although they are sometimes accused of eating the eggs of game birds, this does not seem to be a prevalent habit, and, on the whole, their destruction of ants and insects considered pests by people makes them largely beneficial.

The only park in which Nine-banded Armadillos are native is Hot Springs, and they are often seen there at night. They also occur, from introduced stock, in Everglades, and are sometimes seen along the northern border of that park.

Nine-banded Armadillo: Everglades (I), Hot Springs

Hares, Rabbits, and Pikas

Hares, rabbits, and pikas are widespread throughout the world and consist of about sixty species. This group of mammals, although popularly thought to be rodents, are classified in their own order. Although they resemble rodents in some ways, they not only are fundamentally different but also evidently have a different origin and have been distinct from rodents for some 50 million years. Lagomorphs, as hares, rabbits, and pikas are called, have two pairs of upper incisor teeth, a small pair directly behind the prominent front ones; rodents have only a single pair of upper incisors. Other characteristics of the lagomorphs include hairy-soled feet, and testes that are located in front of the base of the penis, not behind it, as in rodents and all other mammals, except the marsupials. The lagomorphs are divided into two families— the pikas of Asia and western North America, and the hares and rabbits.

PIKAS

Of the fourteen species of pika, only two live in North America, exclusively in the mountainous regions of the west and northwest. Unlike hares and rabbits, pikas are diurnal and vocal, and they have short, rounded ears, no visible tail, hind legs that are not notably longer than the front ones, and eyes that are comparatively small.

The two species in North America are quite similar in appearance and habits and may only represent subspecies of a single species. The Collared Pika is the only one in the Alaskan parks or monuments, whereas the Pika is the only species in the Cascades, Sierra Nevada, and Rocky Mountains. These two species strongly resemble small guinea pigs. They are only 6–8 inches (152–203 millimeters) long and weigh only up to 6 ounces (170 grams). Gray in color, like the rocks over which they scamper, Pikas are active throughout the day and are usually associated with talus slopes and other areas of rock rubble. The Collared Pika ranges as low as sea level, but the Pika is more often found above 8,000 feet (2,438 meters) in mountains, and above timberline to 13,500 feet (4,115 meters) in some places.

Rock rubble provides pikas with armored underground access and relatively little need to burrow extensively, although, in the few places where they live away from rocks, they construct underground burrows. Each animal occupies a territory about 100 feet (30 meters) in diameter, which is defended against incursion from other pikas. The territory is proclaimed vocally, by urine deposits, and by rubbing a cheek gland against rocks to mark them. The pika's more extensive home range, which is shared with others, usually extends from the talus slopes into adjacent meadows, where most food is obtained.

Almost entirely vegetarian, pikas feed on many species of plant, especially grass and sedge. In summer, they commence cutting vegetation and carrying it to their territories. There it is laid out on or under rocks to cure and dry, thus hay is made. Pulling up plants by the roots and carrying them crosswise in its mouth, the pika scurries back to its territory and adds the plants to the growing haystack, taking about a minute for each round trip. It is not unusual to find more than twenty species of plant in a pika's store pile. Haystacks may be almost two feet high and three feet wide. Some pikas move the vegetation to sheltered spots after it has dried; others leave the pile where it was started, adding to it at a rate of almost one inch (25 milimeters) a day. Hay storage areas may be used year after year.

When winter comes, pikas use their stored hay for food, tunneling extensively under the snow to reach their supplies. They may also tunnel through the snow to forage on meadow plants. Although little is known of pikas' winter behavior, territories are probably maintained; the stored food would have to be reserved and protected for the animal that collected it. Their tracks in the snow are seldom seen. Thus it appears that all of their activity in midwinter seems beneath the snow. However, a sign of their

Pika

winter alertness is sometimes heard—their characteristic call coming from beneath the snow.

Pikas breed in April or May, and the two to six young are born one month later. They are naked and blind and weigh one-third of an ounce (9 grams) each. The babies grow rapidly and, by ten days, are toothed, furred, and ambulatory. The adult female may breed again soon after the first litter is born, so the young probably leave the mother's den when the second litter appears, at which time they are themselves a month old. At two months old, pikas are adult size and are gathering their own winter food supply.

Usually the first awareness that a park visitor has of the presence of a pika is hearing its high-pitched "kank" cry. Pikas are quite vocal, their sounds serving not only to alert all other pikas in the area to potential danger but also to announce territorial claims and probably to proclaim social status. The sound has a ventriloqual quality, thus the animal calling may not be readily sighted until it moves. Among the better parks in which to see pikas are Rocky Mountain, Glacier, Yellowstone, Crater Lake, Mount Rainier, Sequoia, and Yosemite.

COTTONTAILS AND BRUSH, MARSH, AND SWAMP RABBITS

Cottontails and brush, marsh, and swamp rabbits are a group consisting of nine North American species, seven of which are found in the parks and monuments discussed in this book. They are generally among the more frequently seen mammals in the parks. They are, despite some different common names, rather similar, with long ears, long hind legs, and a short tail. The four species called cottontail—the Eastern, Desert, Nuttall's, and New England—are remarkably similar in appearance, size, and habits, and two species that occur together often cannot be distinguished from each other in the field. The most widespread of the cottontails is the Eastern, which occurs mainly in the eastern two-thirds of the United States, except for New England. It is about 14—18 inches (356—457 millimeters) long. It is recognized by many people as the familiar suburban cottontail. Overlapping the range of the Eastern Cottontail in the east, and occurring from northern Vermont through the Appalachian Mountains to northwestern Georgia, the New England Cottontail is the same size and so similar in appearance that telling it from the Eastern Cottontail is virtually impossible in the field, and nearly so in the laboratory. The New England species prefers more wooded habitat, rather than brushy, open, and forest-edge areas, which are used by the Eastern Cottontail.

In the west, the two predominant cottontails are the Desert Cottontail and Nuttall's Cottontail. They are slightly smaller than the Eastern and New England species, up to 15 inches (381 millimeters) long, but telling the Desert from Nuttall's can be difficult. Usually the two do not occur together; the Desert Cottontail has a more southerly distribution and prefers lowlands, whereas Nuttall's usually lives at higher elevations and is more likely found in forested areas.

Cottontails are most active at dawn and dusk and are usually seen at these times, frequently along the edges of roads, where they feed on grasses. A great variety of plants are eaten, generally grasses and herbs, but cottontails, especially in winter, also eat bark. In agricultural areas they damage crop trees. Cottontails are not territorial, and the size of their home range varies with the habitat and the species. Eastern Cottontail males usually range over about 6 acres (2.5 hectares), whereas females have a home range of up to 4.7 acres (1.9 hectares). Desert Cottontails may have larger home ranges, averaging up to 15 acres (6 hectares); New England Cottontails have small home ranges of ½—2 acres (0.2—0.8 hectares). Although they are not territorial, cottontails utilize a social dominance hierarchy, and when these rabbits spar, fight, or chase one another, it is usually in relation to the social dominance behavior.

Cottontails breed for most of the year in the southern parts of their range and may have as many as seven litters during this time. However, where the breeding season is extended, the litter size averages smaller (around three) than in the northern areas, where there may be two litters and the number of young averages about six. The babies are born in a nest, which is usually about 6 inches (152 millimeters) deep in the ground and lined with grass, as well as fur torn from the mother's belly. The gestation period is about 4 weeks but varies. Although the females may breed immediately after giving birth, lactation may prevent quick development of the second litter. Newborn cottontails are about 4 inches (102 millimeters) long and weigh about 1.4 ounces (40 grams). They are lightly haired. They are blind when born, but within a week their eyes are open. By two weeks of age, cottontails have generally moved out of the maternal nest and are on their own.

Cottontail mortality is high, and the average life span in the wild is less than a year. Adults and young are preyed upon by a variety of birds and mammals, especially hawks and owls, weasels, foxes, Coyotes, and Bobcats. When not frightened, cottontails generally keep their tail with the white underside down; when alarmed, they raise it, and the white underside alerts other cottontails in the vicinity, and they all dash for the nearest cover. Once in some brush, they stop and lower the tail and effectively "disappear" from view. When less alarmed, cottontails "freeze" standing very still and alert, but dash off if whatever frightened them comes closer. Cottontails are readily seen at dawn and dusk in almost any national park where they occur.

The Brush Rabbit is a short-legged, short-eared, small cottontail, which averages about 10 inches (254 millimeters) in length. The one-inch (25-millimeter) tail has a white underside, as in the other rabbits of this group called cottontails. Because it is more nocturnal than some of the other rabbits and because it rarely travels farther than 30 yards (9 meters) from dense cover, the Brush Rabbit is not as often seen as the other species. Brush Rabbits live in Oregon and California, mainly west of the Cascades and Sierra Nevada, in chaparral and brush, and have a small home range. They breed from December through May in California and a month later in Oregon; they probably have three litters a year. The gestation period is about 27 days, and the usual number of young is three. The development of the babies is rapid, and, by two weeks of age, they leave the nest. As with other cottontails, the mother may breed soon after the litter is born. By the time they are five months old, perhaps even earlier, Brush Rabbits are sexually mature. They are small, secretive animals, with a small home range near dense cover, and thus are not usually seen in the national parks, except in Redwood, where they are the only rabbit and are frequently seen near the beaches.

Marsh Rabbits are found only in the southeastern United States, and the only national parks where they live are Everglades and Biscayne Bay.

They are abundant and easily seen in Everglades. They are dark brown in color and have an inconspicuous grayish underside to their tails, instead of the bright, white underside that is characteristic of the cottontails. As its name implies, this rabbit frequents wetlands and swamps, and nearby thickets, takes readily to water, and swims well. It is a fairly large rabbit, up to 18 inches (457 millimeters) long, and begins its breeding season in February, having several litters of two to five young a year. Overall, its habits are similar to those of the other cottontails, with the exception of its aquatic tendencies. At dawn, Marsh Rabbits are one of the most readily seen mammals in Everglades. A species with similar habits is the Swamp Rabbit. It lives in swampy areas from eastern Texas to Alabama and north along the Mississippi River to southern Illinois and Indiana. It resembles the Marsh Rabbit, but the two do not occur together anywhere, and the only park where Swamp Rabbits might occur is Hot Springs, although that is doubtful.

Brush Rabbit: Kings Canyon (?), North Cascades (?), Redwood, Sequoia, Yosemite; **Marsh Rabbit:** Biscayne, Everglades; **Eastern Cottontail:** Big Bend, Everglades, Great Smoky Mountains, Guadalupe Mountains, Hot Springs, Mammoth Cave, North Cascades (I), Saguaro, Shenandoah, Theodore Roosevelt (?), Voyageurs (?), Wind Cave; **New England Cottontail:** Great Smoky Mountains, Shenandoah; **Nuttall's Cottontail:** Bryce Canyon (?), Canyonlands, Crater Lake, Death Valley, Grand Canyon, Lassen Volcanic, Mesa Verde, North Cascades (?), Rocky Mountain, Theodore Roosevelt (?), Yellowstone, Zion; **Desert Cottontail:** Arches, Badlands, Big Bend, Canyonlands, Capitol Reef, Carlsbad Caverns, Death Valley, Grand Canyon, Guadalupe Mountains, Joshua Tree, Kings Canyon, Mesa Verde, Organ Pipe Cactus, Petrified Forest, Saguaro, Sequoia, Theodore Roosevelt (?), Wind Cave, Zion; **Swamp Rabbit:** Hot Springs (?)

HARES AND JACK RABBITS

Although they are called Jack Rabbits, the American animals that bear this name are actually hares, not rabbits. The difference between hares and rabbits is mainly their modes of reproduction. The offspring of rabbits are born with their eyes closed and are thinly haired; those of hares are born fully haired and with their eyes open, and they can run within hours of their birth.

Jack Rabbit is the name used for several species of long-eared hare that live mainly in the western United States. The shorter-eared, more northern species are called hares. Of the seven species that are native to North America north of Mexico, five are found in the national parks. The Snowshoe

Hare (also called Varying Hare and Snowshoe "Rabbit") is a species that is found from coast to coast, mainly in coniferous forest, and from Alaska to the mountains of central California, northern New Mexico, and North Carolina. It is about the size of a large cottontail, up to 20 inches (508 millimeters) in length. When it occurs with cottontails, it can generally be distinguished by its dark brown summer coat and usually by the absence of a clear white underside to its tail. In winter these hares molt to white hair, from which their common name, Varying Hare, is derived. Snowshoe Hares are creatures of the forest and dense brush and are less likely to be seen than cottontails when the two occur in the same park or monument. Like the Snowshoe Hare, the Northern Hare also lives in Alaska and characteristically molts from brown summer color to white winter pelage. It is much larger, however, than the Snowshoe Hare; it may reach 2 feet (0.6 meters) in length and up to 10 pounds (4.5 kilograms) in weight, compared with the 4 pounds (1.8 kilograms) of the Snowshoe Hare. The Northern Hare is mainly a creature of open tundra.

The open prairies and deserts are the domain of the Black-tailed Jack Rabbit (see plate 5a), a large species with ears 6–7 inches (152–178 millimeters) long. It can be distinguished from the other jack rabbits in its area by its black-tipped ears and a black streak on the top of its tail. It ranges from southeastern Washington to Mexico, and east to southern South Dakota and central Arkansas. The White-tailed Jack Rabbit, as its name implies, has a tail that is white both above and below. Although it is about the same size as the Black-tailed Jack Rabbit, it has slightly shorter ears (5–6 inches; 127–152 millimeters). Its winter coat is rather pale gray, in contrast to its summer one, which is brownish. White-tailed Jack Rabbits prefer open grasslands or sagebrush flatlands for their habitat. In south-central Arizona, and ranging down into Mexico, the Antelope Jack Rabbit is a species that has big ears (7–8 inches; 178–203 millimeters long). It can be distinguished from the Black-tailed Jack Rabbit by the absence of black on its ear tips and by its almost totally white sides.

Snowshoe and Northern Hares are well adapted for the snowy climates where they live; they not only turn white in winter for concealment but also have large, well-haired hind feet, which function as snowshoes to enable them to get about easily on the surface of snow. Their relatively short ears function to conserve heat. In summer they feed on grasses and other green vegetation, but in winter their main source of food is the bark of trees, especially willows and aspens. The gestation period for the Snowshoe Hare is about 5 weeks, and, although the young (usually three) can run and feed within a few hours after birth, they may be nursed for as long as a month. There may be two or three litters in summer. The Northern Hare generally has only a single litter of four to eight young, born in midsummer.

The change of coat color in these species is induced primarily, but not entirely, by the amount of daylight; the coat turns white as the days shorten in the autumn, and it turns brown during an increasing photoperiod in the spring. However, day length is not the entire influence, for, in Olympic

Park, Snowshoe Hares do not change to white in the winter, and, at Mount Rainier, those at high elevations become white, whereas those living below 3,000 feet (914 meters) do not.

Snowshoe Hares undergo cyclical population fluctuations, more or less on a nine-to-eleven-year period. After building to enormous numbers, the populations "crash," so that scarcely a hare can be found the next year. Predators that depend mainly on Snowshoe Hares for food, especially the Lynx, have population fluctuations that follow those of the hare, but one year behind, so that the Lynx population peaks just when the hares are at their lowest and "crashes" the next year. Normally remaining in forest cover, Snowshoe Hares have a small home range of about 10 acres (4 hectares) and are primarily nocturnal. They are less often seen than cottontails, but their tracks in winter snow are evidence of their abundance. One good place to see them is along the beach in Olympic Park, early in the morning.

Jack rabbits are mainly nocturnal and are conspicuous to drivers at night along high plains and desert roads. Their long ears, from which their common name ("Jackass Rabbit") is derived, function to let them lose heat in the warm climates where they live; the southern kinds have longer ears than the northern ones. Long hind legs enable them to leap as much as 17 feet (5 meters) in a single bound and run as fast as 40 miles (64 kilometers) an hour. When pursued, jack rabbits characteristically make every fourth or fifth leap an especially high one, which enables them to get a view over the brush at their pursuers. In summer jack rabbits feed mainly on green vegetation, but in winter they rely on the bark and twigs of woody plants. They rest during the day in shallow depressions under bushes or shrubs and do most of their feeding at night, although it is not unusual to see them, where they are not hunted, foraging late in the afternoon or well into the morning.

The well-furred and open-eyed young are born in a shallow nest after a gestation period that is probably about 5–6 weeks long. Most of the breeding is done in the warmer part of the year, and, in the south, reproduction may continue for ten months of a year, with up to four litters per female and two to four young in a litter. The babies are nursed only at night, probably to avoid attracting predators, and, before they are a month old, they are on their own. Fairly large numbers of these hares may exist in a relatively small area, and the White-tailed Jack Rabbit may undergo sizable population fluctuations. Jack rabbits are a main source of food for Coyotes (as cattle ranchers learned when the jack rabbit populations increased as a result of the elimination of Coyotes) and foxes, as well as avian predators.

Black-tailed Jack Rabbits are readily seen in the southwestern parks and monuments, especially Big Bend, Saguaro, and Organ Pipe Cactus, whereas the White-tailed Jack Rabbit can be found, although less readily, at Yellowstone and Glacier. The Antelope Jack Rabbit occurs only in Organ Pipe Cactus and Saguaro and is less often seen than the Black-tailed Jack Rabbit, which also occurs in those monuments.

Snowshoe Hare: Acadia, Capitol Reef (?), Crater Lake, Denali, Glacier Bay, Glacier, Grand Teton, Isle Royale, Katmai, Lassen Volcanic, Mount Rainier, North Cascades, Olympic, Rocky Mountain, Voyageurs, Yellowstone, Yosemite; **Northern Hare:** Katmai; **White-tailed Jack Rabbit:** Badlands, Bryce Canyon (?), Capitol Reef (?), Glacier, Grand Canyon (?), Grand Teton, Kings Canyon, North Cascades (?), Rocky Mountain, Sequoia, Theodore Roosevelt, Wind Cave, Yellowstone; **Black-tailed Jack Rabbit:** Arches, Big Bend, Canyonlands, Capitol Reef, Carlsbad, Death Valley, Grand Canyon, Guadalupe Mountains, Joshua Tree, Kings Canyon (?), Mesa Verde, Organ Pipe Cactus, Petrified Forest, Saguaro, Sequoia, Yosemite, Zion; **Antelope Jack Rabbit:** Organ Pipe Cactus, Saguaro

Rodents

There are more species of rodent in the world than any other kind of mammal; about 40 percent of all mammals are rodents. They are mostly small animals, the largest in the world being the South American Capybara, an aquatic guinea pig–like animal that may weigh more than 150 pounds (68 kilograms). The largest rodent in North America is the Beaver. Rodents are characterized by their strong, ever-growing, chisellike upper and lower pairs of incisor teeth, and by the large space between the incisors and the molars; no rodent has canine teeth. Although many people think that rabbits and hares are rodents, they are not. They actually have two pairs of upper incisors, and they are a group of mammals that has had a separate and different evolutionary origin from the rodents, despite their resemblances.

Eight of the world's thirty-four families of rodent are native to America north of Mexico, and about 10 percent of the 1,700 species live here. Most of the rodents that are active and visible during the day are discussed in this book, but these compose only about one-third of the North American species and are mainly the larger ones. The remainder fall mainly under the headings of rats, mice, and pocket gophers, and there are more than one hundred species of them.

RATS, MICE, AND POCKET GOPHERS

The North American pocket gophers, a dozen or so species, are mainly in the western and southern states. All are adapted for spending their lives underground and have small eyes and ears, and powerful forelegs and claws. They get their common name from the external, fur-lined cheek pouches (pockets), which they use to carry food or bedding. They are wholly vegetarian in their diet. They may be active day or night, but do not often venture above ground. The usual sign of their presence are the mounds of dirt that they have pushed to the surface. These may be mistaken for mole hills, but gophers and moles do not usually occur in the same areas, and the mounds of gophers are less conical and symmetrical than those of moles because the gopher pushes dirt to one side of the hole when it excavates, whereas the mole pushes the dirt straight up. Another sign of pocket-gopher activity is a ropelike cast of dirt that remains on the ground after snow has melted. In winter, as gophers burrow beneath the snow, they often pack tunnels with excavated soil, and these casts (about 2 inches wide) remain when the snow has melted. They are sometimes called gopher "worms." Because there is so little chance of ever seeing a pocket gopher above ground, these animals are not discussed in this book.

There are about one hundred species of animal that are called rats or mice in America north of Mexico. There is no adequate definition of what is called a mouse or what is called a rat. Although generally the larger species are called rats and the smaller ones mice, there are "rats" that are smaller than "mice." Rats and mice are grouped into a variety of families and subfamilies, including the pocket mice and kangaroo rats, the voles and lemmings, the jumping mice, the white-footed mice, cotton rats, rice rats, and many more. In addition, in many of the parks there are nonnative rats and mice, especially the Norway or Brown Rat and the House Mouse, both of which were unintentionally carried to North America from Europe and which have spread throughout the United States.

Rats and mice vary in size from tiny pocket mice, which have a body only about 2½ inches (64 millimeters) long and weigh a quarter of an ounce (7 grams), to chunky meadow voles, which have 6-inch (152-millimeter) bodies that weigh as much as a quarter of a pound (113 grams) or more.

Each kind of mouse or rat is highly adapted to a particular habitat or environment. Some pocket mice and kangaroo rats are so well adapted to desert life that they may never drink water, obtaining all the moisture they need from the metabolism of their food. Although most rats and mice feed on forms of vegetation (seeds, leaves, grasses, bark, buds), a few, such as grasshopper mice, are mainly insectivorous and may even prey on other species of mouse. Most rats and mice are nocturnal and ground dwellers. They are usually not seen by park visitors during the day, although hikers may sometimes glimpse a jumping mouse bounding away through the grass or see the dark shadow of a meadow vole that is scurrying through a runway in the grass alongside a mountain stream. At night in cabins or around campsites, rats and mice may venture forth, but are more often heard rustling than seen foraging. Identification of these various rats and mice is difficult for the inexperienced, especially in places where similar kinds occur. Because rats and mice are so seldom seen, this book deals mainly with the more visible and diurnal rodents, the tree and ground squirrels (but not the nocturnal flying squirrel), Beavers, Muskrats, and Porcupines, and the relatively large but nocturnal rats—the woodrats—which do frequent campsites and which also make conspicuous nests that are often seen.

Mountain Beaver

MOUNTAIN BEAVER

Described as looking like a tailless Muskrat or a small Woodchuck, the Mountain Beaver is a chunky-bodied rodent about 1–1½ feet (0.3–0.5 meter) long, with a one-inch (25-millimeter) tail. It is generally regarded as the most primitive of all living rodents, and the single living species is put in its own family. It is neither a Beaver nor a close relative of one, nor does it especially inhabit mountains. Its habits are more similar to those of the fossorial pocket gophers than to any other species.

Mountain Beavers occur only in the northwest, from southern British Columbia to northern California, and live in moist situations, usually forest meadows or thickets, often along a stream. There they construct extensive burrows, usually shallow but up to 5 feet (1.5 meters) deep and 4–10 inches (102–254 millimeters) in diameter. The nest chamber, lined with vegetation, is usually in the deepest section. Branches of the burrow contain rooms for storing food and also garbage and feces, and there are several entrances to the burrow system. In winter, Mountain Beavers continue to burrow, piling dirt into snow tunnels, where, when summer comes, the 6-inch (152-millimeter) wide dirt cores are visible on the ground.

Mountain Beavers are entirely vegetarian, harvesting a variety of lush grasses and herbs, but also feeding on the tips of deciduous trees, such as willows and alder, occasionally climbing to obtain them. In summer, a characteristic sign of their presence is the "hay pile," a small stack of cut grasses and leaves laid out on the ground, evidently to wilt and dry. Whether the purpose of these haystacks is to provide winter food, for they are subsequently carried underground, or to serve as dry bedding for the nest, is still a matter of dispute. It seems likely, however, judging from the volume, that much of the haystack is used as winter food. It is known that Mountain Beavers venture forth under snow in the winter, as they do not hibernate, and tunnel their way to the bases of deciduous trees, where they gnaw on the bark and clip twigs.

The breeding season is late in the winter, and the gestation period is about a month. The three to five young are born in the underground nest and are nursed for at least six weeks. By the time they are three months old, in early autumn, they leave to establish their own burrows. They are two years old before breeding. In most places, Mountain Beavers are nocturnal, but can occasionally be seen out at dawn or dusk. They evidently function best at low-light levels, because they are sometimes abroad on overcast days. Their home range seems small, perhaps only about one-third of an acre (0.1 hectare), and populations are small, even though extensive burrowing may give the appearance that a large number of animals are involved.

In many places, Mountain Beavers seem to have a restricted distribution, and colonies fluctuate in size and extent, sometimes disappearing for a few years from places they once occurred and returning later. These animals are seldom seen by park visitors, although sign of them can usually be found in Olympic, Mount Rainier, Crater Lake, and Redwood parks.

Mountain Beaver: Crater Lake, Kings Canyon, Lassen Volcanic, Mount Rainier, North Cascades, Olympic, Redwood, Sequoia, Yosemite

---

SQUIRREL FAMILY

Most members of the squirrel family are diurnal and are therefore among the most frequently seen mammals in the national parks and monuments. More than sixty species of this family live in North America, including animals commonly called chipmunks, marmots, ground squirrels, prairie dogs, tree squirrels, and Woodchucks. Some forty-five species occur in one or more of the parks and monuments, and, except for two species of flying squirrels which are wholly nocturnal, all are active during the day and most are relatively conspicuous.

The squirrel family shows great diversity of habitat preference and specialization, as well as a variety of social structures. Prairie dogs and marmots are colonial and highly social, whereas Red Squirrels and Woodchucks are territorial and intolerant of others of their kind. Most ground squirrels hibernate in cold weather, and some estivate in hot, dry weather; tree squirrels are active throughout the year. Prairie dogs and Woodchucks are excellent diggers, whereas tree squirrels specialize in arboreal life. Tufted ears, bushy tails, and uniform coloration characterize some species; bold stripes or spots and strongly marked tails are usual in others.

Most members of this family have a fairly high reproductive rate to compensate for their high mortality. They usually form the primary food source for many small- and medium-sized carnivores, especially members of the weasel family. All squirrels are primarily plant eaters; some mainly eat grass, others eat seeds and nuts. Many species store food for use at less productive times of the year, or accumulate much fat as an energy supply for hibernation or estivation.

Chipmunks. Fifteen of the twenty-one North American species of chipmunk are found in the parks and monuments featured in this book. Many of the western parks have two or more species, but differentiating the kinds of chipmunk by field observation is difficult if not impossible. Chipmunks are not only hard to identify, even using museum specimens, but their classification fluctuates as scientists use new characters and new techniques to distinguish the various but similar species. Adding to the confusion is the fact that the Golden-mantled Ground Squirrel has coloration and body stripes similar to those of chipmunks, as do some Antelope Squirrels. However, chipmunks can always be identified as the only North American ground squirrels that have facial stripes, usually a single dark stripe that runs through the eye and that is bordered by white stripes. Mammalogists divide chipmunks into eastern and western groups.

Eastern Chipmunk. The single species of Eastern Chipmunk inhabits open forests, hedgerows, forest edge, and rock walls from eastern North Dakota to the East Coast, and in suitable locations as far south as Louisiana. The only park where it occurs together with a western species is Voyageurs. There, the Eastern Chipmunk's slightly larger size, its rust-colored rump, and the absence of side stripes extending to the base of the tail distinguishes it from the Least Chipmunk.

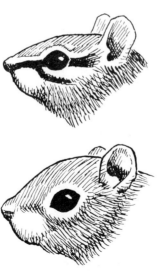

top, *Chipmunk;* bottom, *Gold-mantled Ground Squirrel*

Eastern Chipmunks are ground dwellers, although they climb trees for food. They customarily live in burrows which they excavate themselves, the entrance of which is about 2 inches (51 millimeters) in diameter and is often concealed under a rock or fallen log. No pile of dirt or mound marks the entrance, and the burrow may be 15 feet (4.6 meters) long and 3 feet (0.9 meters) below the surface of the ground. The nest is a chamber about one foot (0.3 meters) in diameter and half as high and is lined with dry leaves. Other hollowed-out chambers serve as storage areas for food, and there are usually several entrances to the burrow.

Seeds and nuts are an essential part of the Eastern Chipmunk's diet; when nuts are ripe in late summer, it begins storing vast quantities in the underground chambers of its den. Using the cheek pouches inside its mouth, each of which has a capacity equivalent to a heaping tablespoon, the chipmunk gathers as many seeds or nuts as it can at a time before scurrying off to cache them. One was seen stuffing thirty-one large corn kernels in its pouches, another carried eight white-oak acorns, and another carried four hickory nuts. In storage, there may be dozen or more quarts of nuts and seeds. Perishable foods, which are eaten on the spot, include berries, mushrooms, and the seeds and fruits of many trees and bushes. These chipmunks also eat a variety of animal matter, including beetles, slugs, snails, butterflies, and even bird eggs and fledglings. However, these items are a comparatively small part of a chipmunk's diet.

Eastern Chipmunks have a relatively small home range, usually less than 2 acres (0.8 hectares) in area. Although other chipmunks may be tolerated in some parts of the home range, chipmunks seem to maintain exclusive territories around their burrows and are not especially sociable. Except when the female has young in her den, chipmunks seem to maintain their den only for their own use. In the northern parts of their range, most Eastern Chipmunks hibernate, but some individuals may be seen above ground even on cold but clear winter days. They may go into deep torpor, with lowered body temperature, breathing, and heart rates for days at a time, but evidently may also awaken, feed, and even venture out into snow before retiring again to the burrow to hibernate some more. In the south, they remain active throughout the year.

Breeding takes place in the spring, usually April, and about four (to as many as eight) hairless, blind babies are born about a month later. Their hair appears in a week and their eyes open in about a month. When about six weeks old and two-thirds the size of an adult, the young first venture forth from the burrow, follow the mother, and begin to sample solid food. By two months, they are adult size. Some may go off on their own. Usually the female breeds again in July or August, and the young of the second litter may spend the winter denned with their mother. These young breed, as do the first litter, in the next spring breeding season, when less than a year old.

Eastern Chipmunks are noisy animals; their two most characteristic sounds are a high "chick-chick" usually accompanied by the twitching of the tail, which is a sign of nervous agitation and an alarm call. They also make a rapid "chip-chip-chip" trill that may be "sung" for minutes at a time, several animals sometimes forming a chorus. The latter is probably used in social communication. Chipmunks have broad food tastes and thus forage around campsites and picnic areas, as well as at viewpoints, where park visitors like to feed them. Feeding of chipmunks, however, is no longer permitted in the parks. The kinds of food offered them are usually unsuitable for storage, and these animals may be unprepared for winter and may not survive it.

In some parks, Eastern Chipmunks may not be abundant because the habitat is not ideal. The best parks in which to see Eastern Chipmunks are Shenandoah, Great Smoky Mountains, and Hot Springs.

Western Chipmunks. The western chipmunks (see plate 2b) are a group that has adapted to a variety of habitats, from dry sagebrush flats, brushy foothills, mountainous coniferous zones, and high talus slopes to cool, moist redwood and hemlock forests, and from sea level to above 13,000 feet (3,962 meters) in the mountains. Except for the Least Chipmunk, all of the western species are found in parks from the Rocky Mountains, eastern New Mexico, and western Texas westward. Distinguishing species of western chipmunk from one another in the field is difficult in some places, impossible in others. Often the altitude and habitat are the best guides to what species is likely to be observed, but in many places two or more species either forage together or not far from one another. Individual, age, and seasonal variation in pelage sometimes results in conflicting identifications. In Voyageurs National Park, the only park in which the Least Chipmunk and the Eastern Chipmunk occur, the former is smaller, has stripes reaching the rump, and lacks a rust-colored rump patch.

Although western chipmunks do most of their foraging on the ground, they climb well and are not averse to climbing high into trees for nuts and seeds and even occasionally for bird eggs. They stuff their cheek pouches with nonperishable nuts or seeds and scurry off to bury them in small depressions in the ground. Many of these caches are not used, and these chipmunks are responsible for much reforestation by burying seeds, especially in burned or lumbered sites. Their predations on insects and insect larvae and pupae are generally considered beneficial to human interests.

The burrow entrance, about 2 inches (51 millimeters) in diameter, is often under a rock or log, and western chipmunks may not make as extensive a burrow excavation as the eastern species, perhaps only 5 feet (1.5 meters) long, and the nest area may only be 5 inches (127 millimeters) in diameter. Like the Eastern Chipmunk, some species of Western Chipmunk store food in burrow chambers and probably use it during waking periods in winter hibernation, as well as in spring, when other food may be scarce. Depending upon the species and the habitat, western chipmunks may hibernate for nearly six months or scarcely at all.

Western chipmunks mate as early as March in warmer areas and as late as May in others; in some places there may be a second litter produced in midsummer. The gestation period is about a month, and the usual number of young is three or four. The babies begin to leave the maternal den when they are a little more than six weeks old, and the early litters, at least, may be on their own by late summer. Although they are preyed upon by carnivores and birds, chipmunks live as long as five years in the wild. Just what the social structures of western chipmunk species are is not well known. Certainly they aggregate in large numbers at viewpoints in many national parks and seem relatively tolerant of one another under these conditions. They probably have home ranges of less than one acre (0.4 hectares), in places where food is relatively abundant. Western chipmunks have an extensive vocal repertoire, and there are differences in the calls of each species.

In most western national parks, it is virtually impossible to avoid seeing some of these chipmunks in summer. In places their populations at visitor sites number in the thousands. Among the better parks in which to see western chipmunks are Bryce Canyon, Sequoia, Yosemite, and Rocky Mountain.

Eastern Chipmunk: Acadia, Great Smoky Mountains, Hot Springs, Mammoth Cave, Shenandoah, Voyageurs; **Alpine Chipmunk:** Kings Canyon, Sequoia, Yosemite; **Least Chipmunk:** Badlands, Bryce Canyon, Capitol Reef (?), Glacier, Grand Canyon, Grand Teton, Mesa Verde, North Cascades (?), Rocky Mountain, Theodore Roosevelt, Voyageurs, Wind Cave, Yellowstone, Zion; **Yellow-pine Chipmunk:** Crater Lake, Glacier, Grand Teton, Lassen Volcanic, Mount Rainier, North Cascades, Olympic, Yellowstone; **Townsend's Chipmunk:** Crater Lake, Lassen Volcanic, Mount Rainier, North Cascades, Olympic, Redwood, Yosemite; **Merriam's Chipmunk:** Joshua Tree, Kings Canyon, Sequoia, Yosemite; **Cliff Chipmunk:** Bryce Canyon, Capitol Reef (?), Grand Canyon, Saguaro, Zion; **Colorado Chipmunk:** Arches, Canyonlands, Capitol Reef, Kings Canyon (?), Mesa Verde, Rocky Mountain (?), Sequoia, Zion; **Red-tailed Chipmunk:** Glacier; **Gray-footed Chipmunk:** Guadalupe; **Long-eared Chipmunk:** Yosemite; **Lodgepole Chipmunk:** Kings Canyon, Lassen Volcanic, Sequoia, Yosemite; **Panamint Chipmunk:** Death Valley; **Uinta Chipmunk:** Bryce Canyon, Grand Canyon, Grand Teton, Rocky Mountain, Yellowstone, Yosemite, Zion

---

Woodchuck and Marmots.   The largest members of the squirrel family—the Woodchuck and marmots—have heavy bodies, short legs and tails, and large, blunt-nosed heads. These rodents live mainly in the mountains of the west and in clearings in eastern woodlands. They are ground dwellers, excavating their own burrows, and they feed largely on grasses and similar vegetation and do not store food. They hibernate in winter, putting on large amounts of fat in the fall to sustain them. Of the six North American species, four are found in U.S. national parks.

Woodchuck.   Also called "Groundhog," the Woodchuck is the only

Woodchuck

member of the marmot group of the squirrel family that inhabits the eastern parks. A grass eater that has prospered by forest-clearing and agriculture, it is a visible and familiar species along many roadsides. About 6 inches (152 millimeters) tall at the shoulder, the Woodchuck may be as long as 20 inches (508 millimeters) and has a tail that is 4–6 inches (102–152 millimeters) long. The usual weight of a male is 5–10 pounds (2.3–4.5 kilograms), and females are a little smaller and lighter. However, their weights vary enormously depending upon the season; they grow very fat in autumn and may lose almost half their total weight by the time they emerge from hibernation in spring. They are generally brownish in color, and their bodies are grizzled with frosted hairs, and their feet are blackish. They often sit on their haunches, peering out over grass.

In nonagricultural areas, Woodchucks usually dig their dens a few feet inside forests at the edge of meadows, but may also den in the middle of cleared fields. The opening of the den, 8–12 inches (203–305 millimeters) in diameter, is conspicuous and usually has a mound of dirt around it. The tunnel slants down to a depth of 3–6 feet (0.9–1.8 meters) and then may extend as much as 45 feet (14 meters), with several nest chambers branching upward from the main tunnel. About 15 inches (381 millimeters) in diameter, the nests are lined with dry leaves and grasses. Other entrances to the burrow are smaller and are not marked by mounds of dirt. They probably serve as emergency exits and entrances. Abandoned dens are used as homes by a variety of other animals.

Woodchucks are solitary and territorial. Their home range, which may be shared with other Woodchucks, is usually 300–600 feet (91–193 meters) in diameter and may cover 2–160 acres (0.8–65 hectares) in area. It is smaller where food is abundant, larger where the animals must travel more for forage. The area close around the den is not shared but is defended from other Woodchucks; it is identified by scent from glands, as well as by the low whistles given by the resident when it emerges from the den.

In summer and fall, Woodchucks become fatter, eventually depositing a blanket of fat that may be a half-inch (13 millimeters) thick over the back and shoulders. As cool weather approaches, the animals seem more lethargic and are seen only on warmer, sunny afternoons. As early as the end of September in the north and later in the south, Woodchucks go into their dens, usually in the woods, to hibernate. In a grass-lined chamber, often plugged with dirt and below the frost line, the animal becomes torpid. The body temperature drops from its usual 99° F. (37° C.) to between 38° F. and 57° F. (3° C. and 14° C.), and the heartbeat slows from more than one hundred beats to four beats per minute. The animal may breathe only once every six minutes. In this state of reduced metabolism, Woodchucks pass the winter. If dug from the burrow at this time, they cannot be awakened for hours. Folklore claims that Woodchucks awaken on 2 February to venture forth and, by the presence or absence of their shadow, predict the extent of winter. This tale had its origin in central Europe in conjunction with Candlemas Day, and with hedgehogs as forecasters. However, over most of their range, Woodchucks emerge from hibernation in late February or early March, at which time they may have lost from one-fourth to one-half of their prehibernation weight. A 10-pound (4.5-kilogram) adult male lost 5.5 pounds (2.5 kilograms) between 14 October and 22 March in its hibernation den in Maryland.

Mating takes place soon after emergence from hibernation, with males wandering widely in search of females. At this time, males may fight one another, and torn ears and scarred bodies are evidence of the battles. A male and female may remain together for only a day or two before separating. The young—usually four or five—are born about 32 days later, are naked and blind, and weigh about one ounce (28 grams). Woodchucks' eyes open when they are a month old, and, at an age of six or seven weeks, they begin to move out of the maternal den. When a little more than two months old, the babies are driven away by the mother and establish their own dens and territories. By hibernation time, the young Woodchucks weigh about 4.4 pounds (2 kilograms), which evidently is fat enough for them to survive the winter, although some may weigh as little as 2.6 pounds

(1.2 kilograms) by spring, having lost more than 40 percent of their autumnal weight. Young animals may breed in their first year. Woodchucks have lived as long as six years in the wild.

Woodchucks make several whistling sounds, some of which are probably used in territory identification, and others when startled, as well as sounds in social contexts between mothers and young. These large rodents are among the more visible species, usually active early and late on hot summer days, and more active throughout the day if it is warm and sunny, as fall approaches. Mammoth Cave National Park is a excellent place to observe Woodchucks and so are places in Great Smoky Mountains National Park.

Marmots. Throughout the western mountains, marmots are an often-seen large rodent. Three species inhabit the national parks—the Yellow-bellied Marmot, from the Rockies to the Sierra Nevada; the Hoary Marmot, from northern Montana and Washington to Alaska; and the Olympic Marmot, restricted to the Olympic Peninsula. Because no two species occur in the same park, nor do any coexist with Woodchucks, the locality serves to identify any of these large-bodied (to 20 inches, 508 millimeters) short-tailed (8 inches, 203 millimeters) animals. As its name implies, a yellow underside is characteristic of the Yellow-bellied Marmot (see plate 6a), whereas a grayish, frosted appearance identifies the Hoary Marmot and its close relative, the Olympic Marmot. Hoary Marmots tend to be the heaviest of the group, sometimes reaching 20 pounds (9 kilograms) in weight.

Unlike the solitary and intolerant Woodchuck, western marmots are much more social, tolerant, and playful. The Hoary and Olympic Marmots are the most social, living in colonies that sometimes consist of dozens of animals. The degree of sociality may be a result of the short summer growing season for the young, as well as limitations of den sites and hibernation sites imposed by the environment. Yellow-bellied Marmots, which may have less than three months to feed in summer, are less than one-fourth their adult weight when their first hibernation season arrives and must overwinter in the hibernation den with their mother. As yearlings, when they are 60 percent their adult weight, they disperse; they may have their first litter when two years old. Hoary and Olympic Marmots, in even more severe climates and with as little as two months of foraging time in which to grow, are only about one-third their adult weight when they are yearlings, and they remain with the mother until they are two years old. By this time, they are 70 percent of their adult weight and are capable of dispersing. They have their first litters when they are three years old.

The Yellow-bellied Marmot colony is essentially a harem that consists of a territorial male and numerous females and their offspring. Other marmot males may have their dens on the periphery of the territory. The average territory size is 1.67 acres (.67 hectares). Hoary and Olympic Marmot colonies may contain more adult males with smaller harems, and the males seem not to defend territories, but are quite tolerant of all colony residents.

Marmots constantly greet one another by sniffing cheeks (there is a gland in each cheek) and arching their tails. The larger the colony, the more greeting and grooming takes place, often several times an hour. Not only do these actions reaffirm the identity of the colony members but they also serve to determine the status of individuals. A marmot from the periphery of the colony is usually chased and shrieked at, and it shows its subordinate status by keeping its tail down or letting itself be groomed. For Yellow-bellied Marmots, at least, the area around the den mouth seems to give status to the resident, as a territory generally does.

Further evidence of the social nature of the western marmots is their usual communal hibernation. A mother and her yearlings spend the winter together, and an adult male generally hibernates with one or two adult females. This sociality may be a result of the limited number of suitable places for hibernation. In the mountains of Colorado, Yellow-bellied Marmots may hibernate for eight months—from early September to May—and animals that weighed 10.6 pounds (4.8 kilograms) emerge in spring weighing half that amount.

Breeding takes place within two weeks of emergence from hibernation and a territorial male Yellow-bellied Marmot mates with all the adult females of his harem. The young are born in the den about 31 days later and weigh about one ounce (28 grams). They are weaned when they are six weeks old and weigh about one pound (453 grams). There are usually four or five young in a litter. During the long summer days, marmots feed on succulent grasses in the mountain meadows in the morning, then usually sunbathe on rock surfaces, and feed again later. By dark, however, they have all returned to their burrows. They also seem to remain underground more when it is cloudy and stormy. Often one animal seems to serve as a sentinel, watching for predatory birds, such as eagles, and mammals, such as foxes, Coyotes, Mountain Lions, Bobcats, Lynxes, domestic dogs, and human beings. The shrill whistle alerts the colony, and the animals scurry to their dens. Alarm whistles are often given even when no danger is evident, and marmots are sometimes locally called "Whistlers."

Marmots require rocky areas adjacent to meadows for their home ranges, and so are usually found around talus slopes and rocky clearings in the mountains. There are several entrances to a burrow, which may extend 14 feet (4.4 meters) into a hillside and have several chambers. There are usually well-marked trails between burrows. Among the better parks in which to see marmots are Rocky Mountain and Yosemite, for the Yellow-bellied Marmot, and Olympic and Denali, for the Olympic and Hoary Marmots.

Woodchuck: Acadia, Great Smoky Mountains, Hot Springs, Mammoth Cave, Shenandoah, Voyageurs; **Yellow-bellied Marmot:** Bryce Canyon, Capitol Reef, Crater Lake, Kings Canyon, Lassen Volcanic, Rocky Mountain, Sequoia, Yosemite, Zion; **Hoary Marmot:** Denali, Glacier Bay, Glacier, Katmai, Mount Rainier, North Cascades; **Olympic Marmot:** Olympic

Antelope Squirrels. Of the four species of antelope squirrel that occur north of Mexico, three are found in national parks or monuments, and no two occur in one park. They are often confused with chipmunks because they too have a white stripe on each side, but, unlike chipmunks, antelope squirrels do not have any facial stripes. The most widespread species is the White-tailed Antelope Squirrel, which is found west and north of the Colorado River in California and Nevada, and east of it in southeastern Utah and northwestern New Mexico. The closely related and very similar Texas Antelope Squirrel inhabits central and southeastern New Mexico and southwestern Texas. Both species have a conspicuous bright white underside to the tail, but the other species, Harris' Ground Squirrel, which lives in Arizona, south and east of the Colorado River, has a tail with a grayish underside.

Antelope squirrels are well adapted to desert life, and the habits of all three species are similar. They are extremely tolerant of desert heat, partly because they can endure a broad range of internal body temperature. An antelope squirrel can have an internal temperature of 110° F. (43.3° C.) before it has to cool itself. It usually does this by seeking shade, going into its burrow, or, as a last resort, by salivating and smearing the resultant moisture over its head and face with its forepaws, deriving a cooling effect from the evaporation. By being able to let its body temperature rise so high before needing to cool it, by utilizing water derived from its metabolism of dry seeds, and by excreting concentrated urine to conserve water, the antelope squirrel survives hot desert conditions. These squirrels obtain moisture by eating green vegetation, including cacti, but can survive for several weeks eating nothing but dry seeds; they drink water when it is available. Another characteristic that may function in their temperature regulation is their habit of running with their tails raised so that they are flat against their backs. Thus the back is shaded, and the white underside reflects solar heat.

In addition to seeds, berries, fruits, and succulent vegetation, the diet of antelope squirrels consists of grasshoppers, grubs, crickets, beetles, and other insects. In captivity they eat meat. Antelope squirrels gather seeds in the cheek pouches inside their mouths and then store the seeds in their burrows. One naturalist counted forty-four mesquite beans stuffed into a squirrel's cheek pouches. In the northern parts of their range, antelope squirrels may hibernate, and stored food can save them if they awaken during winter, or serve to feed them when they emerge in spring.

Although antelope squirrels live in burrows, they are not as proficient at digging as are many other ground squirrels and often utilize the abandoned tunnels of some other species. Because they usually live in sandy soil, digging is not difficult, and each animal may have several burrow systems within its home range. The main den, with an inconspicuous hole for an opening, goes straight down about one foot (0.3 meter) before leveling off to wind around roots and rocks, giving the animal some protection from Badgers, which may try to dig them out. The nest chamber is lined with soft, dry vegetation and hair. In addition to the nest burrow, there are other burrows—also about one foot (0.3 meter) deep, but short—in which antelope squirrels can take refuge from predators, as well as heat, during their food searches.

Perhaps because their food sources are widely scattered, antelope squirrels have large home ranges and are known to roam hundreds of yards (meters) from their nest burrow. A typical home range may encompass a dozen acres (4.9 hectares). Little seems to be known of the breeding habits of these species. Young are born in the spring, and the litters are large, averaging eight; there may be a second litter born later in the year. When the babies are about half their adult size, they emerge from the burrow and play at its edge, but the mother does not seem overly protective of them. Predation by raptorial birds and predatory mamals and other mortality must be high to warrant such a high reproductive rate, for some adults are known to have lived for six years in the wild.

Although not colonial and often sparsely distributed, antelope squirrels are conspicuous inhabitants of some parks and monuments, where their midmorning and afternoon activity allows them to be readily observed. Saguaro and Organ Pipe Cactus are the best parks in which to see Harris' Antelope Squirrel; Big Bend and Guadalupe mountain, the Texas Antelope Squirrel; and Zion, Arches, and Capitol Reef, the White-tailed Antelope Squirrel.

Harris' Antelope Squirrel: Grand Canyon, Organ Pipe Cactus, Saguaro; **White-tailed Antelope Squirrel:** Arches, Canyonlands, Capitol Reef, Death Valley, Grand Canyon, Joshua Tree, Petrified Forest, Zion; **Texas Antelope Squirrel:** Big Bend, Carlsbad Caverns, Guadalupe Mountains

---

Ground Squirrels. There are eighteen species called ground squirrels in America north of Mexico, most of them in the western United States, and fifteen of them live in one or more of the national parks and monuments discussed in this book. Two of the species—the Golden-mantled and Cascade Golden-mantled Ground Squirrel—have a white stripe bordered with black on their sides and resemble large chipmunks. The only other striped species is the Thirteen-lined Ground Squirrel, which, as its name implies, has thirteen dark and light stripes, which run the length of its body; there are pale spots within the dark stripes. Antelope Squirrels, a different group of ground-dwelling squirrel, also have a white stripe on the sides, but these are not bordered with dark stripes. The other ground squirrels are spotted or mottled in color, but can be further identified either by geography or by the length and shape of their tails. Townsend's, Richardson's, Uinta, Belding's, Columbian, Spotted, and Round-tailed Ground Squirrels all have tails that are less than 5 inches (123 millimeters) long; Franklin's and California Ground Squirrels, as well as the Rock Squirrel, have long tails, more than 5 inches (127 millimeters) in length. The Arctic Ground Squirrel, with a tail 3–5½ inches (76–140 millimeters) long, is the only species in Alaska. In its range, the Mexican Ground Squirrel can be distinguished from the similar-looking Spotted Ground Squirrel by having a longer (more than 3½ inches, 89 millimeters) tail. In most places, ground squirrels are easily identified by size, coloration, or habitat.

Arctic, Belding's, Columbian, Richardson's, Townsend's, and Uinta Ground Squirrels. Arctic, Belding's, Columbian, Richardson's, Townsend's, and Uinta ground squirrels are characterized by grayish, somewhat mottled or spotted fur and by relatively short tails. All of the squirrels in this group inhabit meadows or tundra in which there is a relatively short summer season, and they hibernate for long periods. They live in colonies and are conspicuous; they often sit erect. Where two species occur in the same park or monument, they do not occur together; they generally have particular, often altitudinal, habitat separation.

The large Arctic Ground Squirrel, which reaches almost 20 inches (508 millimeters) in total length and weighs an average of about 1½ pounds (0.7 kilograms), is the only ground squirrel in the Alaskan parks and monuments. Arctic Ground Squirrels are colonial, and their den sites and burrow systems are elaborate and used by successive generations. Although the burrows are rarely deeper than 3 feet (0.9 meter), they may be very long; one that was dug out extended for 68 feet (21 meters), traveled through three levels, and had six openings. Nests are lined with dry grass and bits of fur. Burrow systems are located in relatively well-drained places, where the underlying permafrost is deep, rather than close to the ground surface. Permafrost is probably avoided by the shallow 3-foot (0.9-meter) limit to the depth of a burrow.

During the long Arctic days, ground squirrels are active mainly from about 4 A.M. to 9 P.M., and, unlike many other more conspicuous ground squirrels, they crawl close to the ground, through the vegetation, to and from feeding sites. Some of these travels in search of food may be extensive—as much as 1,500 yards (1,372 meters) from the den site—and sometimes necessitate swimming across large puddles formed by melted snow. Grasses and sedges are eaten, and berries and seeds are stuffed into the ground squirrels' cheek pouches and then carried to the burrow, where they are stored for consumption after the animals emerge from hibernation. They also eat beetles, insect larvae and pupae, mushrooms, and carrion. Caribou antlers are gnawed as a source of salts.

By September, Arctic Ground Squirrels are fat; they enter hibernation in late September or early October and remain in torpor for as long as seven months, emerging in April or May. Characteristically, they mate soon after hibernation, and the naked and blind young—usually four to eight in number—are born about 3½ weeks later. By three weeks of age, the babies are weaned and begin feeding rapidly in order to grow and put on fat for the winter. When six weeks old, they are four-fifths of their adult size; when four months of age, they are adult size. At this time, they dig a winter burrow and enter hibernation.

Arctic Ground Squirrels are a prime source of food for many Arctic birds, as well as for wolves, foxes, weasels, and even Grizzly Bears. A common name, "Parka Squirrel," derives from the Eskimos' extensive use of their fur for parka linings. These rodents are readily seen in both Katmai National Park and Denali National Park, and their loud "keek-keek" call is often heard.

Richardson's, Belding's, Uinta, and Columbian Ground Squirrels—all of which are close relatives of the Arctic Ground Squirrel—typically inhabit grassy mountain meadows or high plains, and they are, at least in most places, colonial and have complex social structures. Another characteristic of this group is that its members spend much of their lives in hibernation, some as many as eight-and-a-half months a year, utilizing the extensive amounts of fat that they deposit under the skin for metabolism.

Richardson's Ground Squirrel ranges west from the Minnesota border to Montana, Nevada, and south to central Colorado in the mountains. It stands erect on its hind legs so characteristically that it is usually called "Picket Pin" Ground Squirrel (or Gopher, which it is not) and, when on all fours, flicks its tail; this movement is the origin of another sobriquet, "Flicker-tail." Before agriculture occupied the high prairies where Richardson's Ground Squirrels lived, they were, in places, more numerous than prairie dogs, with estimates of 5,000 ground squirrels per square mile (2.6 square kilometers) not unusual, and as many as 50 in a single acre (0.4 hectare). High populations close to agricultural activities led to the killing of most of these populations, but, in the national parks, Richardson's Ground Squirrels are still numerous.

The natural food of these rodents is mainly grass and seeds, the latter of which are carried in the cheek pouches to storage chambers in the burrow. They also eat insects. In fall, Richardson's Ground Squirrels put on much fat, in addition to storing food for winter hibernation. The main entrance to the burrow is conspicuously marked with a bare mound of dirt, but there is usually another hidden opening. About 3 inches (76 millimeters) deep, the burrow may extend for nearly 50 feet (15 meters), although most are shorter. It contains a grass-lined nest chamber.

Depending upon their location, Richardson's Ground Squirrels come out of hibernation as early as the end of January or as late as April. In some places, when greenage dries in the midsummer heat, they may also estivate. Mating takes place soon after emergence from hibernation, and up to ten babies may be born after a gestation period that is thought to be 3½ weeks. By June, the young ground squirrels are out feeding with their mother and probably disperse soon after to dig their own burrows and to prepare for winter.

These ground squirrels are noisy, and the high-pitched whistle is usually accompanied by a flicking tail. The best park in which to see Richardson's Ground Squirrels is Rocky Mountain.

Belding's Ground Squirrel also has colonial habits, and 560 burrows were once found in one acre (0.4 hectares). Found only in east-central and northeastern California, eastern Oregon, and adjacent Idaho and Nevada, Belding's Ground Squirrels feed mainly on grasses and other greenage, but insects and their larvae are also eaten. They do not seem to store food. In areas, generally at lower elevations, where grass dries in summer, these rodents may commence estivation in mid-July and then stay in the den, where they go directly into hibernation, for an underground torpid existence of eight months. Where the grass remains green, the animals forage longer and may not estivate, and young animals, which have had relatively little time to grow and to deposit fat for hibernation, may remain out until the first snowfall. By the middle of February, some males have emerged from hibernation, burrowing through snow to do so, and by early March, most have emerged. Mating takes place about a week after the females come out, with the males fighting fiercely for them. The females are promiscuous, mating with several males, and the young—three to eight—are born after a 25-day gestation. In July, the young have emerged from the maternal den and are weaned, and they remain out foraging after the adults have started hibernating. The young females generally remain in the area where they were born, but young males disperse to new living sites. The social nature of Belding's Ground Squirrels is mainly tied to closely related animals; they do not fight with one another, they share territories, and they defend one another's dens from nonrelatives. However, yearling males may attack and kill nonrelated baby squirrels, and females whose litters have died may move to a new area, kill babies there, and establish a new home. In addition to predation from hawks, Coyotes, and Badgers, these ground squirrels suffer from severe weather, either during hibernation or on emergence in the spring. Surviving males live three to four years, females to six years. Among the better parks in which to see Belding's Ground Squirrels are Yosemite and Sequoia.

The overall habits of Columbian Ground Squirrels are much like those of Belding's, except that the former store some food in their hibernacula, probably to be used in spring when they emerge. They live in meadows in eastern Washington, northeastern Oregon, Idaho, and northwestern Montana, as well as adjacent Canada. Two to seven babies are born after a 24-day gestation period and, by a month of age, are weaned and ready to dig their own burrows. The only park in which Columbian Ground Squirrels live is Glacier, where they are abundant and readily seen.

The Townsend's Ground Squirrel is also dependent upon green vegetation and emerges early, in January or February, to mate. The five to ten young (rarely as many as fifteen) are born in March, after a 24-day gestation period, and by May they are out of the burrow and starting to eat grass. The adult males fatten quickly, in about 120 days, but the females and young require about 135 days. After this process, they go underground to estivate and hibernate, so that from July to February these mammals may not be visible.

Townsend's Ground Squirrels range from south-central Washington and southern Idaho to southern Nevada and western Utah. Although they live in large aggregations, they may not be as social as some other species in this group; lost young, squeaking loudly, seem not to evoke any response from their mother. The only parks where Townsend's Ground Squirrels occur are Lassen Volcanic and Zion.

The Uinta Ground Squirrel is another long hibernator; the adults start to estivate in midsummer, but the young remain active above ground until September. Their colonies are generally in moist habitats, where there is soft soil for burrowing. They usually have five or six young in May, after a 24-day gestation period. By one month of age, the offspring are out feeding on grass. The range of this species is mainly along the Wyoming-Idaho border, as far south as central Utah. Grand Teton and Yellowstone are the only two parks where these ground squirrels can be seen; they are easy to find and observe there.

Arctic Ground Squirrel: Katmai, Denali; **Richardson's Ground Squirrel:** Glacier, Rocky Mountain; **Belding's Ground Squirrel:** Kings Canyon, Lassen Volcanic, Sequoia, Yosemite; **Columbian Ground Squirrel:** Glacier, North Cascades (?); **Townsend's Ground Squirrel:** Zion; **Uinta Ground Squirrel:** Grand Teton, Yellowstone

---

Rock and California Ground Squirrels. Rock and California Ground Squirrels, which are relatively large, have the general appearance of the arboreal Gray Squirrel. Up to 18 inches (457 millimeters) in total length, of which more than a third is their rather bushy tails, these squirrels are often seen in parks. The only national parks in which the California Ground Squirrel (sometimes called Beechey's Ground Squirrel) is found are in California and Oregon, whereas the Rock Squirrel is an inhabitant of parks in Utah, Colorado, Arizona, New Mexico, and western Texas.

As its name implies, the Rock Squirrel prefers rocky areas, where its grayish or brownish-gray pelage blends in well with the boulders on which it frequently suns itself. Compared with other ground squirrels, Rock Squirrels climb well and often go up into trees in search of nuts, berries, and fruits, which make up most of their diet. Stuffing their cheek pouches with food, they scurry off to store it under rocks, in hollows, or in their underground dens. In many parts of its range, a Rock Squirrel may be active throughout the year, but it does get quite fat in the fall and, in the north, may hibernate, at least for a short time. The gestation period is probably about a month, and usually five young are born in spring; there may be a second litter in August in some areas. Rock Squirrels are not highly social, and it is unusual to see several together, with the exception of a mother and her young. In the hotter parts of their range, Rock Squirrels may estivate in the summer. When alarmed, they emit a piercing whistle. They are fairly long-lived rodents, having survived ten years in captivity.

California Ground Squirrels (see plate 8a) also prefer open, dry country, and they are much more colonial than Rock Squirrels. Burrow systems, some as many as 200 feet (61 meters) long, may be occupied by successive generations of squirrels, although individuals generally each have their own entrance. Runways 3 inches (76 millimeters) wide, from hole to hole and to feeding sites, are usually evident. California Ground Squirrels start their activity at dawn and spend most of their time within

100 feet (30 meters) of their burrows. They spend a few hours feeding, keeping their den in repair, and sun- and dust-bathing, but most of the day is spent underground. Seeds, nuts, berries, fruits, mushrooms, and green vegetation, as well as bird eggs, birds, and insects, are in their diet. In agricultural areas, they are notorious pests, even climbing trees for fruits and nuts.

During midsummer, adult California Ground Squirrels estivate in their burrows, but the young continue to be active above ground. In September, the adults emerge again and feed heavily, putting on fat and, in late October or November, go into hibernation until January, February, or even March. Some of these ground squirrels are, however, seen out at any time, even in winter; they are probably young animals. Mating takes place soon after the adults emerge from hibernation, and, about a month later, five to eight young, naked and blind, are born in the leaf-lined nest. The babies' eyes open at about five weeks, and at two months of age, they venture out of the burrow and play at its entrance. Young California Ground Squirrels remain with their mother until they are about four months old, at which time they leave to find an unoccupied den in the colony. By eight months, they are of adult size and may dig burrows of their own.

California Ground Squirrels are preyed upon by hawks, eagles, snakes, weasels, Badgers, Raccoons, foxes, Bobcats, and Coyotes; these predators often can be seen near ground squirrel colonies. These ground squirrels have a high reproductive potential and relative longevity—at least five years in the wild. They have benefited from agricultural activities and are usually more abundant near farms than in undisturbed situations in the parks. Yosemite National Park is one of the better places to see California Ground Squirrels.

Rock Squirrel: Arches, Big Bend, Bryce Canyon, Canyonlands, Capitol Reef, Carlsbad Caverns, Grand Canyon, Guadalupe Mountains, Mesa Verde, Organ Pipe Cactus, Petrified Forest, Rocky Mountain (?),Saguaro, Zion; **California Ground Squirrel:** Crater Lake, Death Valley, Joshua Tree, Kings Canyon, Lassen Volcanic, Redwood, Sequoia, Yosemite

---

Thirteen-lined and Franklin's Ground Squirrels. Both Thirteen-lined and Franklin's Ground Squirrels have more easterly distributions than other ground squirrels and are basically prairie species; the former is typical of short-grass prairie, the latter, of tall-grass areas.

The Thirteen-lined Ground Squirrel, as a short-grass prairie species, has a range not unlike that of the main Bison population before the herds were killed off, from Utah and Montana to as far east as Indiana and, west of the Mississippi River, as far south as Texas. In the soft soils, it constructs its burrows, with the entrance (2 inches; 51 millimeters in diameter) unmarked by mounds of bare soil. There are generally two types of burrows: a shallow one for summer use, and a deep one that extends below the frost

Thirteen-lined Ground Squirrel

line for hibernation. From this base, a home range extends over about 12 acres (5 hectares) for males, and about 4 acres (1.6 hectares) for females. The male's home range is largest during the spring breeding season, whereas the female's is at its maximum size when she is pregnant and/or lactating.

These ground squirrels are neither as colonial nor as social as many of the western species, but they are not considered territorial, even though some seem to defend specific areas. They do not fight much, however, and altercations that do occur are mainly between males at breeding time.

Insects, especially larvae and grasshoppers, can make up half or more of the diet of Thirteen-lined Ground Squirrels in summer; grass and seeds comprise the remainder. The ground squirrels gain weight rapidly and their weight may increase by 40 percent before they begin estivation at the end of July or early August, in hotter areas such as Texas, and they can remain torpid for seven-and-a-half months. Elsewhere, they may hibernate from October to March. In hibernation, their temperature drops to 34°–37° F. (1°–3° C.), and their heart rate decreases from 200 beats per minute to 4; their breathing rate can scarcely be determined. The entrance to hibernation burrows is plugged, and these animals are solitary when hibernating. By April, in most places, these ground squirrels have emerged and breed soon afterward. The gestation period is 28 days, and eight young are usually produced. In about a month, the babies are weaned and emerge from the burrows; the emergence usually occurs in mid-June, and again in August if there is a second litter.

Thirteen-lined Ground Squirrels are active during the day from 9 A.M. to 5 P.M. in spring and only early and late in the day in hot summer weather. When it is windy or cooler than 50° F. (10° C.), they are usually not active above ground. Although not as conspicuous as some other park ground squirrels, the Thirteen-lined species can usually be observed at the prairie dog towns at Wind Cave and Badlands.

The Franklin's Ground Squirrel is large—16 inches (406 millimeters) in total length—and dark gray. It is among the more secretive species of this group of ground squirrel. Little remains of its original habitat, which was probably tall-grass prairie, from Indiana to North Dakota, and as far south as Kansas and Missouri. The entrance to the Franklin Ground Squirrel's burrow is concealed in tall vegetation, and although it sometimes lives in small colonies of a dozen individuals, it is usually solitary. Its home range is about 100 yards (91 meters) in diameter. Its diet is primarily composed of plant material, although it is known to kill and eat mice and birds, as well as dine on bird eggs.

Franklin's Ground Squirrels put on large amounts of fat in summer, then hibernate (perhaps in groups) from October to May, and mate soon after their emergence. The gestation period for the four to eleven young is 28 days. By early July the young ground squirrels are weaned, and by the end of September they have almost reached adult size. Whether Franklin's Ground Squirrels occur in any national park is questionable; only Theodore Roosevelt Memorial Park is a possible location.

Thirteen-lined Ground Squirrel: Badlands, Glacier, Theodore Roosevelt, Wind Cave; **Franklin's Ground Squirrel:** Theodore Roosevelt (?)

Mexican, Spotted, and Round-tailed Ground Squirrels. Mexican, Spotted, and Round-tailed Ground Squirrels have restricted distributions in the southwestern United States. The Mexican Ground Squirrel is found only in southern Texas and southwestern New Mexico. Its range is overlapped by that of the Spotted Ground Squirrel, which lives in sandy-soiled areas as far north as southern Nebraska and as far west as Arizona. Although both species have whitish spots on their sand-colored backs, the Mexican Ground Squirrel's spots are more distinct and occur clearly in rows, whereas those of the Spotted Ground Squirrel are more random. The Mexican Ground Squirrel's tail is bushier than the Spotted's thin, pencillike appendage. The Round-tailed Ground Squirrel's tail is also pencillike, and although colored like that of the others, this animal has no whitish spotting. In the United States, its range is from southwestern Arizona to southeastern California.

Each of these species prefers sandy soil, usually in arid areas, that are sparsely vegetated. Seeds and green plant parts make up most of their diet, although they are also known to eat insects and carrion. The normal home range for the male is about 7½ acres (3 hectares); the female has a smaller one. Spotted Ground Squirrels hibernate in the northern parts of their range, but may not do so in the south. Some Round-tailed and Mexican Ground Squirrels may be active all winter, others may hibernate or estivate. Mating, which occurs in the spring, is usually about two weeks after hibernation for Spotted Ground Squirrels. The gestation period is probably 28 days, and the number of young varies from four to ten.

Extreme heat is a deterrent to these ground squirrels, and in New Mexico, for example, Spotted Ground Squirrels generally emerge in the morning when the temperature is 68° F. (20° C.), but return underground when it reaches 93° F. (34° C.). In midsummer, there are usually none to be seen at midday.

These ground squirrels seem more secretive and less colonial than most mountain species and are thus less often seen. Big Bend and Guadalupe Mountains are parks where Spotted Ground Squirrels can be seen; Big Bend has the only record of a Mexican Ground Squirrel; and Death Valley, Organ Pipe Cactus, and Saguaro are where Round-tailed Ground Squirrels can be found.

Mexican Ground Squirrel: Big Bend; **Spotted Ground Squirrel:** Big Bend, Grand Canyon, Guadalupe Mountains, Petrified Forest; **Round-tailed Ground Squirrel:** Death Valley, Joshua Tree, Organ Pipe Cactus, Saguaro

---

Golden-mantled Ground Squirrels.   Golden-mantled Ground Squirrels, which are found from the Rocky Mountains to the Cascades and Sierra Nevada, are sprightly, conspicuous animals that are often seen together with and confused with chipmunks. Although both are striped and may have similar coloration, Golden-mantled Ground Squirrels are larger (their body length is about 7 inches; 178 millimeters) and lack facial stripes. Chipmunks are rarely as long as 6 inches (152 millimeters) in body length and always have a dark stripe, bordered with white, on the sides of the face.

Although they are customarily found in pine, spruce, and fir forests, Golden-mantled Ground Squirrels also venture into chaparral and even live above timberline in places. The burrow is the center of a small territory, which may be as much as 100 feet (30 meters) in diameter; however, the larger home range—which extends some 600 feet (183 meters) from the burrow—is shared with other Golden-mantled Ground Squirrels. The entrance to the burrow, about 2 inches (51 millimeters) in diameter, is often under a log, a rock, or between tree roots, and the underground portion, with several chambers, may extend for 10 feet (3 meters). In addition to a leaf- and bark-lined nest chamber, there are storage alcoves into which seeds and nuts are stuffed, and there is usually a well-concealed secondary access hole to the burrow system.

Unlike chipmunks, Golden-mantled Ground Squirrels tend to put on fat in the fall. Depending on local conditions, they may hibernate from October to May. Since food is stored in the burrow, these ground squirrels probably awaken and feed briefly in winter, although this food supply could also serve them when they emerge from hibernation in the spring, if other foods are in short supply. Soon after emergence, the squirrels mate, and two to eight (usually four or five) young are born after a gestation period of about four weeks. They are raised solely by the mother. When the babies are about a quarter of their mother's size, they emerge from the burrow and commence feeding on the great variety of foods that makes up this species' diet. Conifer seeds and nuts, berries and fruit, and seeds of many herbs and shrubs are the diet's main components. Golden Mantled Ground Squirrels also show a proclivity for insects—grasshoppers, flies, and beetles—as well as for carrion and mushrooms. With such diverse tastes, it is little wonder that these rodents congregate around campsites, picnic grounds, and auto turnouts, where they boldly beg for food and search for garbage. When feeding of park animals was permitted, Golden-mantled Ground Squirrels were a major attraction, stuffing their cheek pouches full before scurrying off briefly to unload and then returning again and again for more handouts.

The Cascade Golden-mantled Ground Squirrel, which has a limited distribution in the northern Cascade Mountains, may prove to be only a subspecies, rather than a distinct species, of the Golden-mantled Ground Squirrel; its habits are the same.

In most of the western parks where they occur, these ground squirrels are difficult to miss. Rocky Mountain, Bryce Canyon, Yosemite, and Mount Rainier are among the best parks in which to see them.

Golden-mantled Ground Squirrel: Bryce Canyon, Capitol Reef, Crater Lake, Glacier, Grand Canyon, Grand Teton, Kings Canyon, Lassen Volcanic, Mesa Verde, Redwood (?), Rocky Mountain, Sequoia, Yosemite; **Cascade Golden-mantled Ground Squirrel:** Mount Rainier, North Cascades

---

Prairie Dogs. Four species of prairie dog live in the western United States, from central Kansas and Nebraska to western Utah and Arizona, and from southern Canada to Mexico. The best known and most eastern in distribution is the Black-tailed Prairie Dog (see plate 6b), which is an inhabitant of short-grass prairies. The other three species are white-tailed: the White-tailed Prairie Dog, of northeastern Utah, northwestern Colorado, and central Wyoming; the Utah Prairie Dog, of southwestern Utah; and Gunnison's Prairie Dog, of southeastern Utah, south-central and southwestern Colorado, northwestern New Mexico, and northeastern Arizona. The three white-tailed species are usually inhabitants of mountain meadows 6,000–12,000 feet (1,829–3,658 meters) in elevation. Because they compete with livestock for grass, prairie dogs have been extensively poisoned and now occupy only a small portion of their original range; some of the white-tailed species are endangered.

Black-tailed Prairie Dog. The Black-tailed Prairie Dog, a highly colonial species, had "towns" that once extended for thousands of square miles on the short-grass prairies. Not only were these diurnal ground squirrels themselves conspicuous, but the large, bare mound surrounding the entrances to their burrows readily proclaimed their presence and made them most vulnerable to extirpation. These rodents are excellent and prodigious diggers. The burrow entrance, 6–8 inches (152–203 millimeters) in diameter, descends almost vertically 3–16 feet (0.9–4.9 meters) and narrows to 4–5 inches (102–127 millimeters). A few feet below the entrance is an alcove, where the animal may take momentary refuge while deciding the extent of a threat. At the bottom of the vertical shaft, the tunnel turns horizontal and may extend for another dozen feet (3.7 meters) or so, with branches leading to a nest chamber and other abandoned and plugged nesting sites and toilet chambers. The mound around the entrance, which is bare and packed hard, may be as many as 4 feet (1.2 meters) in diameter and 2 feet (0.6 meters) high. It not only serves as a lookout post but also diverts floodwaters from entering the burrow.

Green plants, almost entirely grasses and forbs, make up the major part of a Black-tailed Prairie Dog's diet. Because these animals do not like to live in places where grass is too tall to see over, the grass around burrows is usually scant from their feeding, and their original distribution on the prairies was closely tied to the grazing activities of Bison and, later, to areas overgrazed by cattle. Prairie dogs will eat insects, especially grasshoppers, in addition to grass. They do not store food, but during the summer they grow enormously fat, and, although they probably do not actually become torpid, as in true hibernation, they spend most of the winter in their underground chambers. On clear days, however, they emerge briefly and feed on any exposed grass.

Black-tailed Prairie Dogs have a highly complex social structure. The essential unit of the social system consists of a male and one to four females and their young, and is called a *coterie*. This family unit is territorial; all of the members help to keep other prairie dogs from the protected zone around the burrow mouth. The average size of a coterie's territory is about three-fourths of an acre (0.3 hectare). The larger social units of a "town" are called *clans* and *precincts*, and although there is territorial antagonism between members of different coteries, all cooperate in mutual warning of danger. Even when feeding, prairie dogs are alert and usually pause every half minute or so to sit up and scan the horizon or sky for signs of danger. In addition, there is usually one animal not feeding at all, who serves as a sentry. At the first sign of danger, a warning is given— a short, double-note chirp—and all prairie dogs that hear it repeat it and race for the safety of a burrow. In addition to the auditory warning, tail-flicking is another part of the repertoire and accompanies the alarm calls.

Within the prairie dog town, vocal communication is loud and frequent, with at least nine different calls, which include chirps, chatters, snarls, and a yipping bark. When meeting, members of coterie touch noses ("kissing"). They also spend much time grooming one another's fur. The nose-rubbing is probably a form of scent identification. Some of the calls are accompanied by characteristic movements. For example, when prairie dogs give alarm barks, they stand erect and then point their muzzles skyward.

Depending upon the locale, mating takes place from late January to March, and usually three to five young, naked and blind, are born after a 28–32 day gestation period. When the young are about six weeks old (a week after their eyes have opened), they start coming above ground. When six months of age, they will be adult size. They may remain with their parents until they are two years old and of breeding age; some may start digging their own burrows when only 10-weeks old, whereas others take over the parental burrow when the adults move out to take up a new residence, usually on the periphery of the town.

Coyotes, foxes, and Badgers prey on prairie dogs, as do hawks, eagles and snakes. One member of the weasel family, the Black-footed Ferret, utilizes prairie dogs as its primary source of food. Because of the extirpation of this rodent from so many places, the ferret is so reduced in numbers that it is now one of the rarest North American mammals.

Black-tailed Prairie Dogs are a popular attraction in the parks where they occur. Badlands, Wind Cave, and Theodore Roosevelt parks, and Devil's Tower National Monument, each have colonies readily visible from the roadsides, and, in these locations, the animals have become quite accustomed to the presence of human beings.

White-tailed Prairie Dogs. The three species of white-tailed prairie dog are more closely related to one another than any is to the Black-tailed Prairie Dog. They are mainly animals of montane meadows, where they feed on succulent grasses. Although their overall habits are essentially like those of the Black-tailed species, their social system is less complex, their primary social group is that of a mother and her young, and they are less territorial. While not firmly proved, it seems likely that white-tailed prairie dogs, at least the northernmost ones, actually hibernate for a while in winter. They have a repertoire of vocalizations that is similar to, but distinct from, not only that of the Black-tailed species, but also from each of the other white-tailed kinds.

Colony sizes of the white-tailed prairie dogs tend to be smaller than those of the black-tailed ones, even in an area of equivalent size, and the average number of young (five) tends to be lower. The gestation period is about 30 days, and the babies emerge from the burrow early in the summer. Their burrows are generally shallower than those of the prairie dwellers, and, in many ways, the life history and behavior of the white-tailed species seem more like that of some ground squirrels than that of the Black-tailed Prairie Dog.

Because of their restricted habitat and limited numbers, all three species of white-tailed prairie dogs have been much reduced in number and distribution by poison campaigns. They occur in a few parks and, even in these, live a precarious existence. Because of this, to avoid disturbing the animals, the park service may not reveal the location of colonies or encourage visitation. Travelers to these parks should abide by these requests in order to enable the colonies to become firmly enough established that their continued existence will be assured and they can be viewed by future generations.

Black-tailed Prairie Dog: Badlands, Carlsbad Caverns (E?), Guadalupe Mountains, Theodore Roosevelt, Wind Cave; **White-tailed Prairie Dog:** Arches (E?), Canyonlands (?); **Utah Prairie Dog:** Bryce Canyon, Capitol Reef (E?), **Gunnison's Prairie Dog:** Canyonlands (?), Grand Canyon (E?), Mesa Verde (E), Petrified Forest

---

Tree Squirrels.   Tree squirrels, as their name implies, are forest dwellers that make their homes in trees and are proficient at climbing and leaping. Unlike those ground squirrels and chipmunks that also live in forests, tree squirrels do not have cheek pouches, are active throughout the year, and have long, bushy tails. None has a striped or mottled back. Although they nest in trees and obtain much of their food there, all tree squirrels also forage on the ground and store food there. North of Mexico, there are eight species of tree squirrel, seven of which are found in one or more parks. They are classified into a number of related subgroups, but in this book will be grouped by similarity of appearance and habits, rather than their actual relationships. Thus eastern and Western Gray Squirrels and the Fox and Arizona Squirrel, which belong to different subgenera, are considered together.

Two other tree-dwelling squirrels are not discussed in this book. These are the two species of flying squirrel—the Southern and the Northern Flying Squirrel—which inhabit North American forests and many national parks and monuments. They are the only nocturnal members of the Squirrel family in North America, and, because they do not venture out until it is fully dark (and because they are small), are rarely seen or identified by park visitors. Flying squirrels are not capable of true, sustained flight, unlike bats, the only true fliers among the mammals. Flying squirrels are gliders, making extensive airborne leaps and gliding, supported by layers of loose skin that extend along each side of their bodies from wrist to ankle. At night, the best a camper may hope for is a glimpse of a flying squirrel against the sky as it glides between two trees, or a red eyeshine in the beam of a flashlight. During the day, flying squirrels that are startled by people emerge from hollows in trees.

Gray and Fox Squirrels.   The Gray Squirrel, Western Gray Squirrel, Arizona Gray Squirrel, and Fox Squirrel are all large tree squirrels that are similar in appearance and habits. Actually, the Arizona Gray Squirrel and the Fox Squirrel are closely related, and the eastern Gray and Western Gray Squirrels are no more closely related to each other than either is to the Fox Squirrel or Arizona Squirrel. An additional complication is that Gray Squirrels and Fox Squirrels from the east have been introduced and become established in the Pacific Coast states. Fortunately, thus far, neither of these species has spread to enter any of the national parks, where they might be confused with native western species.

The eastern Gray Squirrel is a familiar species to many people because it is the common squirrel in many eastern and midwestern city parks and suburbs, as well as, by introduction, some cities on the Pacific Coast. About 15–20 inches (381–508 millimeters) in length, half of which is the bushy tail, and ¾–1½ pounds (0.3–0.7 kilograms) in weight, these squirrels are characteristic of hardwood forests and wooded riverbottoms. Oak, beech, and hickory trees are especially preferred for their homes because their nuts provide so much of a Gray Squirrel's food. The Gray Squirrel generally has a gray back and sides, a brownish head, a white belly, and gray tail that is fringed with white-tipped hairs. Both albinos and melanistic animals are not especially rare, although they are more prevalent in urban areas where natural predation is reduced. In parks where the eastern Gray Squirrel may exist with the Fox Squirrel—Shenandoah, Mammoth, Great Smoky Mountains, Everglades, and Hot Springs—the Fox Squirrel may usually be distinguished by its buffy-tipped tail hairs (in contrast to the Gray Squirrel's white-tipped ones) and larger size; usually the Fox Squirrel's belly is yellow or orange, rather than white.

A Gray Squirrel's home is high in a tree, preferably a hollow where a limb has died or broken off, but often in a globular nest constructed of dead leaves and twigs. In summer, additional leaf nests are constructed for temporary shelter. In general, a home range encompasses .02–7 acres (0.08–2.8 hectares), overlapping the home ranges of other Gray Squirrels. Where food sources are limited, these squirrels may wander widely. In spring, Gray Squirrels feed on buds and tender twigs and lick sap. In summer, fruits and berries make up a major portion of their diet, and when hardwood nuts ripen, they become the main focus of Gray Squirrel activity. Acorns, walnuts, hickory nuts, and butternuts are eagerly cut and dropped or carried to the ground (as chestnuts were in the past). The squirrel carries one off in its mouth, digs a shallow hole in the ground with its paws, and buries it, carefully tamping down the dirt to conceal the cache. Hundreds of nuts are buried, each taking one to three minutes to conceal. When winter comes, buried nuts are the animal's main food source. Squirrels do not remember where they bury nuts (tests indicate that they forget the location within a half hour) and in winter, even when there is snow on the ground, find each nut by a keen sense of smell and dig down to excavate it. Buried nuts not utilized by squirrels are an important reforestation source.

Midwinter is the breeding season for Gray Squirrels, and males can be seen pursuing females in the trees and on the ground, often fighting among themselves for mating privileges. After a gestation period of 6½ weeks, babies, naked, blind, and weighing a half-ounce (14 grams) each, are born in the maternal nest. By five weeks of age, they are fully haired, their eyes and ears have opened, and their incisor teeth have erupted. At this time they weigh 3–4 ounces (85–113 grams). As soon as they are this well developed, the young squirrels begin to explore outside the nest, and the mother begins to wean them; however, they may continue to nurse until they are three months old. In many places, the mother breeds again in early summer, and if this occurs, she will generally leave the original litter and move to another nest to rear the second one. Otherwise the first litter may remain with the mother until late fall. Mother squirrels are extremely protective of young in the nest, even attacking and biting human beings who attempt to handle nestlings. The babies make a piercing squeal, which brings the mother quickly to their defense. The mother will often move nestlings to another den or nest if they have been disturbed, and sometimes for no apparent reason, except perhaps to escape the large number of fleas and other ectoparasites that inhabit the nests.

In autumn, Gray Squirrels put on much fat, although they do not hibernate. They also acquire fresh, thick pelage, which, together with the fat layer, serves as insulation in cold weather. Late litters will usually share a winter nest with their mother, and when temperatures are very cold, they will not go out. However, their abundant tracks in snow and the excavations where they have unearthed nuts are evidence of their winter activity.

Although females with young in the nest are fiercely defensive, Gray Squirrels seem not to be typically territorial. Males evidently maintain a dominance hierarchy, perhaps established during competition for mates. These squirrels, which have a superb sense of smell, leave scent trails and urine marks throughout their pathways, thus identifying themselves, and perhaps their status, to other squirrels in the area.

Under natural conditions, Gray Squirrels are much more wary of people and are less often seen than those that inhabit urban parks and suburbs. They are preyed upon by hawks, snakes, Raccoons, foxes, Coyotes, Bobcats, and weasels and are often infested with parasites. Nevertheless, they may survive six or seven years in the wild and have lived more than fifteen years in captivity. Among the better parks in which to see Gray Squirrels are Hot Springs, Great Smoky Mountains, Mammoth, and Shenandoah.

Western Gray Squirrel. Closely resembling the eastern Gray Squirrel, but larger, the Western Gray squirrel is an inhabitant of oak and mixed oak and pine forest in the Pacific Coast states. Up to 2 feet (610 millimeters) in total length and weighing as much as 2 pounds (0.9 kilograms), it seems to prefer more open forests than does the eastern Gray Squirrel and may be found from sea level to about 8,000 feet (2,438 meters). Their primary habitat, and source of food, is oak forest, and acorns are harvested and buried one by one. The deep snows of the Sierra Nevada limit their winter searches and thereby confine them more to the lower altitudinal zones. The nest is usually high in a tree, either built of twigs and bark or in a hollow trunk, and it is here that the single annual litter of three to five young is born. The gestation period is about 45 days, but, depending upon locale, breeding may take place from January to May. Overall, the habits of the Western Gray Squirrel seem to be quite similar to those of the eastern Gray Squirrel.

Western Gray Squirrels may undergo population fluctuations; in some years they are commonly seen, in others they are scarcely observed. Among the better parks in which to see Western Gray Squirrels are Redwood, Yosemite, and Sequoia.

Arizona Gray Squirrel. Similar in size and coloration to the Western Gray Squirrel, the Arizona Gray Squirrel is actually a kind of Fox Squirrel that lives only in the mountains in southern Arizona, where there are oak and pine woodlands. It also lives in wooded canyons, where there are walnut, cottonwood, and sycamore trees. Its overall habits are similar to those of the other gray squirrels, and it can be found in the mountains of Saguaro National Monument.

Fox Squirrel. The largest of the tree squirrels—the Fox Squirrel—may weigh as much as 3 pounds (1.4 kilograms) and be 28 inches (711 millimeters) long from tip of nose to tip of tail. It occurs over most of the eastern United States, as does the Gray Squirrel; however, unlike the Gray Squirrel, it is absent from the northeast. Fox Squirrels (see plate 2a) are inhabitants of hardwood, open southern conifer, cypress, mangrove, and mixed hardwood-conifer forests and are usually found where there are oaks and hickories. Overall, they seem to prefer sunnier woodlands than the Gray Squirrel does.

The Fox Squirrel spends much of its time foraging on the ground, but nests in a hollow high in a tree or constructs a twig and leaf nest on a branch or in a crotch. Winter nests are quite sturdy, but temporary summer leaf nests are not very substantial and fall apart quickly if deserted or not maintained. Acorns are the Fox Squirrels' main source of food, but a variety of other nuts, seeds, fruits, and berries are eaten, and nuts are buried for winter use. Most of the time, Fox Squirrels wander on the forest floor within 1,000 feet (305 meters) of the den tree, but in the fall, if the population is high and food supplies are low, they may range farther. Usually a home range is 10–40 acres (4–16 hectares) in area.

Fox Squirrels mate in winter, and two to five young are born after a 45-day gestation period. The young are weaned and start to leave the nest when they are about three months of age. Two-year-old and older females usually have a second litter in summer. Except during the mating period, when a male and female may nest together for a while, adult Fox Squirrels are solitary, but females with a subadult litter are sometimes found nesting together and may overwinter in the same den.

There are several color phases of the Fox Squirrels, some similar to those of the Gray Squirrels. Generally, however, where the two occur together, the Fox Squirrel is marked with rust-colored hair at the edges of the tail or on the belly, or has a blackish head. Fox Squirrels are not readily seen in any of the parks where they occur.

Abert's and Kaibab Squirrels. In the central Rocky Mountains, from northern Colorado to central New Mexico, north-central Arizona, and in southeastern Utah, is Abert's Squirrel, a large squirrel with 1½ inch (38 millimeters) tufts of hair on its ear tips in winter. Essentially grayish in body color, a typical Abert's Squirrel has a white belly, which is demarked from the sides by a black band, and a rusty-colored midback. The underside of the tail is white. In some places, especially central Colorado, the reddish middorsal area may be missing and the back is solid gray. North of the Grand Canyon in Arizona, on the Kaibab Plateau, there is an Abert's Squirrel that is markedly different. The belly is black instead of white, and the tail, rather than gray above and white below, is entirely white. This black-bellied, white-tailed squirrel evidently crossed the Colorado River and became isolated on the Kaibab Plateau. It has been regarded as a species distinct from the Abert's Squirrel and is called the Kaibab Squirrel. However, current opinion tends to classify the Kaibab Squirrel as a subspecies, not species, of Abert's Squirrel. Part of the basis for subspecific rank for the Kaibab Squirrel is that its characteristic black belly occasionally appears in Abert's Squirrels on the south side of the Grand Canyon and in Colorado and New Mexico, and there are also a few records of squirrels from the Kaibab Plateau with white bellies. The habits of both kinds of squirrel are the same and are discussed here as those of Abert's Squirrel. These squirrels are sometimes called Tassel-eared Squirrels because their ears become tufted in winter. In summer, their ears are usually untufted.

Ponderosa pines provide home and food for Abert's Squirrels, and their distribution is always in the range of these trees. Bushy nests of pine twigs, 1–3 feet (0.3–1 meter) in diameter, are located 16–89 feet (5–27 meters) up in ponderosa pines, generally in the crotch formed by a branch and the trunk. Several nests may be used, located in the home range, which, in summer, may be 1–4 acres (2.4–9.7 hectares) in area and which is smaller in winter. Ponderosa pines provide most of the Abert's Squirrel's food: buds, seeds, flowers, and inner bark are eaten. These squirrels do not hibernate; pine cones are cut, dropped, and buried individually for additional winter food. In late winter, Abert's Squirrels breed. Several males generally follow a receptive female all day long, the dominant male eventually becoming the one to mate with her. The gestation period is 40–46 days, and there are usually two to four young, which are naked, blind, and weigh one-half ounce (14 grams) each. Their eyes open at about six weeks, and by seven weeks the young squirrels begin to come out of the nest and to eat solid food, but it is not until they are nine weeks of age that they may first descend to the ground; they are weaned a week later.

Populations of Abert's Squirrels seem to fluctuate markedly from year to year, perhaps depending on the abundance of pine seeds the previous year. When populations are low, their home ranges are larger; thus they may seldom be seen. Even when populous, Abert's Squirrels tend not to be as attracted to people and their food as much as are other squirrels. The South Rim of the Grand Canyon is the best place to observe Abert's Squirrels. Kaibab Squirrels are found on the North Rim of the Grand Canyon, but generally are less often seen there than Abert's are on the south side.

Red and Douglas' Squirrels. Red and Douglas' Squirrels are creatures of the coniferous forests and are the smallest tree squirrels in North America, generally less than 14 inches (346 millimeters) in total length—of which 4–6 inches (102–152 millimeters) is the tail—and usually weighing about half a pound (227 grams). The two species are similar in appearance, but do not occur together. Douglas' Squirrel inhabits the forests from British Columbia to the Sierra Nevada in central California, whereas the Red Squirrel is much more widely distributed, from Alaska to as far south as the mountains of Mexico, and in the north-central and northeastern states as far south as the mountains of Georgia. A white belly and rusty-red back characterize the Red Squirrel in summer, but the Douglas' species is more grayish or olive above, and yellowish or orange below. In summer, both species have a dark line along the sides, separating the lighter-colored undersides from the darker back, and in winter each species grows hairy tufts on its ears. Either species may be locally called a "Chickaree."

Red and Douglas' Squirrels are the most territorial of our tree squirrels, and one way in which they define their territory is vocally. The loud chattering of a Red or Douglas' Squirrel is often the first knowledge a park visitor has of its presence. The home range is fairly small, usually about 200 yards (61 meters) in diameter, and within it may be several nests or shelters. Preferred are dens in hollow trees, where the nest is usually lined with soft vegetation and fur, but also constructed are leaf-and-twig nests, about 1½ feet (0.5 meter) in width, near the top of a tree, against the trunk. A third type of nest is underground, often beneath a stump, log, or rock, descending a foot or more below the surface, where, in addition to a dry, soft, lined nest chamber, there are a number of underground trails in soft duff leading to storage rooms and pits.

Over their wide area of distribution, Red and Douglas' Squirrels have diverse sources of food. Conifer cones and hardwood nuts are their main diet, and these squirrels, in contrast to other tree squirrels, make large caches of cones and nuts in the ground. The concentration of this source of winter food is possible because of the territorial and aggressive behavior of these squirrels; such behavior prevents the loss of their entire hoard to another squirrel. In addition to storing cones and nuts, Red and Douglas' Squirrels feed on seeds and berries, and they harvest mushrooms, which they dry on a branch before storing. They even seem able to eat species that are poisonous to human beings. Bird eggs, snails, insects and larvae, and fledglings are also part of their diet.

Red and Douglas' Squirrels generally have favored feeding sites, such as a branch or a stump, which can readily be identified by the debris of cone scales, nut husks, and cores of stripped cones that litter the ground for 10 feet (3 meters) and constitute a pile 3 feet (0.9 meters) high. This midden serves as a garbage heap as well as a food storage place, for it will be honeycombed with excavations where a squirrel has dug to retrieve pine cones and nuts, and it may even contain an underground nest. These squirrels spend much of their time on the ground and are active from dawn to dusk; in winter, they burrow under snow to reach their caches. There are even a few reports of these squirrels, unlike any of the other tree squirrels, harvesting nuts at night.

Late winter is the breeding season, depending somewhat on the latitude and altitude, and at this time males chase females and one another and fight for breeding privileges. The gestation period is 38–42 days, and the usual litter size is about six. The babies' eyes open in about four weeks, and they may be nursed by the mother for three months. In some areas, a second litter is born in August or September, and these youngsters will overwinter with their mother.

Because they are noisy, conspicuous, and widely distributed, Red and Douglas' Squirrels are familiar to most park visitors. Among the better parks in which to see Red Squirrels are Acadia, Katmai, Rocky Mountain, and Yellowstone. Douglas' Squirrels can best be seen at North Cascades, Olympic, Sequoia, and Yosemite.

Red-bellied Squirrel. In Biscayne National Park, on Elliott Key, there is a population of Red-bellied Squirrels, a species native to Mexico which was introduced in this area in the 1930s. About 20 inches (500 millimeters) in total length, of which half is the tail, these squirrels are gray, with orange-colored bellies. There are variable patches of red to brown on the neck or rump, or on the shoulders. Melanism is prevalent in this population, and most of the animals are black. This species can readily be seen during the day on Elliott Key.

Gray Squirrel: Acadia (I?), Everglades, Great Smoky Mountains, Hot Springs, Mammoth Cave, Shenandoah; **Western Gray Squirrel:** Crater Lake (E?), Kings Canyon, Lassen Volcanic, North Cascades (?),Redwood, Sequoia, Yosemite; **Abert's Squirrel:** Canyonlands (?), Grand Canyon, Mesa Verde, Rocky Mountain, Saguaro (I); **Fox Squirrel:** Badlands (?),Everglades, Great Smoky Mountains, Hot Springs (?), Mammoth Cave (E?), North Cascades (?I), Shenandoah, Voyageurs, Wind Cave (?); **Arizona Gray Squirrel:** Saguaro; **Red Squirrel:** Acadia, Bryce, Capitol Reef (?), Denali, Glacier Bay, Glacier, Grand Canyon, Grand Teton, Great Smoky Mountains, Isle Royale, Katmai, Mesa Verde, North Cascades, Rocky Mountains, Shenandoah, Theodore Roosevelt (?), Voyageurs, Wind Cave, Yellowstone, Zion; **Douglas' Squirrel:** Crater Lake, Kings Canyon, Lassen Volcanic, Mount Rainier, North Cascades, Olympic, Redwoods, Sequoia, Yosemite; **Red-bellied Squirrel:** Biscayne (I)

BEAVER

Few mammals have had as much impact on American history and the environment as the Beaver. The largest of the North American rodents, the Beaver is about 4 feet (1.2 meters) long, including its one-foot (0.3-meter) flattened, scaly tail, and weighs up to 100 pounds (45 kilograms). Beavers were originally found throughout North America, as far north as the trees in the Arctic and as far south as the Mexican border and were only absent from peninsular Florida and some areas of the desert southwest. The search for their valuable skins led early explorers westward, and, by 1900, Beavers were virtually extinct over the main portions of their original

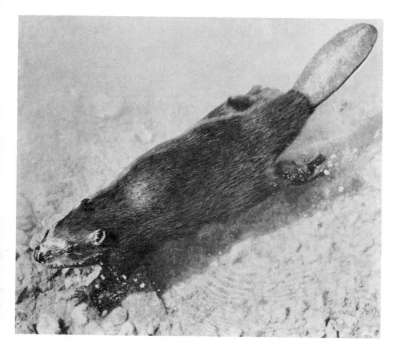

Young Beaver

range. Fortunately, they have survived and have been successfully reintroduced into places where they once occurred. Because they are mainly
nocturnal, Beavers are not often seen, but signs of their presence along
rivers, streams, and lakes are conspicuous and unmistakable.

Although most mammals alter their environment, in one way or another,
to make it more suitable for their existence, only the Beaver does so on a
major scale. Essentially, these aquatic rodents create an environment that
contains water of sufficient depth not to freeze to the bottom of the pond
in winter and providing a medium for transportation and storage of food
and protection for the animals as well as their homes. Stimulated by the
sound of rushing water, they dam small streams and rivers. Beavers sit on
their haunches and chip away at trees until they fell them. They use their
large, orange-colored incisor teeth (which grow continually). Trees and
branches, together with mud and stones, are used to construct a dam
which will flood an area to provide a pond for the Beavers. Generally, within
the pond, a lodge of mud and sticks will be constructed. The domed lodge
usually rises several feet above the water level and may be 4 or 5 feet (1.2
or 1.5 meters) in diameter; lodges 7 feet (2.1 meters) high and more than
30 feet (9 meters) in diameter have been found. Within the lodge is a
large above-water room, the entrance to which is underwater. The outside
plaster of mud and sticks is usually sufficient to deter most predators that
might swim or wade to the lodge in summer; the Beaver can readily escape,

in any event, into the water. In winter, when the mud on the lodge's exterior has frozen, it is impervious to attack.

Although the lodge may be the sole home of a colony of beavers, there also may be a den, with an underwater entrance, on the shore. Beavers that live on major rivers, such as the Rio Grande in Big Bend National Park, do not build lodges but live only in bank dens.

The flooded area produced by the dam around the lodge is the sole property of a single family of Beavers, consisting of an adult male and female, and litters of Beaver kits up to two years of age. The pond is a territory, which is defended by the owners from incursions by other Beavers. Scent glands are used to mark the territory. Generally, the Beaver urinates on a mound of mud or a pile of twigs; thus the castoreum scent, which is from a special gland, identifies its property.

The main food of Beavers is the bark of trees, especially deciduous ones, such as aspens and willows. In summer, Beavers feed more on aquatic vegetation, and, in fall, they fell trees and carry branches into the pond to store, implanted in the mud, underwater. In winter, when the top of the pond is frozen, Beavers can swim under the ice to retrieve these sticks and eat the bark.

Although it is generally thought that Beavers eat only bark from deciduous trees, it is evident in many places that, although they may prefer deciduous trees, they will also gnaw and eat the bark from various kinds of conifers. A stroll along the shore, wherever there are Beaver colonies will generally reveal stumps, felled trees, and scarred wood, where these animals have dined. According to folklore, Beavers know how to fell trees so that they always fall toward water. However, there is ample evidence that this is not so, and there are even instances in which Beavers have been killed by the trees they were cutting. Usually it is one Beaver, working alone, that cuts a tree; a willow that is 5 inches (127 millimeters) thick can be cut in less than three minutes.

Beavers are mainly monogamous, one male and female maintaining a territory for life, but some cases of infidelity are known. The mating season is during the last half of winter, and the gestation period lasts for four months. Baby Beavers, usually three or four, are born in the den and are fully furred, have incisor teeth and their eyes are partially open. They weigh ¾–1¾ pounds (340–680 grams) and nurse for three months, although they can eat solid food earlier. At the time the young are born, the adult male Beaver and any yearlings or other young present are evicted from the lodge or main den and take up residence elsewhere in the territory. Some baby Beavers are able to swim as soon as they are born, and all can swim by a week of age. Tired babies are sometimes ferried on the mother's back.

A Beaver family is a cooperative one, and when dams wash out, all will respond to repair them rapidly and effectively. Adult animals are dominant over the youngsters, and there is some evidence indicating that the adult female of a colony is the dominant individual. The warning slap of a tail on the surface of the water receives more attention when performed by an adult female than by a male. The tail slap usually sends all the animals diving underwater, where they can remain for as long as fifteen minutes without coming up for air. When swimming, Beavers propel themselves mainly with their hind feet and use the tail as a rudder. They do not carry mud or sticks on the tail, as legend has it. Beavers are, however, as folklore tells us, industrious animals.

In most places, Beavers are strictly nocturnal and therefore not often seen. However, Alaska and other northern parts of their range are the best places in which to observe Beavers, because there is little darkness on summer nights, and thus these animals are active at any time. Elsewhere, in some of the parks where Beavers have not been disturbed, they may be active while it is daylight or early dusk; observers must remain still and quiet. The warning slap of a tail on the water surface is often heard at night, when campers are alongside a Beaver pond. Dams, lodges, and cuttings are evident in most parks where these animals occur. Katmai, Mount Rainier, and Rocky Mountain national parks are ones where there is a good chance of seeing Beavers.

Beaver: Acadia (RI), Arches, Big Bend, Bryce Canyon, Canyonlands, Capitol Reef, Carlsbad Caverns (E), Crater Lake (E), Denali, Glacier, Grand Canyon, Grand Teton, Great Smoky Mountains, Hot Springs (?), Isle Royale, Katmai, Mammoth Cave (RI), Mesa Verde, Mount Rainier, North Cascades, Olympic, Redwood, Rocky Mountain, Sequoia (I), Shenandoah (RI), Theodore Roosevelt, Voyageurs, Wind Cave (E), Yellowstone, Yosemite (I), Zion

---

WOODRATS

All nine of the species of the attractive rodents called woodrats which inhabit America north of Mexico are found in one or more of the national parks or monuments. Also called "Pack Rats" or "Trade Rats," Woodrats live mainly in the more arid lands of the southwest, but are also known from the southern states and occur northward into New York in the Appalachian Mountains, as well as into Canada in the west. Unlike the Black and Norway Rats, which are not native to North America and to which many people have an aversion, Woodrats are handsome gray or brownish animals with soft fur, white bellies and feet, large black eyes, and a furred tail. One distinctive characteristic of woodrats is that they construct large houses, usually of twigs and sticks and various local dried debris, including pieces of cactus and dried cow dung. These nests, depending upon the locale in which they are built, may be on the ground, under or in a rock crevice, or even up in a tree. The collection of items for the nest is diverse, and something that catches the animal's attention will often cause it to drop whatever it might be carrying at the time and pick up a new item. Campers who have discovered that their keys, coins, or cooking utensils have disappeared and that a twig or nut has been left in exchange will

Bushy-tailed Woodrat

understand why these woodrats are called "Trade Rats," and should seek out the nearest woodrat nest if they want to retrieve their property.

Woodrats have a head and body length of about 6–9 inches (152–229 millimeters) and usually a tail of about the same length or slightly shorter. They weigh 4–20 ounces (113–567 grams), depending upon the species, the Bushy-tailed Woodrat being the largest.

Woodrats are strictly nocturnal and therefore rarely seen by park visitors, even though they may be abundant in places, sometimes even living in tourist cabins. Identification of the species can be difficult in places where more than one species occurs, although habitat and altitude may be a guide. Seeds, beans, nuts, and even cactus pulp make up the diet of some woodrats in the arid southwest, whereas in California, the leaves, seeds, nuts, and fruits of a variety of trees and shrubs are the main source of food. Woodrats do not hibernate but instead store food, sometimes a bushel or more, in their dens. These rodents have favored feeding sites, generally evident from debris of seed husks and bits of cuttings on and beneath stumps and rocks, where they have paused to dine.

There is much variation in the kind and content of nests, not only between the different species but also between individuals. In desert areas, cactus joints will often be used to armor the outside of the den, whereas in other places the nest is mainly a pile of debris, rather shapeless, wedged into a rock crevice. Some species show a preference for vertical crevices; others build under horizontal rock overhangs. A woodrat house is usually the home of a single individual, even though it may contain two or more soft, lined nests and have a number of entrances. The area around the house is defended against other woodrats, but a wider foraging area—the home range—may be shared. Usually the home range is fairly small, generally less than 5 acres (2 hectares), and some individuals wander no more than a diameter of 82 feet (25 meters) around their dens. Males travel more widely than females, especially during the breeding season, when they seek mates.

The gestation period of most woodrats is about 35 days, and the litters are generally small, from two to four. In some places, woodrats breed at any time of the year and have several litters; in others, there is a restricted breeding season. The naked and blind young weigh about a half-ounce (14 grams) each. By the time they are three weeks old, their eyes have opened. Weaning may start at the fourth week. When they are two months of age, young woodrats may be on their own. Females are capable of reproducing when as young as six months, although most do not reach puberty until the spring of the year after their birth. Most woodrats live less than a year in the wild, but some have been known to live more than five years in captivity.

Woodrats occur in most of the national parks and monuments, but are rarely seen because they are small and nocturnal. Evidence of their presence is usually their elaborate stick-and-debris nests.

Eastern Woodrat: Hot Springs (?), Mammoth Cave, Shenandoah; **Southern Plains Woodrat:** Big Bend, Carlsbad Caverns; **White-throated Woodrat:** Big Bend, Canyonlands, Capitol Reef, Carlsbad Caverns, Grand Canyon, Guadalupe Mountains, Joshua Tree, Mesa Verde (?), Organ Pipe Cactus, Petrified Forest, Saguaro; **Desert Woodrat:** Arches, Bryce Canyon, Canyonlands, Capitol Reef, Death Valley, Grand Canyon, Joshua Tree, Organ Pipe Cactus, Saguaro, Sequoia, Zion; **Arizona Woodrat:** Grand Canyon; **Stephen's Woodrat:** Grand Canyon; **Mexican Woodrat:** Big Bend, Canyonlands, Grand Canyon, Guadalupe Mountains, Mesa Verde, Rocky Mountain, Saguaro; **Dusky-footed Woodrat:** Death Valley, Joshua Tree, Kings Canyon, Redwood, Sequoia, Yosemite; **Bushy-tailed Woodrat:** Arches, Bryce Canyon, Canyonlands, Capitol Reef, Crater Lake, Death Valley, Glacier, Grand Canyon, Grand Teton, Lassen Volcanic, Mesa Verde, Mount Rainier, North Cascades, Olympic, Redwood (?), Rocky Mountain, Sequoia, Theodore Roosevelt, Wind Cave, Yellowstone, Yosemite, Zion

---

MUSKRAT

Widely distributed throughout most of North America, the aquatic Muskrat lives in streams, rivers, lakes, ponds, and marshes, where there is a fair growth of aquatic vegetation. About a foot long (0.3 meters), with a tail of approximately the same length, the Muskrat weighs up to 4 pounds (1.8 kilograms) and is much smaller than the Beaver, which may inhabit some of the same sites. In addition to the size difference, the Muskrat does not have the broad, top-to-bottom flattened tail that is characteristic of the Beaver, but a relatively slender tail, which is flattened from side to

Muskrat

side. Beavers and Muskrats are often found together because Muskrats seem to take advantage of the environment created by the larger rodent and to set up housekeeping, with no objection from the Beaver, in its pond. Unlike Beavers, which feed mainly on bark, Muskrats prefer softer food, such as sedges, cattail stems, grasses, reeds, rushes, and floating pondweeds. Muskrats are also known, occasionally, to eat clams, crayfish, and even turtles, but they do so either when there is little other food available for them or when the populations of these animals are enormous. Sign of the Muskrat's presence in a pond is related to its food habits and is usually a few blades of cut sedges or cattail stalks afloat.

Muskrats build conical houses of sedges, cattails, and reeds (rather than of sticks, as the Beaver does), which are usually 2–3 feet (0.6–0.9 meter) in diameter and about the same height above the water. There is a nest chamber in the center, led to by an underwater entrance. Muskrats generally feed close to their lodge or at a feeding station, usually within a diameter of 50 feet (15 meters), and so have a relatively small home range. In some places they may also live in bank burrows, which also have underwater entrances.

Ordinarily, Muskrats construct small feeding platforms in a pond, and on these mats of vegetation do most of their eating, carrying the food to it. Feeding platforms are sometimes roofed over, forming another house that is similar to, but smaller than, the main nest. Muskrats are nocturnal but, in places where they are not molested, can often be seen swimming at dusk. They do not hibernate in winter. Winter is a hard time for Muskrats, because they do not store food and must feed on whatever they can find in the way of vegetation. They are adept swimmers and use their hind feet for propulsion and their tail mainly as a rudder. They can remain underwater for as long as 20 minutes. Because they can close their lips behind their incisor teeth, they are able to snip underwater plants and, in winter, under the ice, will gnaw open holes in ice in order to lead to a feeding chamber on the surface.

In winter, several Muskrats sometimes stay together in the same den, but these animals are primarily solitary and, as the breeding season approaches, become highly territorial, fighting to defend the area around their nest. Marking of the territory is done with the inch-long (25-millimeters) musk glands on the animal's abdomen. Both males and females defend territories vigorously, and some of the fights result in the death of an animal, although subordinate individuals are usually driven out of an area and thus become more vulnerable to the numerous predators of Muskrats. The gestation period varies from as few as 25 to as many as 30 days, and the number of young produced is usually between four and eight. In the southern parts of the range, these rodents may produce five or six litters a year, but in the north, only one or two. The babies are naked, blind, and weigh about three-fourths of an ounce (21 grams). By the time they are two weeks old, their eyes are open, they are furred, and they can also swim well. The babies are weaned when a month old and are soon evicted from the nest by the mother. They can breed when about a year old, or younger, if they were in a litter born late in the year. Muskrats may live three or four years in the wild.

Populations of Muskrats fluctuate greatly, as these animals are dependent upon many environmental factors, such as flooding and drought, severity of the winter, and abundance of food. When the populations are high, there is much fighting, and a good deal of the emigration is of young and subordinate animals leaving in search of new, less populated places. Emigration often occurs in the fall or early winter, at which time Muskrats can sometimes be seen wandering far from their normal haunts. These animals are not commonly seen by park visitors, but if visitors are quiet and patient at still ponds away from places where there are many people, muskrats can usually be seen swimming at dusk.

ROUND-TAILED MUSKRAT

A much smaller version of the Muskrat, and one that lacks the compressed tail, is the Round-tailed Muskrat, or Florida Water Rat. Slightly over half the size of a Muskrat and weighing up to only three-quarters of a pound (0.3 kilograms), this miniature version of a Muskrat occurs only in southeastern Georgia and Florida, which is outside the range of the Muskrat. Its small, conical houses are seldom more than 2 feet (0.6 meters) in diameter, and its general habits are quite similar to those of its larger relative. They live in shallower environments, however, preferring areas where the depth of water is 6–18 inches (150–460 millimeters) and where the aquatic vegetation is dense. Round-tailed Muskrats also undergo great population fluctuations, varying from as many as 100 per acre (250 per hectare) to 20 per acre (50 per hectare) in as short a time as from the beginning of March to early May. Only one animal inhabits a house. Round-tailed Muskrats do not range widely, usually feeding within 33 feet (10 meters) of home.

In the warm environments where they live, Round-tailed Muskrats breed throughout the year, producing relatively small litters, which average two animals. The gestation period is 26–29 days, and at birth the babies are 3½ inches (90 millimeters) long and weigh 0.4 ounce (12 grams). By the time they are three weeks of age, they have fur, teeth, open eyes, and are weaned; by the age of three months, they are sexually mature.

Round-tailed Muskrats are primarily nocturnal and, in the dense marshes where they live, are seldom seen. Their houses and runways can be found readily. The only park where they occur is Everglades.

Muskrat: Acadia, Arches, Canyonlands, Capitol Reef, Carlsbad Caverns (E?), Crater Lake (?!), Denali, Glacier, Grand Teton, Great Smoky Mountains, Isle Royale, Katmai, Lassen Volcanic, Mammoth Cave, Mesa Verde, Mount Rainier (E), North Cascades, Olympic, Redwood (RI?), Rocky Mountain, Shenandoah, Theodore Roosevelt, Voyageurs, Wind Cave (E), Yellowstone, Zion; **Round-tailed Muskrat:** Everglades

---

PORCUPINE

The United States' only quilled mammal, the Porcupine, is second in size only to the Beaver among North American rodents, reaching a weight of as much as 35 pounds (17 kilograms), a total length of up to 35 inches (889 millimeters), and a height at the shoulder of 12 inches (305 millimeters). Although primarily creatures of coniferous forests, Porcupines also inhabit brushy areas in the southwest and are found from coast to coast in the northern United States, as far south as Virginia in the east and Mexico in the west. Although they seem clumsy, they climb well, and most of their winter food is the rich, inner layers underneath the bark of trees. Debarking activities scar trees and may kill them, either by girdling or by exposing them to infection. Porcupines also eat buds and twigs of a variety of trees and bushes, and they will even eat prickly pear fruits in desert areas.

Quills, which are stiff, hollow, modified hairs, cover the animal's back and tail. Contrary to popular belief, Porcupines cannot throw their quills, although the quills are so loosely attached to the skin that some may shake loose when the animal is running or when it flails its tail about in self-defense. The tips of the quills are covered with tiny barbs, and quills that

Porcupine

are imbedded in the flesh of an attacker cannot readily be pulled out. If the quills are not removed, however, muscle contractions work them deeper into the flesh, and quill tips are known to have eventually passed completely through a person's calf, coming out under the skin on the other side of the leg. Animals that have preyed upon Porcupines and that have many quills imbedded in their faces are sometimes found dead. However, a few species—Fishers, Mountain Lions, and Bobcats—seem able to kill and eat Porcupines, including some of their quills, without disastrous consequences. These predators probably manage, in some way, to attack the Porcupine's belly, which lacks quills.

In summer, Porcupines are solitary, except for mothers with young, but in winter it is not unusual to find several denning together in a cave or up in a tree. In late fall, males begin to seek out females, and courtship is elaborate, because it is important for the female to be receptive. During copulation, the male sits upright and does not lean over the female, who must turn her quilled tail aside. The gestation period is long, about seven months, and the single baby is about 10 inches long (254 millimeters) and weighs about one pound (0.5 kilograms) at birth. It is covered with hair and has soft quills, which harden within a half hour. With its open eyes and well-formed teeth, a newborn Porcupine is capable of eating solid food immediately, although it will generally nurse, at least in captivity, for four or five months, at which time it may weigh 3–4 pounds (1.4–1.8 kilograms).

Porcupines are inoffensive animals that never attack. They are active mainly at night and are sometimes noisy, making a loud grunting sound as they move along. They seem to be attracted to salty objects, and campers have discovered them gnawing on ax handles, canoe paddles, gunstocks, and other sweat-stained wooden objects. They also chew on plywood—as some people learn when they discover that their boats have been gnawed. Porcupines are active in winter, and their trails through snow are often evident. They are known to have lived as long as ten years in captivity.

Encountering Porcupines is largely a matter of chance, but they are frequent visitors to campsites at night. Among the better parks and monuments in which to find them are Katmai, Yellowstone, and Glacier.

Porcupine: Acadia, Arches, Badlands, Big Bend, Bryce, Canyonlands, Capitol Reef, Carlsbad Caverns, Crater Lake, Death Valley, Denali, Glacier Bay, Glacier, Grand Canyon, Grand Teton, Guadalupe, Katmai, Kings Canyon, Lassen Volcanic, Mesa Verde, Mount Rainier, North Cascades (?), Olympic, Organ Pipe Cactus (?), Petrified Forest, Redwood, Rocky Mountain, Saguaro, Sequoia, Theodore Roosevelt, Voyageurs, Wind Cave, Yellowstone, Yosemite, Zion

---

Whales, Dolphins, and Porpoises

These wholly marine mammals—whales, dolphins, and porpoises—are divided into two groups, the toothed whales (which include dolphins and porpoises) and the baleen whales. Some forty-four species are known to inhabit North American waters, but many of these are quite rare. Seeing whales is largely a matter of chance, although there is some predictability for sighting the Gray Whale. Identifying whales is also difficult in many instances, because little is seen of the animal at sea, sometimes only 2–3 feet (0.6–0.9 meter) of an animal that may be 60 or 70 feet (18 or 21 meters) long.

Of the ten species of beaked whale, only one, the Goose-beaked Whale, has been recorded from a national park, Redwood. Up to 28 feet (8.5 meters) long, grayish-black in color, and with a small dorsal fin, Goose-beaked Whales are rarely seen, and, like most of the beaked-whale group, little is known of their habits. They are believed to feed mainly on small- and medium-sized squids, and most information about these whales has come from studying occasional individuals that have been found dead on beaches.

Sperm Whales are the largest of the toothed whales; the males reach 60 feet (18 meters) in length, the females reach half that size. They may weigh more than 50 tons (45,360 kilograms). The huge, square snout and the absence of a dorsal fin characterize this species. Sperm Whales which are essentially deep-water animals, are highly social, usually traveling in pods that consist of an adult male and a harem of females and their offspring. They are sometimes seen from Olympic or Redwood parks but not regularly, and sometimes these whales, for reasons still unknown, strand themselves on beaches, where their own great weight, no longer supported by water, suffocates them. Sperm Whales, which feed mainly on squid, can dive very deep and remain underwater for more than an hour before surfacing for air. When Sperm Whales "blow," the direction of the spout is forward, at an angle of almost 45 degrees.

Dolphins and porpoises are small whales and are the ones that are more frequently seen in coastal waters. Many people use the terms *dolphin* and *porpoise* interchangeably, but they apply to similar, but distinctively different, groups of mammals. Those properly called porpoises have teeth that are expanded at the tip so that they look like miniature lollipops, whereas dolphins have conical teeth. Dolphins have a distinct "beak," protruding upper and lower jaws, set off by a bulbous forehead, while porpoises are not beaked, and the slope from the top of the head to the tip of the snout is even. Some scientists have adopted the habit of calling all of these animals porpoises, because there are fish that are called dolphins, but doing so has caused even more confusion. Another complication in the use of common names for the dolphin group of toothed whales is that some of the larger ones are called whales; for example, Killer Whales and Pilot Whales are so named, and the latter are also called Blackfish.

Although almost any of the twenty species of dolphin and porpoise of American waters may be seen on the coasts, only ten of the species have been recorded from national park beaches or waters. On the Atlantic side, the Bottle-nosed Dolphin, a grayish animal up to 12 feet (3.7 meters) long, is not at all rare in the waters around Everglades and Biscayne national parks. This is the same species of dolphin that is usually kept to perform in captivity and is well known from television shows. On the Pacific Coast, it and a similar species are less frequently seen.

In the waters of the Pacific Coast parks, Pilot Whales are perhaps the most frequently seen dolphin. These black, melon-headed, 20–28 foot (6–8.5 meters) animals often swim close to shore in pods of several dozen. Pilot Whales can be seen from most of the coastal national parks—Everglades and Biscayne, in the Atlantic; and Redwood, Olympic, and Channel Islands, in the Pacific. Probably the best location in which to see them is between Anacapa Island of Channel Islands National Park and the mainland. Distinguishing between the two species—the Common Pilot Whale and the Short-finned Pilot Whale—is difficult in the field.

Another fairly common dolphin along the Pacific Coast is the Pacific White-sided Dolphin, which, as the name implies, has a whitish strip along each side, and a belly that is also white. Up to 9 feet (2.7 meters) long, these dolphins sometimes travel in very large groups and, because they feed on small fish, such as anchovies, they sometimes swim fairly close to shore. They are often seen around the Channel Islands and also from the beaches at Redwood and Olympic parks.

Most of the other dolphins are not seen in any of the park waters with regularity. The largest member of the dolphin family, the Killer Whale, which is up to 30 feet (9 meters) in length, is a wide-ranging species that can be found almost anywhere. It is readily identifiable by its huge dorsal fin and sharply demarked black and white body. It usually travels in pods of five to as many as thirty-five or forty animals, and it is one of the more conspicuous and readily identifiable species of whale. They are regularly seen in waters around the Alaskan parks, as well as Olympic. In addition to fish, Killer Whales feed on baleen whales, sea birds, seals, and sea lions. When Killer Whales are around a sea lion colony, the sea lions usually make haste to get out of the water and onto the shore, where they are safe.

Among the other dolphins that are sometimes seen in the waters of national parks are: Grampus, which are up to 14 feet (4.3 meters) long and have dark gray, light-streaked bodies); Northern Right Whale Dolphins which are up to 8 feet (2.4 meters) long, black, and lack a dorsal fin; and False Killer Whales, which are up to 18 feet (5.5 meters) long, black, with a sharply recurved dorsal fin and a nonbulging head.

Both species of porpoise, the Common and Dall's, are found in park and monument waters, sometimes quite frequently. They are small animals, reaching about 6 feet (1.8 meters) in length. They are characterized by their triangular dorsal fin and, unlike the Bottle-nosed and other dolphins, they lack a pronounced snout. The Common Porpoise is black, with a white belly and a pale area that extends up its sides. It occurs in both the Atlantic and Pacific oceans, and the only Atlantic park in which it is reported is Acadia. It is frequently seen in Glacier Bay and Olympic parks. The Dall's Porpoise is also black, but it has a distinct white patch, which extends up its sides below the dorsal fin. It is only found in the Pacific, mainly in the cooler, northern waters. It is a frequently seen species at Glacier Bay.

The baleen whales, in contrast to the dolphins and porpoises, have no teeth. They feed by filtering water through the plates of whalebone (baleen) that hang from their upper jaws. Five of the nine species of North American baleen whales are sometimes seen in park waters, most of them sporadically.

The largest of all whales and, indeed, the largest mammal that is known to have ever lived, is the Blue Whale. It can reach over 100 feet (30 meters) in length and weigh as much as 150 tons (136,000 kilograms). As a result of commercial whaling activities, the numbers of Blue Whales are so reduced that there is concern for the survival of the species. Because Blue Whales are mainly deep-water animals, they are not often seen close to shore. Channel Islands and Olympic are the only two parks where they have been reported with certainty. As do other baleen whales, Blue Whales feed on small (3–4 inch; 76–102 millimeters) crustaceans, which float in the water. In the summer, they feed in high-latitude waters, where there is a rich bloom of this krill, but by the polar winter, Blue Whales have migrated to warmer waters to bear their young and to breed. At birth, Blue Whales are about 24 feet (7.3 meters) long and lack a thick layer of fat, which insulates the adults from cold water; this is a reason for the annual migration. There is little food in the warm water for Blue Whales, and they may not feed during the months they are there. The Blue Whale is difficult to identify; it is quite similar to the Fin Whale, except in size. However, its slate-gray color is characteristic, and it lacks white on its baleen and lower jaw; the Fin Whale is black and is usually white on its baleen and the left side of the lower jaw. In addition, the underside of the Fin Whale's flukes are white; those of the Blue Whale are grayish.

Fin Whales, which can reach 70 feet (21 meters) in length, are the large whales that are more likely to be seen in park waters, because they are more numerous than Blue Whales. Similar in appearance and habits, except as previously noted, Fin Whales, like their relatives the Blue Whales, usually travel in small pods of four to six animals, and are the only baleen whales that are seen sometimes in an Atlantic park, Acadia. On the West Coast, they are sometimes seen around the Channel Islands, as well as off Redwood National Park and Glacier Bay National Park.

The Minke Whale is the least seen of this group of Whales, the Rorquals. It is a small species, about 30 feet (9 meters) long, which has a whitish baleen and a white band across the upper side of each flipper. It does swim in coastal waters; it feeds more on small fish than do the other rorquals.

Although sighting whales is generally considered a matter of chance, there is one whale that can be viewed with certainty. This is the Gray Whale, a medium-sized baleen whale, which can reach 45 feet (13.7 meters) in length and about 35 tons (31,752 kilograms) in weight. It is grayish-black in color and is blotched with pale spots; it does not have a dorsal fin. Gray Whales are relatively slow swimmers, usually traveling at 4–6 miles (7–10 kilometers) an hour, and they migrate each year from the northern Pacific seas to the coasts of northern Mexico, a trip that can be as long as 6,800 miles (11,000 kilometers). The journey southward in autumn and northward in spring is made close to shore, with the whales moving alone or in loose groups of up to a dozen or so animals. During the long northern days, the whales feed constantly, putting on the layers of fat that are necessary to sustain them on migration and during the birthing and breeding period. Their northern feeding grounds are in shallow water, and they obtain their food, mainly invertebrates, by stirring the mud with their snouts and then engulfing the water, mud, and organisms that float up.

In autumn, Gray Whales start southward, and by December they pass the California coast, at which time they are so readily visible that there are organized boat tours, especially from San Diego, to view the whales. The whales can be viewed from the shores of Cabrillo National Monument in San Diego. They continue to the shallow lagoons of southern Baja California, where they spend the winter. Soon after the arrival of the pregnant mothers, which lead the migration, the calves are born. They are about 15 feet (4.5 meters) long at birth. Mating also takes place during the winter.

The gestation period for a Gray Whale is a little less than a year. Females that have just given birth do not mate, but do so the following year, producing a baby every other year.

By February, some of the whales have started northward once more, again swimming fairly close to the coast, and, by the end of March, all have left the wintering grounds. Characteristically, they remain submerged for about four minutes, and then swim along the surface and blow five or six times before diving again. When "sounding," they go down almost vertically, raising their flukes high out of the water.

The Gray Whale is one of the finest examples of a species that has benefited from protection. Between 1850 and 1925, virtually all Gray Whales were killed by whalers, and as few as 300 were thought to have survived. Finally protected by international agreement, they slowly began to increase in numbers, and, by 1960, an estimated 6,000 Gray Whales made the annual migration. By 1980, the population of Gray Whales was thought to exceed the original size of the population before the animals were killed off by whalers.

During migration, Gray Whales can usually be seen off the shores of the Channel Islands, Redwood, and Olympic parks, in addition to Cabrillo National Monument.

Another species that has suffered greatly from whaling activities is the Hump-backed Whale. Up to 50 feet (15 meters) in length, with a black body, it is characterized by its very long, narrow flippers, which are white beneath and have a row of knobs along their front border and an uneven rear edge. They also have knobs on their snouts. Hump-backed Whales derive their name from the deep arches that their bodies make when they dive. They have a small dorsal fin, and the underside of the flukes is white; the flukes also have irregularly scalloped rear edges. Hump-backed Whales seem to stay in deeper waters for most of the year, but, when the winter calving season comes, they tend to seek shallower waters. Their gestation period is about a year, and, unlike most of the other baleen whales, which only nurse their young for about six months, Hump-backed mothers do so for a year.

Recent research has determined that Humpbacks have an extensive repertoire of sounds that they form into "songs," repeating refrains over and over again and altering them. The role of the Humpback's song is uncertain, but it is believed to be part of a courtship pattern. Individual whales seem to have their own singing styles. When courting, Humpbacks may leap out of the water (breach), flail their tails above the surface, and embrace one another with their long flippers.

Glacier Bay is one of the few places where there is some certainty of seeing Hump-backed Whales during the summer, because several usually occupy the area starting in late June. However, it now seems that the extensive tour-boat traffic in the park's waters may be disturbing them and future opportunities to see Humpbacks may diminish unless steps are taken to consider the whales' needs by limiting boat traffic.

Goose-beaked Whale: Redwood; **Sperm Whale:** Olympic, Redwood; **Bottle-nosed Dolphin:** Biscayne, Channel Islands, Everglades; **Northern Right-whale Dolphin:** Channel Islands; **Pacific White-sided Dolphin:** Channel Islands, Olympic, Redwood; **Killer Whale:** Channel Islands, Glacier Bay; **Grampus:** Channel Islands, Redwood; **False Killer Whale:** Channel Islands, Redwood; **Common Pilot Whale:** Acadia, Everglades, Olympic; **Short-finned Pilot Whale:** Channel Islands; **Harbor Porpoise:** Acadia, Glacier Bay, Katmai, Olympic, Redwood; **Dall's Porpoise:** Channel Islands, Glacier Bay, Redwood; **Gray Whale:** Channel Islands, Olympic; **Fin Whale:** Acadia, Channel Islands, Glacier Bay, Redwood; **Minke Whale:** Acadia, Channel Islands, Glacier Bay; **Blue Whale:** Channel Islands; **Hump-backed Whale:** Acadia, Channel Islands, Glacier Bay, Redwood

Carnivores

The seven families of the Order Carnivora that are found in the national parks are: dog family (foxes, wolves, Coyote); bear family; raccoon family (Raccoon, Ringtail, Coati); weasel family (Ermine, Weasels, Marten, Fisher, ferrets, skunks, Badger, Wolverine, otters, Mink); sea lion family (fur seals and sea lions); hair seal family (Elephant Seal and Harbor Seal); cat family (Mountain Lion, Bobcat, Lynx). All of these animals are characterized by their large, sabrelike canine teeth, and many of them have molars and premolars that are specialized for shearing flesh. Although all of these animals are adapted to some extent to feeding on animal matter, many—including bears, Raccoons, Coyotes, and foxes—are omnivorous and eat fruit, berries, and other vegetation, in addition to meat. Seals, sea lions, and River Otters are specialized for eating fish, and the Sea Otter eats mainly invertebrates.

More than any other group of mammals, the carnivores have suffered from human exploitation. Because wolves, Coyotes, foxes, bears, and Mountain Lions have taken advantage of human agriculture to feed on livestock, they have been relentlessly hunted and trapped, with the result that wolves, bears, and Mountain Lions have been extirpated from most of their original range. National parks, in fact, are among the few places in the United States where some of these species persist. Carnivores also are characterized by their fine, often dense coats of hair, and foxes, Mink, Marten, Ermine, Badger, skunks, Bobcats, and otters have been extensively trapped for their fur. Sea Otters and some of the fur seals were nearly extinct before trade in their pelts was controlled. The Black-footed Ferret is now one of the rarest North American mammals because its main source of food, the prairie dog, has been eliminated from so many places in the West.

Most of the land-dwelling species of carnivores are nocturnal or secretive in habit and thus are not usually seen. In the national parks, however, where they are not harassed, daytime views of Coyotes, foxes, bears, Martens, and otters are more likely. Skunks and Raccoons are frequent visitors around campsites in the evening and are a pleasant diversion.

COYOTE, GRAY WOLF, AND RED WOLF

The Coyote, the Gray Wolf, and the Red Wolf are three species of North American wild dog that have been greatly affected by human activity. The Gray Wolf is nearly extinct in the lower United States, although it prevails in Alaska, and the Red Wolf's situation is likewise precarious in the southern areas, where it still remains. However, the Coyote, a species that was once intended for total elimination by the U.S. government, has maintained itself and spread to portions of the country where it may never have lived in the past. All three of these wild species are similar in appearance and conformation; they differ mainly in size. Where two occur together, it may be extremely difficult to distinguish one from another. Because there seems to have been extensive hybridization between Red Wolves and Coyotes, telling these two apart is virtually impossible in the field and complicated and difficult in the laboratory. Of the three, the Coyote is generally the smallest, but there is overlap in size with the others. Coloration can also be similar. To add to the confusion, Coyotes not only resemble some breeds of domestic dog but also have interbred with them; the resulting hybrids are called *Coy-dogs*. Coyotes and Gray Wolves are generally intolerant of one another, with the Coyotes disappearing where wolves are abundant. Visitors to the national parks usually see Coyotes rather than wolves, which are rare.

Coyote.   Although their original range was mainly west of the Mississippi River, Coyotes may now occur not only in all of the eastern states, but throughout the west, as far north in Alaska as 70° latitude. They are found in deserts, mountains, forests, and seashore, and they are among the more adaptable and opportunistic American mammals. They have a body that is 3–4½ feet (1–1.3 meters) long, with a bushy tail that is another 16 inches (406 millimeters). Males are generally larger than females and usually weigh 18–44 pounds (8–20 kilograms); the largest Coyote that was recorded weighed 75 pounds (34 kilograms) and measured 5¼ feet (1.6 meters) from tip of nose to tip of tail. Those from the more northern parts of the range tend to be larger and heavier than those from the southern parts. The color is also variable, although most Coyotes are brownish- to reddish-gray, with those from higher and cooler places tending to be grayer and blacker, those of the deserts, tanner and redder. The belly and throat are pale in coloration, regardless of location. The tail may be tipped with black hairs.

Coyotes are not as social as wolves and are usually seen alone or in pairs. Pairs may occur only during the breeding season. The few sightings of packs of coyotes were believed to be of a female and her almost grown

Coyote

young, perhaps also with an adult male. They are wide-ranging animals, males having a home range that may be about 16 square miles (42 square kilometers), and they are known to make long journeys of up to 100 miles (160 kilometers). The home range of the females is smaller, averaging 4 square miles (10 square kilometers), and because there seems to be no overlap in the home range of females (there is in those of males), it may be that these areas are actually territories. Ordinarily, in the course of a night, a Coyote travels about 2½ miles (4 kilometers). Coyotes can run as fast as 30 miles (48 kilometers) an hour and can trot at 20 miles (32 kilometers) an hour. Although they may be active during daylight, especially early in the morning, Coyotes generally emerge from their dens at sunset and do most of their hunting at night.

One of the reasons for the Coyote's survival is the wide scope of foods that it can and does eat. One student found that Coyotes ate fifty-six species of animals, twenty-eight of plants, and some other items, including old shoes and tin cans. However, their main source of food is meat, which usually composes 90 percent of their diet. In captivity, Coyotes require a daily ration of 1⅓ pounds (600 grams) of meat a day; in the wild, they can eat more or less, depending upon opportunities. Deer, elk, rabbits, hares, and various rodents (rats, mice, ground squirrels, and prairie dogs), as well as ground-nesting birds, amphibians, reptiles, crayfish, and insects make up the bulk of the meat diet. Where there is livestock, Coyotes may eat sheep and poultry. Their plant foods are mainly berries and fruits. Most of the meat of large animals that they eat is in the form of carrion. In a case in which Coyotes had supposedly preyed on livestock, serious investigation determined that many of the animals that the Coyotes fed on had actually died of other causes, including attacks by domestic dogs. Elsewhere, it was discovered that Coyote populations could be reduced by picking up all dead animals before the Coyotes could utilize them as food. In most places, rabbits and hares make up almost half the coyotes' diet, rodents about a third, carrion a quarter, and birds, deer, and fruit, the remainder. In preying on large mammals, such as deer, Coyotes have little success in killing healthy adults, and most of their kills are of the young, old, and sick individuals.

Unlike domestic dogs and wolves, Coyotes have a very restricted breeding season, generally from January to March. The courtship is not so highly ritualized that hybridization with domestic dogs, Gray Wolves, and Red Wolves cannot occur; but in the wild, usually only female Coyotes mate with domestic dogs. The five to six pups, which are born after a gestation period of about 63 days, are blind and helpless. Their birth site, which is in a den that can be in a hillside, under a log, or under rocks, may be dug out to a length of 25 feet (7.5 meters). The babies weigh 9–10 ounces (250–75 grams) at birth and gain about three-fourths of a pound (0.3 kilograms) a week for the first eight weeks of their lives. Their eyes open when they are two weeks old, and soon after they may begin to come out of the den to play. At three weeks of age, they begin to eat semisolid food, which is regurgitated for them by the mother. Male Coyotes also help in raising the litter; they hunt and bring back food for the mother, and they perhaps even regurgitate food for the young. At about six weeks old, the pups are weaned, but it is not until they are six to nine months of age that they start to disperse, sometimes as much as 100 miles (160 kilometers), to establish their own home ranges. At nine months of age, they are about the same weight as an adult. The same den may be used year after year. Although not necessarily monogamous, the same pairs of adult Coyotes are known to mate year after year. Some of the young females may breed in their first year, but most do not. Although Coyotes have lived as long as eighteen years in captivity, the life span in the wild is usually six to eight years.

Coyotes communicate in many ways. They have numerous facial movements (in which their ears, lips, and teeth are used) that are indicative of mood and behavior, and they make at least eleven different kinds of vocal signal. Their evening howls, the precise purpose of which is not clearly understood, are characteristic evening sounds in Coyote country. Coyotes do not seem to use scent glands to mark as much as wolves do, and this may reflect their somewhat less complex social behavior.

Coyotes occur in many western national parks and, in some of them, are seen fairly readily. Among the better parks in which to observe Coyotes are Wind Cave, Badlands, Organ Pipe Cactus, Saguaro, and Yellowstone.

Gray Wolf. Gray Wolves are the largest of the wild dogs; a male may weigh up to 175 pounds (79 kilograms) and measure 6½ feet (2 meters) from the tip of its nose to the end of its tail. However, some males are smaller, and the usual range of measurements is as low as 45 pounds (20 kilograms) and 5 feet (15 meters) in length. The maximum weight for females is 120 pounds (54 kilograms) and the maximum length is 6 feet (1.8 meters). Wolves once ranged over the entire northern hemisphere, from the northernmost land in the Arctic to as far south in North America as central Mexico, and to central India in Eurasia. They have now been exterminated over about 95 percent of their original range, and they persist mainly in areas where there is little domestic livestock and little human activity. In the United States, except for Alaska, Gray Wolves are known to exist only at a few places along the Canadian Border, and they sometimes cross into Arizona from populations in Mexico.

Although called Gray Wolves (partly to distinguish them from the species called Red Wolves), these wolves may range in color from all white (though not albino) to all black. They are usually grayish or brownish, looking rather like domestic dogs, such as German Shepherds or Huskies. Compared with Red Wolves and Coyotes, Gray Wolves are larger and heavier, although it may sometimes be quite difficult to tell the species apart in the field. Gray Wolves also have shorter ears and broader snouts than Coyotes. The Red Wolf is intermediate in size and shape between Gray Wolves and Coyotes. In the past, the two kinds of wolf probably did not occur together; there are some scientists who believe that the Gray and Red Wolves are only

Gray Wolf

subspecies, not full species. Gray Wolves also interbreed with domestic dogs, and, in fact, the domestic dog had its origin from the Gray Wolf more than 12,000 years ago.

The Gray Wolf is a wide-ranging species that can travel for long periods at a rate of 5 miles (8 kilometers) an hour and can run as fast as 43 miles (70 kilometers) an hour. In the Arctic, the Gray Wolves' dense fur protects them so well that they have no difficulty in hunting at temperatures of −40° F. (−40° C.). Their normal home range may vary from 50 square miles (130 square kilometers) to, in Alaska, 5,000 square miles (13,000 square kilometers), and within it an individual may travel as much as 45 miles (72 kilometers) a day. When wolves in Alaska follow migrating Caribou, they may move more than 100 miles (161 kilometers) to keep up with them, and return again in the fall to the home site when Caribou move back from the tundra to the trees. The large home range is now thought to be a territory, the exclusive property of a single pack of wolves. The pack is an extended family, which usually consists of about five to eight animals, but packs comprised of up to thirty-six have been reported. Packs are dominated by one adult male, and there is a linear hierarchy, separate for males and females, of dominance within the pack. It now seems that, in a pack, the pups produced are the products only of the dominant male and dominant female. Lone wolves from other packs are chased from the territory, sometimes even killed, and mating within and between the subordinates of a pack is suppressed by the dominant animals.

Mating between the dominants may start in January in the southern parts of the range and as late as April in the north. The dominant ("alpha") male courts the dominant female, sometimes for weeks, and the pups are born after a gestation period of 63 days. Usually there are six young, born in a den that is an excavation in the ground and that may be used by wolves for many years. The pups are blind at birth, but their eyes open in two weeks. They are weaned when five weeks old. At two months, the pups are moved from the natal den site to a new location with a shallow nest, and they play outside there for several weeks before being moved to still another place, sometimes as much as 5 miles (8 kilometers) away. It is during this period, the first five months of their lives, that the young wolves learn the various behaviors and ties that are involved in the social life of a wolf pack. After being weaned, they feed on regurgitated food and then are led to nearby carcasses to feed. By autumn the young animals, who may weigh 60 pounds (27 kilograms), join the rest of the pack in hunting. The pups do not become physiologically mature enough to breed until their third year and, for social reasons, may not breed until they are older. In the wild, wolves have lived for sixteen years, but ten is usually considered old for a Gray Wolf.

Gray Wolves are meat-eaters, and originally they were the main predators of the larger North American mammals—Moose, deer, Elk, Caribou, Mountain Sheep, Mountain Goats, and Bison. The smallest mammal that is a regular part of a wolf's diet is the Beaver. In the north, wolves are still major predators of Moose and Caribou, but their depredations are aimed mainly at the weaker, younger, or older animals, not at the prime ones. A pack of fifteen wolves was successful in killing an adult Moose only once in twelve attempts, and studies indicate that a wolf pack kills an average of one deer per wolf every eighteen days, or one Moose per wolf every forty-five days. Although a wolf's stomach can hold up to 20 pounds (9 kilograms) of meat, a usual meal is much less, generally 5½–14 pounds (2.5–6.3 kilograms) per day.

One of the places where Gray Wolves have been intensively and steadily studied is Isle Royale National Park. For many years, it seemed that the wolves and their main prey, Moose, had achieved an equilibrium, with about six hundred Moose and twenty wolves. Wolves ate mainly Moose in the winter and Beaver in the summer. For a long time, a single pack had ranged over the entire island, but recently the pack split, and the island was divided into two territories. Because these wolves, when there was a single pack, were producing young only from the dominant female, the population remained fairly stable. When there were two packs, the reproductive rate on the island doubled. Then there were more splits, so that by 1980 there seemed to be five packs composed of more than fifty wolves. At the same time, the Moose population was declining, perhaps not so much because of the increased wolf population but from environmental changes that had reduced the amount of forage for Moose, as well as a decline in the number of Beavers. It will be interesting to see what occurs on Isle Royale in the future, but one must remember that the island represents a small and narrow ecosystem, which is not representative of the normal conditions for wolves or Moose on the mainland.

Wolves maintain their dominance hierarchy within the pack through a variety of facial and bodily movements and also utilize scent marks to delimit their territories and to identify territorial borders for other wolves. They also seem to mark trails within the territory, probably for use of their own pack, by urinating along them. Howling seems to serve both as a call for the pack to assemble, as well as to inform other wolves of the presence of the pack in the territory. Howling by one pack often sets off howling from another.

Observing wolves is very much a matter of chance, and a survey indicated that on Isle Royale, only one visitor in one thousand sees a wolf. On the island, wolves seem to have a good idea of just where people are and avoid them. In autumn, when the visitors have left the island, the wolves use the hiking trails extensively as their pathways. The only other park where there is a reasonable chance of seeing wolves is Denali.

Red Wolf. The Red Wolf is an enigmatic species that falls between the Gray Wolf and the Coyote in size, coloration, and social behavior. Because it has been exterminated from so many parts of its range, and because hybridization with the Coyotes seems to have occurred, there are few places where the true Red Wolf may persist. Originally, the species ranged over the Gulf Coast area, from eastern Texas to and throughout Florida, and it may have occurred northward, in the humid areas along the

Mississippi River drainage to southern Indiana. But now, only the coastal flatlands of eastern Texas and of western Louisiana may be the remaining stronghold of Red Wolves.

The Red Wolf is slightly larger than a Coyote and may overlap a small Gray Wolf in measurements and weight. Melanism seems to be a characteristic of this species, but not of the Coyote. Its social structure also seems more complex than that of the Coyote, but less so than that of the Gray Wolf. Mated pairs evidently mark a territory, and the few packs that have been seen are small, perhaps representing only a pair and their litter. Like Coyotes and unlike Gray Wolves, Red Wolves seem mainly to prey on small mammals, especially rabbits and rodents, and, even where deer are abundant, there are few reports of Red Wolves killing them.

Red Wolves probably mate in late winter, and the gestation period is thought to be 61 or 62 days. The average size of a litter is seven, but most pups seem to die, probably of disease, before six months old. This is one of the current problems in the survival of the species. Those pups that do survive are not ready to breed until their third year. Another factor in the survival of Red Wolves is that they have been easy to trap and poison (not being as cautious as Coyotes) and human beings have been primarily responsible for the decline in their numbers.

Over much of what was once the range of the Red Wolf, except for the coastal section mentioned previously, the wolflike animals now present are believed to be hybrids of Red Wolves and Coyotes. Since there are so few Red Wolves left in this range, eventually the animals present will be essentially Coyotes, with a few Red Wolf genes.

Red Wolves probably once occurred near Hot Springs National Park, but now any doglike wild animal in the region is likely to be a Coyote–Red Wolf hybrid, if not a Coyote.

Coyote: Arches, Badlands, Big Bend, Bryce Canyon, Canyonlands, Capitol Reef, Carlsbad Caverns, Crater Lake, Death Valley, Denali, Glacier Bay, Glacier, Grand Canyon, Grand Teton, Guadalupe Mountains, Hot Springs (?), Isle Royale (E), Joshua Tree, Kings Canyon, Lassen Volcanic, Mesa Verde, Mount Rainier, North Cascades, Olympic, Organ Pipe Cactus, Petrified Forest, Redwood, Rocky Mountain, Saguaro, Sequoia, Theodore Roosevelt, Voyageurs, Wind Cave, Yellowstone, Yosemite, Zion; **Gray Wolf:** Badlands (E), Big Bend (E?), Bryce Canyon, Capitol Reef (E), Carlsbad Caverns (E), Denali, Glacier Bay, Glacier, Grand Canyon (E), Grand Teton (E?), Great Smoky Mountains (E), Guadalupe (E), Isle Royale, Katmai, Kings Canyon (E), Mesa Verde (E?), Mt. Rainier (E), North Cascades, Olympic (E), Organ Pipe Cactus, Rocky Mountain (E?), Saguaro (E), Sequoia (E), Shenandoah (E), Theodore Roosevelt (E), Voyageurs, Wind Cave (E), Yellowstone (?), Yosemite (E), Zion (E); **Red Wolf:** Hot Springs (E)

FOXES

Of the six kinds of fox in America north of Mexico, five are found in the parks and monuments discussed in this book; only the Arctic Fox, a species that lives near coastlines on the tundra, is not. The most widespread and also the most readily recognized of the species is the Red Fox, which ranges from northern Alaska to as far south as the mountains of southern California, as well as through most of the midwestern and eastern deciduous forest areas. Although Red Foxes have a number of color phases— cross (with a black cross over the shoulders), silver (black hairs tipped with silver), and intermediate variations between these—all members of this species have white tail-tips, which readily identifies them. In most areas, the typical color phase consists of a reddish-yellow body and tail, a white belly, and black feet. The remaining four species of fox are characterized by their black-tipped tails. Two are small, sand-colored species that cannot be told apart in the field: the Kit Fox and the Swift Fox. They inhabit deserts and open plains in the southwest from southern California to eastern Oregon, and the plains areas from eastern Texas to southern Canada. Both of these small foxes have proportionately large ears. The final two species, which also have black-tipped tails, are called Gray Foxes. One species, the Insular Gray Fox, is found only on the Channel Islands of California; the other, the Gray Fox, inhabits the United States east of central North Dakota and is also found in the south and southwest, and from the Pacific Coastal states to northern Oregon. As its name implies, the Gray Fox is gray. It is grizzled and has a black strip down the top of the tail, which ends with black hairs. It has rust-colored hair along the belly and around the neck, and a white throat.

Red Fox. The Red Fox (see plate 5b) is a handsome species that, unfortunately, is less often seen than its abundance would indicate. Its bushy coat makes it appear much larger than it really is. It weighs up to 15 pounds (6.8 kilograms) and reaches 3½ feet (one meter) in total length. A trotting Red Fox seems to float above the ground, its cylindrical bushy tail trailing behind. Although they are mainly active at night, Red Foxes can sometimes be seen abroad early in the morning and even at midday, but only where they are not harassed by people.

Red Foxes are creatures of the forest edge, and although they may inhabit dense woods in a few places, they prefer sunny glades and adjacent fields. They do much of their hunting in grassy areas, especially along streams. There they search for rodents, particularly meadow mice, which they can, because of their keen sense of hearing, detect moving in their runways; they are then captured with a four-footed pounce. Rats and mice, tree and ground squirrels, rabbits and hares, and quail and other ground-nesting birds make up much of their diet, especially in winter. In summer, they dine on insects, fruits, and berries, in addition to meat. Red Foxes characteristically cache excess food, covering a small collection of mice with dirt or leaves for a later meal; in winter, they hide such food under snow.

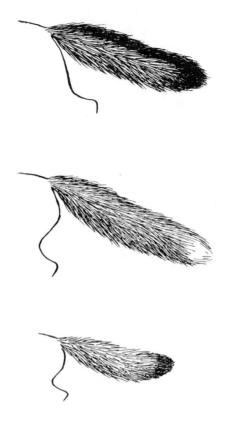

Fox Tails: top, *Gray Fox;* center, *Red Fox;* bottom, *Kit Fox*

Although much maligned by poultry raisers, Red Foxes are considered beneficial to agricultural enterprise because they control rodents. The usual size of a Red Fox's home range is up to 2 square miles (5 square kilometers), and, in the course of a single night's hunting, it may travel 5 miles (8 kilometers). Within the home range is the den, usually a burrow made over from the original owner, perhaps a Woodchuck, and it is often conspicuously marked by a pile of dirt outside it.

Red Foxes use their dens for raising their young, and it is there that seven or eight kits are born about mid-March, after a 53-day gestation. At birth, the kits, which weigh about 3½ ounces (100 grams), have dark brown fur, white-tipped tails, and closed eyes. Although their eyes open when they are about nine days old, it is not until they are five weeks old that Red Foxes start to come out of the den. Both the male and female help care for the young, the male generally hunting and bringing back food for the female at first, and later, when the kits are about eight weeks of age, bringing whole animals back to the young foxes, for them to practice tearing apart. It is usually three months before the kits start to follow their parents to learn hunting techniques, but by autumn, the young foxes move out of the parental den and disperse to start lives on their own. At this time, the male and female may separate, but, in many instances, the same pair mate again in January and use the same den to raise another litter. In winter, these well-insulated animals often sleep out in the open, even in snow. Their bushy tails curl around their bodies and cover their feet and noses.

Although not readily seen during the day in the parks where they occur, Red Foxes are sometimes encountered along roads at night or are seen hunting in open areas. Finding a den in the spring, when kits are present, generally rewards a patient watcher.

Kit and Swift Foxes.   The Kit and Swift Foxes, which are virtually indistinguishable from one another in the field, are the smallest foxes in the United States; they weigh a maximum of 6 ½ pounds (2.9 kilograms) and stand only about 12 inches (305 millimeters) tall at the shoulder. They are characterized by their brownish or grayish body coloration, black-tipped tails, black patches on the sides of their muzzles, and comparatively large ears. Whether these two kinds of Fox actually represent different species or geographic populations of a single species is still a matter of dispute among mammalogists, and even the known distribution of the two kinds is open to question. According to the latest studies, the Swift Fox originally ranged east of the Rockies from southern Canada to northern Texas, to as far as the western border of Minnesota. The Kit Fox lives mainly in the arid basin-and-range intermountain area, as well as desert regions from southwestern Texas to California, and as far north as southeastern Oregon. In at least one place where the two kinds of Fox come in contact, hybrids are known to exist, a fact that lends some validity to the idea that they are but a single species.

Relatively little is known of the habits of the Swift Fox, although they are generally assumed to be quite similar to those of the Kit Fox. Swift Foxes disappeared rapidly from their range with the approach of agriculture and the alteration of the original prairie. In most parts of their range, they were presumed extinct, but, strangely enough, starting in the mid-1950s, they began to reappear in places where they had not been seen for fifty or sixty years. This trend seems to be continuing, even in the national parks, where sightings of this diminutive fox have increased.

Kit Foxes were never in quite so endangered a position as the Swift Foxes, although populations in California and Oregon are greatly reduced, and Kit Foxes have been more thoroughly studied. Both species are strictly nocturnal, and their food supply is composed almost entirely of nocturnal rodents and lagomorphs. In some places, jack rabbits make up more than 90 percent of their diet, whereas in other places, their diet consists almost exclusively of kangaroo rats. Quite probably, these foxes feed on whatever small mammal is most readily captured in the area. Even though ground squirrels or prairie dogs may be abundant, they are diurnal and thus seem not to be exploited as food.

The den is usually made over from some other animal's diggings, and the entrance is small enough to exclude Badgers or Bobcats from entering. There are generally one or more additional entrances to the burrow system, which may be 4 feet (1.3 meters) deep and extend for as much as 20 feet (6 meters). Within the home range, there are several dens, and vacant ones are occupied by ground squirrels, lizards, and owls. Ordinarily, Kit Foxes range over 640–1,380 acres (260–520 hectares), but seldom wander more than 2 miles (3 kilometers) from their den. Because several groups of Kit Fox utilize the same hunting area, it is thought that they may not maintain territories, at least not for hunting.

Although Kit Foxes have often been alleged to be monogamous, mating for life, this seems not to be the usual case. Polygamy has been noted in several instances, and, in one area, only one of seven males mated with the same female in two successive years. The gestation period is not known, but is thought to be about 50 days, and the litters are generally born in February or March. Four or five pups are usually born. They are covered with short hair, are blind, and weigh 1.4 ounces (40 grams). During the time that the mother is nursing the pups, she rarely leaves them, relying on the male fox to hunt and provide food for her. This food is brought back whole, and there is no evidence that the male regurgitates food for the female, unlike some other members of the dog family. The pups are a month old before they emerge from the burrow, and, when three months old, they may begin to hunt with their parents. By five months, the young foxes have reached adult weight, but they remain with the parents until autumn, when they disperse to establish their own home ranges, usually far from their parents. One or more pups may remain with a parent through the winter. Although Kit and Swift Foxes have lived twelve years in captivity, they probably do not live more than seven or eight years in the wild.

Kit and Swift Foxes do not seem especially fearful of people, and, in places, have become quite friendly. Although it is still unusual to see Swift Foxes, Kit Foxes can best be observed at Organ Pipe Cactus National Monument.

Gray Fox. Found throughout the eastern United States and in the south to the Pacific Coast of California and Oregon, the Gray Fox is an inhabitant of forest and dense brush country. It is the only fox that has any climbing ability and occasionally ascends trees, especially sloping or well-branched ones, for refuge or even for food. About the same size as Red Foxes, Gray Foxes are less often seen because they lack the former's running stamina and so stay closer to hiding places and stay in cover. Strictly nocturnal, they prey on any rodents, hares, or rabbits that come their way. They sometimes can be seen along roads at night feeding on dead animals. In summer, they also seem to feed on berries and fruit.

The den of a Gray Fox, which is usually better hidden than that of a Red Fox, is often beneath a log or under rocks. Unlike the Red Fox, the Gray Fox uses its den throughout the year. The mating season is in winter. The gestation period is about 53 days; three to five is the normal litter size. At birth, the babies weigh about 4 ounces (113 grams), are dark brown in color, and are blind. Their eyes open in about eleven days. When the pups are five weeks old, they start to come out of the den, and, at three months, they begin to follow the adults on their nightly hunts. Although they are capable of fending for themselves when four months old, the kits usually remain with the adults for a few more months before dispersing. They often establish their own home ranges fairly close to those of their parents. Although normally smaller, home ranges of Gray Foxes may be as much as 10 square miles (26 square kilometers), and young foxes that disperse have been known to move 50 miles (80 kilometers).

Because of their preference for cover, Gray Foxes are not usually pests in agricultural areas, but likewise are not often seen in national parks. They respond to artificial game calls, such as the imitation of a rabbit's distress cry, and, in chaparral country, one way to see them is to simulate this cry. Canyonlands, Arches, and Capitol Reef are the best national parks in which to see Gray Foxes at night.

Insular Gray Fox. On some of the Channel Islands, there is a short-legged Gray Fox that is believed to be a species distinct from the one on the mainland; it is called the Insular Gray Fox. Its habits are the same as those foxes on the mainland. Of the islands in Channel Islands National Park, this species inhabits San Miguel, Santa Rosa, and Santa Cruz, and it can often be encountered during the day on the former, where it is numerous.

Red Fox: Acadia, Badlands (?), Bryce Canyon, Canyonlands, Capitol Reef, Crater Lake, Denali, Everglades, Glacier Bay, Glacier, Grand Canyon (?), Grand Teton, Great Smoky Mountains (I?), Hot Springs, Isle Royale, Katmai, Kings Canyon, Lassen Volcanic, Mammoth Cave, Mesa Verde, Mount Rainier, North Cascades, Rocky Mountain, Sequoia, Shenandoah, Theodore Roosevelt, Voyageurs, Wind Cave (?), Yellowstone, Yosemite (?), Zion; **Kit Fox:** Arches (?), Big Bend (E?), Capitol Reef (?), Carlsbad Caverns, Death Valley, Grand Canyon, Guadalupe Mountains, Joshua Tree, Organ Pipe, Petrified Forest, Saguaro (?), Zion; **Swift Fox:** Arches (?), Badlands, Canyonlands, Theodore Roosevelt, Wind Cave (?); **Gray Fox:** Arches, Big Bend, Bryce Canyon, Canyonlands, Capitol Reef, Carlsbad Caverns, Crater Lake, Death Valley, Everglades, Grand Canyon, Great Smoky Mountains, Guadalupe Mountains, Hot Springs, Joshua Tree, Kings Canyon, Lassen Volcanic, Mammoth Cave, Mesa Verde, Organ Pipe Cactus, Petrified Forest, Redwood, Rocky Mountain (?), Saguaro, Sequoia, Shenandoah, Voyageurs, Yosemite, Zion; **Insular Gray Fox:** Channel Islands

---

BEARS

Of the three species of bear in North America, only one, the Polar Bear, is not found in any of the national parks or monuments discussed in this book. Polar Bears are animals of the Arctic coastline and ice and do not

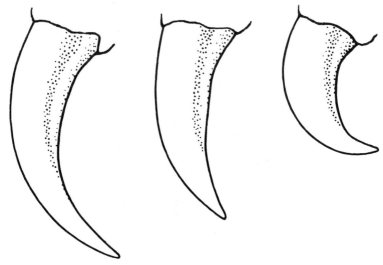
Bear Claws: left, *Grizzly;* center, *Brown;* right, *Black*

range inland to any great extent. The other two species are the Black Bear and the Grizzly or Brown Bear. Black Bears once roamed throughout the eastern deciduous forest, but now persist mainly in the mountain areas, the Gulf Coast, and southern Florida. In the west, they are also inhabitants of the mountain forest from southwest Texas to as far north as Alaska, and in the coastal state mountains to central California. Although their color may vary, in the east they are generally black, but some western animals may be shades of cinnamon and may have the same coloration as the Grizzly or Brown Bear. However, the Black Bear can generally be distinguished from the Grizzly or Brown Bear by a face that is not dished, but rather straight in profile; the absence of a marked hump on its shoulders, and its smaller size—it is no more than 3 feet (0.9 meter) tall at the shoulder. It also has much shorter claws.

The Grizzly and Brown Bears were formerly considered separate species, and were subdivided into a great number of categories. It now seems that these bears are all members of a single but highly variable species, the Brown Bear. The Brown Bear also occurs in the Old World. A popular distinction is still made between the Brown Bear and the Grizzly Bear because there is such a difference in size between the extremes of each group. Essentially, the Brown Bear is a coastal race that, perhaps because of its rich diet of salmon, is huge—the largest of the living carnivorous land mammals—whereas the Grizzly Bear, mainly an animal of open tundra and prairie, is somewhat smaller. There are, however, gradations in size and other characteristics between Brown and Grizzly Bears between Alaska's tundra and its coast. Brown Bears are reputed to weigh up to 1,500 pounds (680 kilograms), whereas Grizzlies weigh only as much as 850 pounds (386 kilograms). Either kind may be variable in color, from pale yellowish to dark brown. White tips to the hairs, "grizzling", led to the

top, *Profile of Black Bear;* bottom, *Profile of Grizzly Bear*

common name for the Grizzly Bear. Since Brown and Grizzly Bears are technically considered subspecies of the same species, they do not occur together, and locality is the best guide for establishing which one is being observed.

Black Bear.   Although it is the smallest of the North American bears, the Black Bear is often the largest mammal in many of the places where it occurs. Reaching weights of up to 600 pounds (272 kilograms) and lengths of 6 feet (1.8 meters), Black Bears are powerful and potentially dangerous animals, despite their seemingly comic antics. They are om-nivorous. In spring, when they emerge from their winter dens, they feed on grasses and sedges, buds, leaves, and bark. They unearth and eat bulbs. Using their paws, Black Bears shove ripe berries and their leaves into their mouths. Nuts, especially acorns, are eaten in vast quantities in the fall. Throughout these seasons, the bears also seek insects and ants, turning over rocks and logs to find them; the eggs and fledglings of ground-nesting birds, and mice, rats, chipmunks, ground squirrels, marmots, and Wood-chucks are dug out and devoured. Black Bears kill and eat fawns of deer and Elk, or any other small mammal that they can overcome, and carrion makes up a major portion of their meat diet. When fish, especially salmon, are in abundance, Black Bears catch and gorge on them. The ability to utilize so many resources for food has been a major factor in the Black Bear's ability to persist in many areas where more restrictive feeders, such as wolves, have not been able to survive. The intensive feeding in summer

Black Bear

enables these bears to put on huge layers of fat, which are sometimes as thick as 5 inches (127 millimeters) and weigh more than 200 pounds (91 kilograms). The bears use this form of stored energy during the winter.

The Black Bear is not a true hibernator. Its body temperature does not drop more than 10° F. (5° C.) below its normal level, and the animal is not in the deep torpor characteristic of true hibernators, such as marmots or ground squirrels, and can awaken instantly. The general metabolic rate is high enough so that the cubs' prenatal development and birth takes place during this period of winter dormancy, but no waste is passed out of the digestive tract, which remains empty. The extent of winter dormancy varies with the locality, and Black Bears in Florida and other warm parts of the range may only be idle for a few days during the coldest part of the winter. In the north, however, the bears generally seek out a sheltered place, often under the logs of a deadfall, or in root masses, hollow trees, and caves, to spend the winter. On warm, sunny winter days, they sometimes emerge and walk around for a few hours.

Although mating takes place in midsummer—the only time of the year that adult males and females associate with one another—the embryo's development halts at an early stage and it does not become attached to the wall of the uterus until late fall, when development resumes. Six or seven months after mating, usually two to three cubs are born in the winter den. Compared to adult Black Bears, the cubs are remarkably small, weighing 8 ounces (227 grams). They are born blind and almost naked, and it is two months before they are developed enough—weighing about 5 pounds

(2.3 kilograms)—to leave the den with the mother. The mother is fiercely protective of her cubs, as is well known, and many of the injuries to human beings caused by Black Bears in the national parks are the result of mothers' attempts to protect their young. Throughout the summer, the mother bear teaches the cubs how to find food. She usually takes them into her winter den with her; Black Bears breed every other year. In the spring, the cubs may be off on their own, or they sometimes stay with the mother through their second summer. However, they find their own winter dens in their second year. Black Bears may live as long as fifteen years in the wild and have lived twice that long in captivity.

Except for females with their cubs, Black Bears are solitary animals. They are primarily nocturnal but are frequently seen out during the day. An individual bear roams over an area of about 10 square miles (26 square kilometers), using brush, caves, or the hollows under uprooted trees, as sleeping sites. Black Bears are adept at climbing, and cubs are usually sent into trees for protection. Although they seem slow moving, bears can walk rapidly, and when they gallop, they can run as fast as 30 miles (48 kilometers) an hour, which is faster than a human being can run.

As adaptable and intelligent animals, Black Bears quickly learned to utilize people's garbage for food and to solicit handouts from tourists in the national parks. This eventually led to a number of severe accidents to park visitors. In recent years, garbage sites have been effectively bear-proofed, and the feeding of bears has been effectively discouraged. One result has been a decline in the number of injuries to people, and another has been the dispersal of the bears away from centers that are populated by people. Black Bears were once a major attraction in some national parks, but there have been few recent observations. In some parks, Black Bears have learned to take advantage of the increasing number of back-packers and to forage near their campsites. This has become enough of a problem that some parks now limit the places where back-packers may spend the night and also to require the use of special equipment to keep food and other items out of the reach of bears. Among the parks with large populations of bears, even though the animals may not often be seen, are Shenandoah, Great Smoky Mountains, Yellowstone, Glacier, Sequoia, and Yosemite.

Grizzly and Brown Bears.　Grizzly (plate 1b) and Brown Bears, although the same species, differ somewhat in their feeding habits, although their lives are similar in most other respects. This species ranges from the great plains of the western United States to the coasts of Alaska, and thus there are local variations in habits, but unless specifically mentioned, the habits accounted here are common to both kinds of bear.

Where salmon run up-river, Brown Bears fish for their food, biting the salmon as they swim by and, rarely, if ever, swatting them out of the water. Fish are characteristic of the Brown Bear's summer diet, but in many places where Grizzlies occur, there are no opportunities for fishing, and carrion—deer, Moose, Elk—provides a major portion of their diet. Grizzly Bears were predators of Bison when they were numerous; these bears were perhaps the only wild animals that could kill an adult of this species unaided. Ground squirrels are laboriously dug out, and Grizzlies do not disdain even a chipmunk. When berries ripen, all of the bears gorge on them. The major part of the total diet of these bears is vegetation. They are far more active during the day than is their smaller relative the Black Bear.

Although they lead mainly solitary lives, these bears may congregate around food sources, such as a good fishing spot along a stream. At these times, a dominance hierarchy prevails, with the subordinate animals readily giving way to the dominant, which is usually the largest and strongest, and the animals avoid close encounters with one another. Status in the hierarchy is usually determined by fights between males during the breeding season, and these may occasionally lead to the death of one of the males. More commonly, one animal flees and is thenceforth subordinate to the other when they meet later in the year. The home range of a Grizzly Bear varies with the locality and the food sources available, and it is not unusual that individuals wander over 50 miles (80 kilometers). Ordinarily, a male utilizes about 100 square miles (260 square kilometers), whereas the home range of females in the same region is much smaller, averaging 30 square miles (78 square kilometers). Brown Bears may have smaller home ranges than Grizzlies; their food sources are more concentrated.

After fattening all summer and fall, these bears sleep through the winter. They dig out a den site, usually on a hillside, or clear and fix up a previously used one where they live from October to April. They are not in deep torpor; their body temperature is only slightly lowered. Temperature records reveal that the temperature in the den, even without a bear present, remains within two or three degrees of freezing. This is because of the insulating value of a lot of snow. During this winter dormancy, bears do not defecate, having eaten material that effectively plugs the colon. When they emerge, they feed heavily on grass and whatever other green vegetation may have sprouted and search for carrion.

Mating takes place in the summer. Female Grizzlies and Brown Bears are usually at least six years old before they have their first litter. The litter, which is generally composed of two cubs, is born in the winter den about seven months after mating. As is usual with American bears, there is a delay in the implantation of the embryo into the walls of the uterus, so that the actual period of development is probably only two or three months. When the cubs emerge from the maternal winter den in the spring, they are carefully guarded by the mother, who tries to avoid contact with other Grizzlies. The cubs remain with the mother the next winter and may be nursed for more than two years. Female Grizzlies do not usually have litters more often than every three years, and the cubs may remain with the mother each winter for three years, although they usually disperse sooner.

Brown and Grizzly Bears seem more active during the day than Black Bears, but that may be because there is so much daylight during the summer in northern latitudes. It is known that they also move and feed at night. Like most bears, they have an excellent sense of smell and good hearing, but relatively poor eyesight. In places where they are common, Brown and Grizzly Bears use the same trails year after year, making deep-rutted parallel gullies in the ground.

In recent years, the status of Grizzly Bears in two of the national parks where they occur—Glacier and Yellowstone—has become controversial. There have been several instances of campers being killed in unprovoked attacks during the night. The cessation of feeding the bears in the parks led to the dispersal of Grizzlies, mainly to the back country, and to a diminution of their numbers. Undisturbed Grizzlies show no fear of human beings. Fortunately, park administrators were strong enough to prevent action on calls for the elimination of Grizzlies from the parks, and present policy is to minimize encounters between Grizzlies and people by closing trails that lead to places where Grizzlies are known to be active. Grizzly Bears can best be seen in Denali National Park and, although with less certainty, in Yellowstone and Glacier. Brown Bears can usually be seen when salmon are running, starting late in June, in Katmai National Park, and also in the northern parts of Glacier Bay.

Black Bear: Acadia, Arches, Big Bend, Bryce Canyon, Capitol Reef, Carlsbad Caverns (?), Crater Lake, Everglades (E?), Glacier Bay, Glacier, Grand Canyon, Grand Teton, Great Smoky Mountains, Guadalupe, Hot Springs (E), Kings Canyon, Lassen Volcanic, Mammoth Cave (E), Mesa Verde, Mount Rainier, North Cascades, Olympic, Redwood, Rocky Mountain, Saguaro, Sequoia, Shenandoah, Theodore Roosevelt (E), Voyageurs, Wind Cave (E), Yellowstone, Yosemite, Zion; **Brown Bear:** Glacier Bay, Katmai; **Grizzly Bear:** Bryce Canyon (E), Capitol Reef (E), Carlsbad Caverns (E), Denali, Glacier Bay, Glacier, Grand Canyon (E), Grand Teton, Guadalupe Mountains (E), Lassen Volcanic (E), North Cascades, Rocky Mountain (E), Sequoia (E), Theodore Roosevelt (E), Wind Cave (E), Yellowstone, Yosemite (E), Zion (E)

FUR SEALS AND SEA LIONS

Fur seals and sea lions are a group of mammals well adapted to life in the sea, although they must come onto shore to breed and bear their young. They are technically known as "eared seals." The presence of ears distinguishes them from the earless or "hair seals," which have no external ears, just a hole where the opening of the ear is. The eared seals have very small but conspicuous external ears. A more obvious difference between the eared and hair seals is the hind feet. Both have flippers, but the eared seals can rotate their hind flippers forward and use them to walk in a quadrupedal way; hair seals cannot turn their flippers forward and must drag them or elevate them when they move on land, progressing mostly on their bellies when out of the water.

At one time, both the eared seals and the hair seals were placed in a single group, whose members were all characterized by being aquatic and having limbs that were modified flippers. The distinction of the eared seals from the hair seals was noted, but both were thought to have diverged from a common ancestor. Recent studies of the ancestry of these animals indicate that the eared seals had their origin from animals that may also have been ancestors of the bears, whereas the hair seals and the otters may share the same ancestors.

Eared seals are grouped into several divisions. Members of the sea lion group are characterized by their relatively blunt snouts, as well as short, coarse guard hairs over relatively little underfur. In sea lions, the first digit of the foreflipper is longer than the second, and the outer digits of the hind flippers are longer then the inner ones. In fur seals, the first digit of the foreflipper is either the same length as or shorter than the second, and the digits of the hind feet are about equal in length. The fur seals also have coarse guard hairs, but these hairs cover a dense, fine underfur. Their snouts are more pointed.

All four species of fur seal and sea lion in North American waters occur in one or more of the parks of monuments discussed in this book. All of the parks and monuments in which they occur border on the Pacific Ocean. All four of these eared seals can at times be found off the California coast. The two fur seals—the Northern Fur Seal and the Guadalupe Fur Seal—may occur together in Channel Islands National Park, usually in considerably smaller numbers than the great herds of California and Northern Sea Lions that breed there. The Guadalupe Fur Seal can be distinguished from the black-headed Northern Fur Seal by its silver-gray head and its generally smaller size. The two species of sea lion also occur on the Channel Islands but can easily be distinguished by size and color. The Northern Sea Lion (sometimes called Steller's Sea Lion) is very large—males reach 10 feet (3 meters) in length and 2,000 pounds (907 kilograms) in weight, whereas California Sea Lion males rarely exceed 8 feet (2.4 meters) in length and 600 pounds (272 kilograms) in weight. The Northern Sea Lion is yellowish-brown; the California Sea Lion is brown to black. There are usually only a few Northern Sea Lions on the Channel Islands, as compared with thousands of California Sea Lions.

Northern Fur Seal.   Also called the Alaskan Fur Seal, the Northern Fur Seal is the species most desired in the fur trade because it has a dense and velvety underfur. The males, which reach lengths of 7¼ feet (2.2 meters) and weights of almost 600 pounds (272 kilograms), are considerably larger than the 4½-foot (1.4-meter), 135-pound (61-kilogram) females. Although these fur seals spend most of the year at sea, usually alone or in small groups of three or four, they start to assemble, in late May and early June, on the beaches of the Pribilof Islands of Alaska by the hundreds of thousands. First come the males, who fight to establish small territories into which they attempt to herd the females upon their arrival a few weeks later. The females usually reach the islands from mid-

Northern Fur Seals

June to mid-July. They come ashore and give birth within a day or two. A week later, they mate with the bull in whose territory they are. Some bulls may have fifty to one hundred females in their harem. The pups, weighing about 10 pounds (4.5 kilograms) at birth, are nursed with the mother's rich milk, but after she has mated, nursing may occur at intervals of as much as a week. With so many fur seals in one place, the mother may have to travel more than 100 miles (161 kilometers) in several days to obtain enough fish for her sustenance and milk production. During the two main months of the breeding season, the harem-master bulls do not eat but fight to maintain their territories, and mate. Thus they lose much weight and are greatly weakened. In addition to the hazards of infrequent nursing, the baby seals are often in danger of being crushed by bulls moving through the harem and fighting among themselves, as well as illness from infections, especially parasitic ones, which take a high toll of young, particularly when the population of fur seals is high.

Although they can probably swim at birth, young fur seals do not enter the water until they are at least a month old. At three months old, they are weaned. By November, all of the animals have left the Pribilofs, some of them to travel almost 6,000 miles (9,656 kilometers) to spend the winter off the California coast. Fur seals have few enemies, except for human beings and Killer Whales. They may live twenty or thirty years. However, mortality is high and half of the pups die before they are one year old, and three-quarters die before they are three. During the winter months, these fur seals swim farther off shore, feeding mainly on squid and fish. They can swim 17 miles (27 kilometers) an hour and dive to more than 200 feet (61 meters). It is mainly in the winter that any of these seals are seen as far south as California.

At one time, the Northern Fur Seal was in danger of extinction, because the animals were killed for their fur when concentrated on land during the breeding season, as well as at sea during the rest of the year. In 1911, by international agreement among the countries engaged in exploiting these animals, the management and harvesting of the fur seals was entrusted to the United States, and, in return for a percentage of the skins, the other countries agreed to cease pelagic (open-sea) hunting. At the time of the treaty, only about 200,000 fur seals remained, but with appropriate management, the herds have been built up to more than a million. Harvest of young males, mainly three to five years old, provides the fur for the trade. Because of the territorial nature of these fur seals, their limited breeding ground, and the population-dependent high mortality, the loss of 50,000 or so bachelor males each year should not affect the reproductive potential of the Pribilof herds.

Northern Fur Seals are not abundant in any national parks or monuments, but sometimes can be seen in the bays of Glacier Bay, on the coast of Redwood National Park, and at the Channel Islands National Park.

Guadalupe Fur Seal. Smaller and considerably lighter in weight than the Northern Fur Seal, the Guadalupe Fur Seal (also called Townsend's Fur Seal) was thought to be extinct at the beginning of the twentieth century. A small breeding population was discovered and then protected on Guadalupe Island, on the Pacific Coast of Baja California, Mexico. From this nucleus, the population has grown to about one thousand animals and has spread as far north as the Channel Islands. However, this fur seal still breeds only on Guadalupe Island. In June and July, the males form small harems in its rocky seaside caves; the pups are born; and mating takes place. Little is known of the habits of the Guadalupe Fur Seal, but they are probably similiar to those of the Northern Fur Seal. The only national park where Guadalupe Fur Seals can be seen is Channel Islands, where each summer a few males reside on San Miguel.

Northern Sea Lion. Ranging from the Aleutian Islands of Alaska to the Channel Islands of California is the Northern Sea Lion, the largest American sea lion. Males may exceed a ton (907 kilograms) in weight, whereas the females are one-third as heavy, at maximum. It is the only sea lion in this area that is sandy in coloration. It breeds throughout its range. Bulls come ashore in May and begin to establish territories when the females arrive two weeks later; the bulls assemble them into harems consisting of ten to twenty members. The pups are born within two days after the females beach, and mating takes place soon after. As also occurs in the Northern Fur Seal, the implantation of the embryo in the uterine wall is delayed for several months. Although female Northern Sea Lions are sexually mature when they are three years old, males do not mature until they are about six, and they may not actually be large and strong enough to establish a territory until they are nine or ten.

Northern Sea Lions: Harem Bull with Cows

Northern Sea Lions feed on many kinds of fish and are often accused of eating fish that are desired by commercial fishermen. They do eat such fish, but they eat larger quantities of other fish, including lampreys, which are not utilized by human beings or which compete with commercially valuable fish. Most of their fishing is done at night. They can dive to depths of 500 feet (152 meters).

This is the species that was first called a "sea lion" by the German naturalist George W. Steller, who described the animal and noted its mane and roar in 1741. Northern Sea Lions can be seen in Channel Islands, Redwood, and Olympic national parks.

California Sea Lion. The California Sea Lion, which is the familiar "trained seal" of circuses, is a common species along the coast of California, especially in Channel Islands National Park. Although the females may be only one foot (0.3 meter) shorter than the male's 7 foot (2.1 meters) length, they usually weigh only about one-third of his 600 pounds (272 kilograms). The adult male has a different profile; it tends to have a high forehead, because of the high crest that develops on the skull. California Sea Lions, which are inhabitants of coastal waters, are rarely found more than 10 miles (16 kilometers) offshore. They are swift swimmers, reaching speeds of 25 miles (40 kilometers) an hour at times, and they can dive more than 400 feet (122 meters), remaining underwater for as long as 20 minutes. They feed on a variety of fish, as well as squid, octopus, and other marine creatures. Although they are thought to eat large quantities of salmon, studies have shown that this is untrue and that they prefer squid to fish. However, sometimes they damage salmon nets.

California Sea Lions are gregarious and playful, and the females remain in large numbers around the breeding grounds throughout the year, whereas males tend to form their own groups and migrate northward for some of the year. Starting in May or June, male sea lions come ashore to establish territories, which they proclaim with their characteristic loud, honking bark. They fight for mating rights with the females that remain within the bounds of their defended areas. Mating takes place in June, shortly after a pup has been born. Newborn pups are about 2½ feet (0.8 meters) long and usually weigh about 12 pounds (5.4 kilograms). They first take to the water when 10 days old. The pup may nurse for a year, but is usually weaned by five months, and it can eat solid food long before. Females can breed when three or four years old; males are not capable of breeding until they are five and, because of the need to fight for a territory, are usually several years older before they first mate.

Sea Lions are often seen floating on the surface of the water, with their flippers raised. (They gain or lose heat through the surfaces of their flippers.) They are intelligent animals and have been trained to perform, although it may take a year before an animal is prepared enough to appear in public. A Sea Lion's career as a performer may last for a dozen years.

Although at times some California Sea Lions may appear off the coast of any of the Pacific national parks, Channel Islands is the only one with a resident population; the animals can readily be seen in the waters and hauled out on the beaches, sunning themselves.

Northern Fur Seal: Channel Islands, Glacier Bay, Redwood; **Guadalupe Fur Seal:** Channel Islands; **Northern Sea Lion:** Channel Islands, Glacier Bay, Katmai, Olympic, Redwood; **California Sea Lion:** Channel Islands, Olympic (?), Redwood

---

RINGTAIL, RACCOON, AND COATI

The Ringtail, the Raccoon, and the Coati are three species that are procyonids, belonging to what is popularly called the Raccoon family. It is a group that includes animals as diverse as the arboreal, fruit-eating Kinkajou and perhaps even the bearlike Giant Panda. (Current opinion is that the Giant Panda is from the bear lineage, rather than the Raccoon stock.) Although the Raccoon, Ringtail, and Coati are not closely related, they do share the common characteristics of a ringed tail and a mask across the eyes. They are readily distinguished from one another. The Ringtail, or Bassariscus, has a tail as long as its head and body, and the tail is clearly and distinctly marked with black and white bands. The Coati also has a long tail, but it is neither bushy nor very distinctively ringed.

Young Ringtails

The Raccoon's tail is relatively shorter, and the rings are either brown, or black and yellowish, rather than white, as are the Ringtail's. The Coati also has a very long, pointed snout. The Raccoon occurs over most of the United States, except in some higher western elevations, whereas the Coati and Ringtail are southwestern in distribution; the Coati is found only from central Arizona to as far east as southern Texas along the Rio Grande, and the Ringtail is found only from eastern Texas to southern Oregon.

Ringtail. Up to 2½ feet (0.8 meters) long, half of which is the bushy tail, the Ringtail is a slender-bodied animal that weighs only about 2 pounds (0.9 kilograms). Its preferred habitat is broken, rocky areas, including cave mouths. Because of its presence around western mines, where it was valued for its rodent-killing abilities, it earned the nickname "Miner's Cat." Ringtails are wholly nocturnal and are never seen abroad during the day. Although the usual den is under rocks, in a shallow cave, or in a tree hollow, Ringtails sometimes take up residence in abandoned cabins. They line their sleeping sites with grass or leaves. They sleep during the day, and at night they hunt for mice, rats, rabbits, birds, insects, and lizards. They also eat fruits and berries. Ringtails are excellent climbers. They seem to live as pairs for much, if not all, of the year. The breeding season is in the spring, and the gestation period is about two months. There are three to four blind, white-haired babies to a litter. It is a little over a month before their eyes open. They begin to leave the den when two months old; they depart from it in the autumn. The male Ringtail helps in raising the young by bringing food to them during the time they remain in the den.

Encounters with Ringtails are usually rare, although these animals may be fairly common in many national parks. They probably do not have a large home range, but their excellent hearing (and probably good night vision) lets them easily avoid human contact. In some places, however, they forage around campsites and even in hotels. The likeliest location in which to see a Ringtail is at Phantom Ranch in the Grand Canyon.

Raccoon. The Raccoon is one of the more familiar species of North American mammal. The black mask across its eyes seems to confirm its reputation as a "bandit"; it often raids suburban garbage cans. Widespread and abundant, Raccoons may get by on a home range as small as 10 acres (5 hectares) in a suburban area or require up to 12,500 acres (5,000 hectares) where food supplies are dispersed, such as on North Dakota prairies. Males usually have a home range that is two or three times as large as that of females, but the normal nightly movements of either sex are generally of relatively short length, usually about a quarter of a mile (0.4 kilometers).

Often a Raccoon's den is in a hollow tree, but Raccoons also use fallen logs, dry culverts, rock crevices, ground burrows of skunks or Woodchucks, and, not uncommonly, the attics of buildings that are either occupied or deserted. Although a Raccoon has a primary den site within its home range, it also has other resting and hiding places within it and may remain in these places for several days during trips through its feeding grounds. Raccoons seem not to be notably territorial, although there is some indication that males may be, unlike females.

Raccoons are omnivores. They do not actually kill many birds or rodents, except those that are injured or otherwise incapable of escaping—fledglings, for example. Raccoons often (but not always) live along streams and marshes, where they find the foods that compose a major portion of their diets: crayfish, crabs, frogs, turtles, baby muskrats, and nestling birds. They also eat almost any kind of berry, seed, nut, and fruit that is ripe and, in agricultural areas, are fond of corn and other crops. One legend about Raccoons is that they always wash their food. Although it is true that Raccoons carry food to water, where they manipulate it before eating it, especially in zoos, they do not always do so in the wild. Compared with other carnivores, Raccoons have a large number of nerves associated with touch in their forepaws and a high degree of development of the brain areas associated with touch. It was thought that the "washing" enabled the animal to detect debris that may be mixed with the food, but it now seems that the animals merely derive pleasure from the sensation. Raccoons with no access to water nevertheless often "wash" the material between their paws.

The breeding season varies with the locality, although it is generally in winter. A male moves into the nest of a receptive female for a week or so at this time, but there is no permanent liaison, and afterward the male moves out and seeks other females. After a gestation period of 63 days, four or five young are usually born in the nest, which is generally up in a

Raccoon

hollow tree. At birth, the babies are blind and weigh 2 ounces (57 grams). At twenty days their eyes open, at forty days they are weaned. At this time, they begin to come out of the nest and soon travel with the mother during her nightly hunts. If food is plentiful, a young Raccoon can be on its own by the time it is four months old. Often, however, the young ones overwinter with their mother and disperse in the spring. In spring, both the males and females, as yearlings, may be able to breed.

Raccoons are capable of putting on large amounts of fat; the northern ones need much fat in order to survive the winter. Raccoons in the colder areas are larger than those in the warm, southern ones. They are also much heavier because of their large amounts of fat; some have reached almost 50 pounds (23 kilograms) in weight. They do not hibernate, although, when it is extremely cold, they may remain in the den for several days, and, over the course of the winter, they may lose half the weight they had started with the previous fall.

Even though Raccoons are primarily nocturnal, it is not unusual to see them out in the morning or, especially, early in the evening, when it is still light. In many of the parks, they are common around campgrounds and picnic areas, where they are usually seen foraging during the night. They can usually be seen in any of the parks where they occur; Mammoth Cave and Yosemite are two of the better ones in which to see them.

Coati. The Coati, a long-nosed, slender-tailed relative of the Raccoon, is a species that has rather recently invaded the southwestern United States from Mexico. Up to 4 feet (1.2 meters) in total length, and weighing

15–25 pounds (7–11 kilograms), Coatis inhabit canyons and wooded areas, usually near streams, in Arizona, New Mexico, and along the Rio Grande in Texas. They are much more social than the other members of the Raccoon family, and packs of Coatis may number up to thirty or forty, although most packs are smaller. Unlike their relatives, they tend to be active during daylight hours. They range over an area, rooting in the ground with their gristly noses to unearth grubs, lizards, and roots. They also climb into trees to obtain fruits, which are also a large part of their diet. When a band of Coatis is on the move, the members usually hold their long, faintly ringed tails erect. They probably den among rocks, but little is known of their habits. Their gestation period is about 75 days, and they have four to six young, which are usually born early in the summer. When foraging, Coatis can be quite noisy, rustling leaves, overturning rocks, and making a variety of sounds, including chirping squeals.

The only park where Coatis are known to be resident is Saguaro National Monument, although transient Coatis have been reported both from Big Bend and Organ Pipe Cactus.

Ringtail: Arches, Big Bend, Bryce Canyon, Canyonlands, Carlsbad Caverns, Death Valley, Grand Canyon, Guadalupe, Joshua Tree, Kings Canyon, Mesa Verde, Organ Pipe Cactus, Redwood, Rocky Mountain (?), Saguaro, Sequoia, Yosemite, Zion; **Raccoon:** Acadia, Arches, Badlands (?), Big Bend, Biscayne, Capitol Reef (?), Carlsbad Caverns, Everglades, Glacier (I?), Grand Canyon, Great Smoky Mountains, Guadalupe, Hot Springs, Joshua Tree (?I), Kings Canyon, Lassen Volcanic, Mammoth Cave, Mesa Verde, Mount Rainier, North Cascades, Olympic, Organ Pipe Cactus, Redwood, Rocky Mountain, Saguaro, Sequoia, Shenandoah, Theodore Roosevelt, Voyageurs, Wind Cave, Yosemite, Zion; **Coati:** Big Bend, Organ Pipe, Saguaro

---

WEASEL FAMILY

Although called the Weasel family, this group contains a number of diverse species, ranging from the marine Sea Otter, the aquatic River Otter, the burrowing Badger, the northern Wolverine, the skunks, Mink, ferrets, Marten and Fisher, as well as animals that we call weasels. Most have well-developed anal scent glands; those of the skunks are the best developed. In the Americas, the majority of the representatives of this family are northern. All of the species in North America are represented in one or more of the national parks. The majority of these animals are hunters of other mammals, but some skunks are largely insectivorous, and the Sea Otter feeds primarily on sea urchins, clams, and other bottom-dwelling invertebrates. Most members of the weasel family are nocturnal and, even where abundant, may seldom be seen. The skunks are an exception and, secure in their defense, are often observed wandering brazenly around campsites looking for food.

Marten and Fisher.   The Marten and the Fisher differ from most other members of the Weasel family in their dentition (they have thirty-eight teeth, as does the Wolverine, rather than the thirty-two to thirty-six teeth of the other members). The Marten, which might be confused with a Mink, is usually larger and has a distinctive yellow or orange patch on the chest. The much larger Fisher is not likely to be confused with any other North American mammal by virtue of its size—it is almost 3 feet (0.9 meter) in total length—and by its dark, almost black, coloration, with grizzled, white-tipped hairs. Both species, which are primarily forest inhabitants, have their main range in the coniferous forests of Canada and Alaska and reach the lower United States only in the central states bordering Canada, in New England; and in the Rocky Mountains, Cascades, and Sierra Nevada.

Marten.   The handsome Marten, rich brown in color, is mainly nocturnal but, more than many other members of the weasel family, is sometimes seen out during the day, hunting rodents in forest clearings. It is an excellent climber, and although it generally makes its den under rocks on the ground, or in fallen logs, it often hunts in trees for squirrels and birds. Over much of its range, Red or Douglas' Squirrels are a primary source of food, but chipmunks, mice, birds, and eggs are also eaten, and seeds and berries make up much of the plant-food portion of their diet. Martens are solitary. They are also probably territorial, and they use their scent glands to delimit their territories. Males may utilize about one square mile (2.6 kilometers) as their home range, while females evidently require only about one-quarter of a square mile. Observers who chirp like birds can sometimes bring Martens close to them.

Martens breed in midsummer, and although the young are born about nine months later, in spring, the implantation of the embryo is delayed until late fall, thus the period of actual development is relatively short. The two to four babies weigh only one ounce (28 grams) each and are covered with fine hair. They are born blind and their eyes do not open until thirty-five to forty days. They are weaned about two weeks later. By three months old, Martens have reached adult weight—about 2 pounds (0.9 kilograms) for males and 1.5 pounds (0.7 kilograms) for females—and move off to establish their own home ranges. The females probably breed when they are a little more than a year old.

Tracks of Marten can often be seen in winter snow, but seeing the animal itself in any of the parks where it occurs is mainly a matter of chance.

Fisher.   Up to 3.3 feet (1 meter) long, of which one-third is its bushy tail, the Fisher is one of the rarely seen animals of U.S. parks and monuments. Fishers live mainly in forest and are usually active only at night. They may den in a hollow tree or on the ground beneath rocks. They feed mainly on Snowshoe Hares and Porcupines. They are one of the few mammals that prey regularly on the Porcupine and seem able to kill that prickly species by flipping it onto its back and attacking its belly, which lacks quills. Some skill is evidently required for such attacks, and Fishers are known to have died from the wounds made by quills. Fishers also eat almost any other small mammal they can catch, as well as some plant material. However, fishing is not a notable habit of theirs, despite their common name.

Fishers roam widely, and a home range of a male, which is larger than that of a female, may be as much as 10 square miles (26 square kilometers). They are solitary, tolerating one another only during the brief mating season, which is in the spring. Like the Marten and some other weasels, Fishers have delayed implantation of the embryo, and the one to four babies are born almost a year after mating. Much more remains to be learned of the habits of Fishers. Over much of their range, they have been greatly reduced in number, if not actually wiped out, for their valuable fur. In the parks, they are rarely seen.

Marten: Crater Lake, Glacier Bay, Glacier, Grand Teton, Isle Royale (?), Kings Canyon, Lassen Volcanic, Mount Rainier, North Cascades, Olympic, Redwood, Rocky Mountain, Sequoia, Voyageurs (E?) Yellowstone, Yosemite; **Fisher:** Acadia, Crater Lake, Glacier, Kings Canyon (?), Mount Rainier (E), North Cascades, Olympic, Redwood (?), Sequoia, Shenandoah (E), Voyageurs, Yellowstone (?), Yosemite

---

Ermine, Weasels, Ferret, and Mink. The five North American species that comprise the Ermine, weasel, ferret, and Mink group are among the most carnivorous and skillful hunters of the weasel family and are characterized by their thirty-four teeth, which are well adapted for a carnivorous diet. Skunks and Badgers, which are also in this family, have the same number of teeth, but their teeth are not so highly specialized for such a diet. None of these animals are commonly seen, and one, the Black-footed Ferret, is among the rarest North American mammals; it is nearly extinct. The Ermine (also called Short-tailed Weasel) and the Least Weasel are both diminutive species; the former is 6–9 inches (152–229 millimeters) in body length, and the latter is about 6 inches (152 millimeters) in body length, but they can be told apart by the tip of the tail. Ermine have a black tip; Least Weasels do not. The tail of the Least Weasel is rarely more than 1½ inches (38 millimeters) long; that of the Ermine is usually more than 2 inches (51 millimeters) long. Over most of their ranges, both of these weasels turn white in winter (although the Ermine maintains its black tail-tip), but there are exceptional places where they do not. Both Ermine and Least Weasels occur in Alaska and the north-central states, but only the Ermine lives in New England, the Pacific coastal states, and the Rocky Mountains. The Long-tailed Weasel, which has a 9-inch (229-millimeter) body and a 6-inch (152-millimeter) tail, is more southerly in its distribution. It is found in all of the lower forty-eight states. Its larger size and black-tipped tail help to distinguish it. In the arid southwest, Long-tailed Weasels are sometimes pale in color and have a dark mask over the eyes and are thus confused with the Black-footed Ferret. In the northern United States, Long-tailed Weasels turn white in winter; in the south and some specific localities elsewhere, they remain the same color year-round.

Black-footed Ferrets are pale yellowish in color, but have black masks over their eyes and also have black feet and black-tipped tails. Their original range was essentially the high plains east of the Rockies, as well as the arid zones where prairie dogs occurred—as far south as southern New Mexico and adjacent Arizona and Texas. Their present distribution is entirely associated with prairie dog towns, and there have been few observations of ferrets in the past decades. The Mink, about 2 feet (0.6 meters) long, of which one-third is the tail, is characterized by its uniform dark brown coloration, with the only touch of white on the chin. It is almost always found near water—streams and lakes—and occurs through most of North America, except for the arid southwest from southern California to eastern Texas.

For the most part, all of these kinds of weasel are nocturnal and relatively secretive. The possible exception is the Long-tailed Weasel, which is sometimes abroad during the day, hunting mice in meadows. All of these animals, which have long necks, cylindrical bodies, and short legs, can usually fit down the same hole as their prey.

Ermine. In its white winter form, with its black-tipped tail, the Ermine has long been a symbol of European royalty. Its fur has bedecked the robes of kings and queens. It is wholly carnivorous, feeding mainly on mice and rats, birds, and young rabbits, and even pursues its prey up into a tree, although it prefers to hunt on the ground. Ermine live in small dens, usually under a log or in a brush pile. They mate in the winter, and the four to eight young are born about nine or ten months later, in the spring. As usual with so many of this family, the implantation of the embryos is delayed. There is some evidence that the males may help in raising the young by bringing food to the den. In places where these weasels may be abundant, there are as many as twenty in one square mile (2.6 square kilometers), and the usual home range is less than 50 acres (20 hectares). Because Ermine are short-legged, small, and extremely fast, a flash of fur is the most that the majority of viewers see of an Ermine.

Least Weasel. With a total length of less than 8 inches (203 millimeters) and a weight of only 2 ounces (57 grams), the Least Weasel is not only the smallest member of its family, but is also the smallest of the mammals in the order called Carnivora. (The largest of such mammals, the Brown Bear, may weigh 12,000 times as much.) The main food of the Least Weasel is mice, especially meadow mice, which it pursues into its burrows. It is active mostly at night. It may range over an area of about 2 acres (0.8 hectares), preferring brushy regions and meadows. It is not particularly abundant anywhere.

One main difference between the Least Weasel and the other weasels in North America is that it seems not to have delayed implantation, so that it usually has several litters a year. They can be born during any month, usually in a fur-lined nest, which may once have been the home of a mouse.

These little weasels are savage killers, using their pointed canine teeth to bite their prey in the skull. When they kill more than they can use at one time, they will cache the surplus near the den. They have the reputation of sucking blood from their prey, but this seems untrue, even though they usually have blood on their snouts after they have made a kill. Voyageurs and Theodore Roosevelt are two parks where these weasels occur, and they may also be present in Shenandoah, but are very rarely seen anywhere.

Long-tailed Weasel.   The Long-tailed Weasel, which has a black-tipped tail about half the length of its 10-inch (254-millimeter) body, is large compared with the Ermine and Least Weasel. If a weasel is seen, it is usually this one, because it occurs in many habitats and is sometimes active during the day. Chipmunks, ground squirrels, mice, rats, tree squirrels, and even rabbits are among its main prey, and when it hears the distress cry of one of these species, it runs boldly to the location of the prey. These weasels have sometimes run up the legs of people holding squeaking birds. Long-tailed Weasels can sometimes be lured from their dens, which are usually under a log or the burrow of some other mammal, by imitating a chirping bird.

Long-tailed Weasels, which probably have a home range of less than 50 acres (20 hectares), usually mate in midsummer. The four to six young are born in spring, the implantation of the embryo in the uterus being delayed. The blind and almost bare young grow quickly; their eyes open in five weeks, but, by the time they are eight weeks old, the young are almost adult size and leave the maternal den. Young females may breed when only three or four months old, but males breed when they are more than one year of age.

Although they occur in many of the national parks, Long-tailed Weasels are not often seen. However, chance sightings are usually made during the day, in an isolated meadow, where the animal is hunting.

Black-footed Ferret.   The original range of the Black-footed Ferret coincided with the distribution of prairie dogs, and evidence from fossils suggests that these two species have long been in association. Larger than most of the other weasels, and about the size of a Mink, the Black-footed Ferret is about 20 inches (508 millimeters) long, including its 5-inch (27-millimeter) tail, and weighs a little more than 2 pounds (0.9 kilograms). As a result of the extensive poisoning campaigns against prairie dogs, most of these ferrets disappeared (as did the rodents), and, at present, the status of the species is precarious. These ferrets were probably never abundant, and perhaps only a single pair occupied a given prairie dog colony. Because they are nocturnal and also spend so much of their time underground, they are difficult to study; much remains to be learned of their way of life.

Most of a Black-footed Ferret's food consists of prairie dogs, although they probably also eat ground squirrels and mice that inhabit the prairie

Black-footed Ferret

dog towns. They seem at a disadvantage when trying to kill prairie dogs above ground, so most of the kills are made in the burrows, probably at night, when the rodent is at a disadvantage. On the ground, prairie dogs are aggressive toward ferrets and can drive them away. The ferret lives in a prairie dog burrow and probably moves from place to place in a town much of the year. Mating takes place in late March or early April, and the young, usually four or five, are born after a 42–45 day gestation period. Their rate of development is not known, but by July, when they emerge from the maternal den, they are almost as large as adults. In the fall, the young leave the prairie dog town and may move considerable distances. They are sometimes killed by automobiles during their travel, and the scarcity of prairie dog colonies is a factor that may be responsible for the long distances the young travel.

Black-footed Ferrets may exist in Wind Cave and Badlands national parks, as well as any of the other places where prairie dogs occur, but their presence is uncertain, and they are not likely to be seen. The most likely time in which to see them is midsummer, when the young have emerged from the den and sometimes play at the entrance during the day. Some of the Long-tailed Weasels in the southern part of the ferrets' range have white and dark markings on their face and pale body coloration that strongly resembles that of the Black-footed Ferret, but they lack black feet. Any sightings of Black-footed Ferrets should be reported to park rangers.

Mink.   Highly valued for their fur, Mink are widespread in North America, but are generally found in forested areas near water. The great number of color varieties that appear in trade furs are the result of genetic manipulation of captive mink, for in the wild these 2-foot (610 millimeter)-long animals are generally uniformly dark brown, with a white patch on the chin. Mink are active mainly at night, when they prowl shorelines in search of mice and rats, frogs and crayfish, and birds and also prey on fish and muskrats. Mink are good swimmers and divers. A hole in the ice in winter sometimes marks the place where a Mink has hunted underwater, digging in the mud for frogs or crayfish. They are solitary animals, and the males, at least, are intolerant of one another and fight on contact. Territories are marked by scent, and Mink have well-developed anal glands. They emit an odor that is considered by many to be more fetid than that of a skunk. Their territory tends to run along the shores of a watercourse and may extend for more than one mile (1.6 kilometers).

Mink mate in midwinter, and the males are promiscuous. The young are born in spring, after a gestation period that varies according to the latitude because implantation may be delayed. The Mink's eyes open when it is three-and-a-half weeks old, and Mink are weaned at a month-and-a-half. The young remain with the mother through the summer, but in autumn leave to establish their own territories. Both males and females can breed during their first year. In captivity, Mink have lived ten years.

A Mink's home is a den, often in a bank, which may be excavated by the animal itself, or taken over from some other species. Mink sometimes set up housekeeping in a Muskrat house or in a Beaver lodge that has been abandoned, but they do not remain permanently in any of their den sites; they move around in their territories at frequent intervals.

Mink are not often seen in the parks, although they occur in many of them. Fishermen are most likely to see them and, on a few occasions, have been surprised to find a Mink stealing catch that they had laid out on a bank.

Ermine: Capitol Reef, Crater Lake, Denali, Glacier Bay, Glacier, Grand Teton, Isle Royale, Katmai, Kings Canyon, Lassen Volcanic, Mount Rainier, North Cascades, Olympic, Redwood, Rocky Mountain, Sequoia, Voyageurs, Yellowstone, Yosemite; **Least Weasel:** Denali, Glacier Bay, Glacier, Grand Canyon, Shenandoah (?), Theodore Roosevelt, Voyageurs; **Long-tailed Weasel:** Acadia, Badlands (?), Big Bend, Bryce Canyon, Canyonlands (?), Carlsbad Caverns, Crater Lake, Everglades, Glacier, Grand Teton, Great Smoky Mountains, Guadalupe Mountains, Kings Canyon, Lassen Volcanic, Mammoth Cave, Mesa Verde, Mount Rainier, North Cascades, Olympic, Redwood, Rocky Mountain, Sequoia, Shenandoah, Theodore Roosevelt, Voyageurs, Wind Cave, Yellowstone, Yosemite, Zion; **Black-footed Ferret:** Badlands, Theodore Roosevelt (E?), Wind Cave; **Mink:** Acadia, Badlands (?), Capitol Reef, Crater Lake (E?), Denali, Everglades, Glacier Bay, Glacier, Grand Teton, Great Smoky Mountains, Hot Springs (?), Isle Royale, Katmai, Lassen Volcanic, Mammoth Cave, Mesa Verde, Mount Rainier, North Cascades, Redwood, Rocky Mountain, Theodore Roosevelt, Voyageurs, Wind Cave (E), Yellowstone, Yosemite

Wolverine

Wolverine. Large and powerful, the Wolverine is a species of the northern forests and tundra of both hemispheres. It is dark brown in color and is marked with a broad yellowish band on each side; these bands usually join over the rump. About 3½ feet (1 meter) long, including the 10-inch (254-millimeter) tail, a Wolverine may weigh as much as 60 pounds (27.5 kilograms). It is strong enough to kill deer, Elk, or Moose that are many times its weight if they are hampered by snow and cannot escape. Wolverines have been known to drive away bears.

Because the Wolverine is a species of the tundra and coniferous forest, there are few places, other than in the western mountains, that it exists south of Canada. Wolverines utilize a vast area for their hunting, sometimes as much as 741,000 acres (300,000 hectares)—which is more than 100 square miles (259 square kilometers). Over this area, a male maintains a territory, marked with scent, from which he excludes other males, but within which two or three females are tolerated. A Wolverine den is usually on the ground, or under brush, rocks, or any sheltered place, but the animal is often away from it, patrolling its vast hunting range along a regular route. The Wolverine's diet consists of almost anything that is available. Carrion seems preferred, but in winter, when deep snow encumbers large mammals, such as Moose, Wolverines kill them. Food not immediately eaten is marked with a foul-smelling scent and returned to for later meals. In summer, birds, eggs, and berries are consumed, in addition to carrion. Trappers in the northlands hate Wolverines because they rob animals from trap lines and invade unoccupied cabins, destroying much food and property.

The mating season is probably extended throughout the warmer months, but implantation of the embryo is delayed so that the young, usually two to four, are born in the spring. The babies nurse for about two months and remain with the mother for two years before she drives them from the territory to fend for themselves. They do not reach sexual maturity until four years old and may live more than fifteen years.

Wolverines climb well, and their broad feet enable them to travel across the surface of snow quite readily. They also swim without hesitation. They are adept at avoiding human beings and are difficult to trap. In the national parks, they are rare and seldom seen. Tracks in the snow are usually the only signs of their presence. At times, Wolverines were seen with some regularity in Glacier National Park, but otherwise the best opportunities to see them—and any chance is slight—are in the Alaskan parks and monuments.

Wolverine: Bryce Canyon (E), Crater Lake, Denali, Glacier Bay, Glacier, Grand Teton, Katmai, Kings Canyon, Mount Rainier (E), North Cascades, Rocky Mountain (?), Sequoia, Voyageurs (E), Yellowstone, Yosemite

---

Badger.   A little over 2 feet (0.6 meters) long, including the stubby 5-inch (127-millimeter) tail, Badgers are flat-bodied animals that appear wider than taller. The Badger is characterized by the white streak down the middle of the face and head (the "Badge"). The rest of the body is grizzled tan or gray, and the feet are black. Where they are not harassed by human beings, Badgers may be active during the day, waddling along in search of rodents. However, they roam at night in most places. They are powerful diggers and are said to be able to bury themselves faster than a man with a shovel can dig.

Badgers are grassland and desert animals and live mainly in the western states, but also in the north-central region as far east as Ohio. In the home range, they find their main food: ground squirrels, prairie dogs, pocket gophers, rats, and mice. The home range is large; it extends more than 1,730 acres (700 hectares) in summer but is much smaller in winter. Within this area, the Badger may dig a new sleeping place for itself almost daily, thus an area where a Badger lives is easily recognized by the numerous excavations. Prey is dug out of its burrow, rather than captured on the ground surface. In addition to rodents, Badgers feed on insects, reptiles, and even birds. Excess meat may be stored underground.

In winter, in the northern parts of the range where the ground may be frozen, Badgers are much less active and forage less widely, probably living on accumulated fat. They do not go into true hibernation and remain underground most of the time. Wintertime, and when their young are small, are the only times that Badgers remain in one den for any length of time. They mate in late summer, and implantation of the embryo into the uterine wall is delayed for six weeks or so. In spring, usually two babies, well-furred but blind, are born in a grass-lined nest deep in a burrow. When they are six weeks old and their eyes have been open for awhile, the young are weaned, but they may remain with the mother for the rest of the summer

Badger

before dispersing. Young females are capable of breeding during the first autumn of their lives, but most do not do so until a year later. Except during the brief mating period and when mothers are with their young, Badgers lead solitary lives; they are probably territorial. In captivity, Badgers have lived for thirteen years, but their life span is probably shorter in the wild.

Although Badgers occur in many of the western national parks, they are not commonly seen. The best place to look for them is around prairie dog towns and ground squirrel colonies, early in the morning, especially if there is much fresh sign of Badger excavations.

Badger: Arches, Badlands, Big Bend, Bryce Canyon, Canyonlands, Capitol Reef, Carlsbad Caverns, Crater Lake, Death Valley, Glacier, Grand Canyon, Grand Teton, Guadalupe, Joshua Tree, Kings Canyon, Lassen Volcanic, Mesa Verde, North Cascades (?), Organ Pipe Cactus, Petrified Forest, Rocky Mountain, Saguaro, Sequoia, Theodore Roosevelt, Voyageurs, Wind Cave, Yellowstone, Yosemite

Skunks. Although there are some six species of skunks listed as occurring in America north of Mexico, they fall into three groups and probably consist of only four species. The smallest is the Spotted Skunk, a weasellike species that is marked with broken white stripes and spots on a black body and is characterized by a triangular or square patch of white on its nose. Spotted Skunks are usually less than 2 feet (0.6 meter) long and weigh less than 2 pounds (0.9 kilogram). They live over most of the United States, except the northeastern region, and, in the eastern part of their range, except Florida, are not common. In the west, Spotted Skunks prefer rocky areas at lower elevations. The Striped Skunk, which occurs throughout the United States, except for Alaska, is the "Common Skunk" that is familiar to most people. Its tail, like that of the Spotted Skunk, is composed of black and white hairs and is a conspicuous warning sign when raised. The Striped Skunk has a narrow strip of white down the middle of the nose, and white V-shaped stripes, joined at the nape, down the back. Closely related to the Striped Skunk, but living only in the southwest from southern Arizona to western Texas, is the Hooded Skunk. Like the Striped Skunk, this species has a narrow strip of white down its nose. However, it has two color phases of body pattern. In one, the back is black, and the white stripes, often narrow, are on the sides of the body and do not join at the nape. In the other phase, the back is covered with mixed black and white hairs and no V-pattern is formed. The last kind of skunk found in the United States is the Hog-nosed Skunk, also a southwestern species; it ranges from eastern Texas to western Arizona and as far north as southeastern Colorado. It has no white patches or stripes on the nose. It has a solid white back and an all-white tail. It derives its name from its long, bare-tipped nose.

All skunks are equipped with a pair of glands, each about the size of a large grape, located on either side of the anus. These contain musk, which can be sprayed through a pair of nipples located just inside the animal's anus. When threatened, a skunk usually tries to run away, raising its conspicuous tail at the same time. If cornered, it still keeps its tail raised, and it may stamp its feet and form its body into a "U," so that both its head and rear are facing the threat. If the threat intensifies, the animal can spray its musk 12 feet (3.7 meters) or more. The musk, aside from its unpleasant odor, is irritating if it gets into the eyes, causing temporary blindness. Skunk musk can be removed from clothing by washing it in a diluted solution of ammonia (a cup of household ammonia to a bucket of water).

Spotted Skunk.   No other North American mammal has a color pattern quite like that of the Spotted Skunk. This weasel-shaped animal is the most carnivorous of the skunks, as well as the most agile. It is an excellent climber and may even, at times, make its den in trees. In the prairies, Spotted Skunks have a home range that may be about a quarter of a square mile (64 hectares), but in the broken-rock arroyos of the west the home range is probably smaller. In winter these skunks feed mainly on rodents, especially rats and mice, but in summer they seem to prefer insects, fruits, and berries. Their home is usually on the ground, under rocks, in brush, or almost any dry, dark place. They are probably not territorial, at least not for most of the year, because they are often found denning together, especially in winter.

Spotted Skunks generally have four or five young in a litter, born in the spring. In the southeast, there may be a second litter late in the year. The breeding season is, for the eastern ones, late winter; however, in the northwest, the mating season may be in the fall and implantation of the embryos is delayed until late winter, when active gestation begins. The babies, which weigh 0.3 ounce (9 grams) at birth, are cared for by the mother alone. They are nursed for two months and attain adult size when they are four months old. During the summer, they follow the mother on her nightly hunting trips, but may disperse in the fall. Groups of these skunks in winter dens may represent a female and her young.

One unusual habit of these skunks is "handstand" behavior. When threatened, sometimes these little skunks rise up on their forefeet, displaying their full back pattern and white-tipped tail, and, arching their back, they can spray an attacker while balanced this way. Spotted Skunks are quite rare in the eastern parks, although Mammoth Cave might be the best place to find some at night. In some western parks, they are common around campgrounds; Phantom Ranch in the Grand Canyon is a good place to find them, and they are often seen at campgrounds in Olympic National Park.

Striped Skunk. Striped Skunks are so secure in their noxious defense that they wander almost anywhere without special caution. For this reason, they are one of the animals most frequently killed on highways, and their carcasses are a familiar sight there in the morning. Their safety requires a degree of intelligence and learning ability on the part of their predators, therefore they have prospered around human developments and are one of the more familiar animals in suburban areas. These skunks are omnivorous. In summer, their diet is composed mainly of insects and their larvae, as well as fruits, vegetables, and berries. In spring, especially, they eat mice, fledglings, and eggs. They dine on carrion at any time of the year. Unlike Spotted Skunks, Striped Skunks cannot climb, and they den on the ground, often under buildings or in the abandoned burrows of other mammals, beneath rocks, or under brush.

In the fall, Striped Skunks become very fat, sometimes weighing as much as 15 pounds (6.8 kilograms). During cold weather, sometimes as many as ten together enter a den, where they remain during the coldest part of the winter. There are usually more females than males in these aggregations, and, if the weather is extremely cold, the animals may remain underground for weeks or even months. They do not hibernate but live off the accumulated fat they have deposited during the summer and fall. Breeding takes place toward the end of February, when promiscuous males roam widely in search of females. The gestation period is about 2 months, but there may be a slight delay before the embryos are implanted. The newborn skunks are naked, but marked with pink skin where white hairs will grow, and with blue-black pigmentation where the black hairs will develop. They are blind and weigh 1.2 ounces (33 grams). Although there may be up to ten in a litter, the usual number is four or five. When the young skunks are about three weeks old, their eyes open, but even before this they have evidently been able to produce the species' characteristic musk.

Young skunks can eat solid food when they are a little more than five

Striped Skunk

weeks old and are weaned soon thereafter. During the summer, they follow their mother on hunting trips, roaming the fields each night, and by fall they may weigh 6 pounds (2.7 kilograms). Dispersal may take place at this time, or else the young may overwinter with the mother. Yearlings may breed the first year of their lives.

In summer, skunks usually forage within 1,000 yards (914 meters) of their home den, but in winter they remain mostly in the den or around it until, as the breeding season approaches, the males begin to wander widely again. One young skunk whose den was covered with snow is known to have remained there for eighty-seven days, evidently without harm.

Because they are omnivorous and bold, Striped Skunks are common visitors to campsites at night. A skunk can usually be chased from a campground by using a combination of caution and perseverance, as its tendency is to escape rather than to spray. Sudden movements should be avoided, as well as loud noises that might scare the skunk. Maintaining a distance of about 15 feet (4.6 meters), and no closer, usually forces the animal out of an area. Skunks are sometimes major carriers of rabies, and, if for no other reason, close contact with skunks should be avoided. Striped Skunks can usually be seen in any of the parks where they occur, generally around campgrounds or along roadsides at night.

Hooded Skunk. The Hooded Skunk, a close relative of the Striped Skunk, lives only in the arid southwest. Little is known of its habits. The Hooded Skunk is nocturnal and not frequently seen, but it is thought that its habits resemble those of the Striped Skunk. Its tail is longer than that of the Striped Skunk; it is usually at least half its total length.

Hog-nosed Skunk. The Hog-nosed Skunk is the only skunk in the United States that has a solid white back as well as an all-white tail. It is a large species, often reaching 2½ feet (0.8 meters) in length, and its hair is quite coarse, compared to that of other skunks. The eastern Texas Hog-nosed Skunk has been considered a separate species from those skunks to the west, but it seems likely that they are the same. The nose of this skunk is piglike and naked on top for about one inch (25 millimeters); the skunk uses its nose, as well as its long fore claws, to root in the ground and unearth insects and grubs, which make up much of its diet. Hog-nosed Skunks seem to prefer rocky areas, where there is much for them to turn over and root under. They are generally solitary. Their homes are usually under rocks or roots, and these skunks are rarely seen out during the day. Little is known of their habits, but they probably have fewer young (perhaps two to four) than do Striped Skunks. Carlsbad and Guadalupe Mountains are the best parks in which to see Hog-nosed Skunks.

Spotted Skunk: Arches, Big Bend, Bryce Canyon, Canyonlands, Capitol Reef, Carlsbad Caverns, Channel Islands, Death Valley, Everglades (?), Grand Canyon, Great Smoky Mountains, Guadalupe Mountains, Hot Springs, Joshua Tree, Kings Canyon, Lassen Volcanic, Mammoth Cave, Mesa Verde, Mount Rainier, North Cascades, Olympic, Organ Pipe Cactus, Petrified Forest, Redwood, Rocky Mountain (?), Saguaro, Sequoia, Shenandoah, Theodore Roosevelt, Wind Cave (?), Yosemite, Zion; **Striped Skunk:** Acadia, Arches, Badlands, Big Bend, Bryce Canyon, Canyonlands, Capitol Reef, Carlsbad, Crater Lake, Everglades, Glacier, Grand Canyon, Grand Teton, Great Smoky Mountains, Guadalupe Mountains, Hot Springs, Joshua Tree, Kings Canyon (E?), Lassen Volcanic, Mammoth Cave, Mesa Verde, North Cascades, Organ Pipe Cactus, Petrified Forest, Redwood, Rocky Mountain, Saguaro, Sequoia, Shenandoah, Voyageurs, Wind Cave, Yellowstone, Yosemite, Zion; **Hooded Skunk:** Big Bend, Organ Pipe Cactus (?), Saguaro; **Hog-nosed Skunk:** Big Bend, Carlsbad Caverns, Grand Canyon (?), Guadalupe Mountains, Organ Pipe Cactus, Saguaro

---

River Otter. Up to 50 inches (1,270 millimeters) long, including the 18-inch (457-millimeter), thick-based tail, the River Otter is one of the largest, but not the heaviest, members of the Weasel family. A large Wolverine outweighs it, because River Otters rarely weigh more than 30 pounds (13.6 kilograms); the marine Sea Otter is larger and heavier. River Otters are not found in the extreme southwestern United States, but they may be found along lakes and watercourses throughout all of the other

River Otter

states, including Alaska. Their thick, sleek fur is valued in the fur trade, and they have been extirpated by trapping from many areas where they once occurred, and, in other places, the effects of pollution on their food supply has eliminated them.

River Otters are completely at home in water, where they are swift and agile swimmers. They can propel themselves with their webbed feet, but also move sinuously, using their muscular tails as sculls. In winter, they have no qualms about swimming under ice, evidently finding sufficient air trapped beneath it for occasional breathing. They also travel well on land, sometimes going long distances to get from one stream to another, but they move best on snow, sometimes sliding 20 feet (6 meters) or more after a running start. These otters are playful, and in winter one of their delights is sliding down a snow-covered bank into a hole in the ice and returning again and again for the sheer enjoyment of the slide.

River Otters are more social than many other weasel family members, and the male, after a brief eviction from the den just after the young are born, remains with the female to help raise the litter. The den is usually in a riverbank and has an underwater entrance. Sometimes an abandoned Beaver lodge is used. There are usually two young, born blind but furred, which are nursed for four months. By early winter, they may be large enough to disperse and find their own homesites. Mating takes place within a few weeks after the young are born, and implantation of the embryos is delayed, so that the total time from mating to birth may be nearly a year, or as little as nine months.

River Otters, which are social animals, are usually seen in pairs. They may maintain territories, marked by scent from anal glands, along many miles of shoreline. Although in most places they are nocturnal, they may be out in the daytime in areas where they are undisturbed. They obtain most of their food from the water, eating primarily fish, frogs, aquatic invertebrates, such as crayfish, and some birds and small mammals. Although fishermen think that they deplete game fish stocks, River Otters evidently feed more on the slower-moving species, which they can catch easily.

Seeing River Otters is largely a matter of chance, but they occur in many of the national parks, and there are good opportunities to see them in Everglades and Olympic.

River Otter: Acadia, Badlands (?), Capitol Reef (E), Crater Lake (?), Denali, Everglades, Glacier Bay, Glacier, Grand Canyon, Grand Teton, Great Smoky Mountains, Isle Royale, Katmai, Lassen Volcanic, Mammoth Cave (E), Mount Rainier (E), North Cascades, Olympic, Redwood, Rocky Mountain (?), Shenandoah (?), Theodore Roosevelt, Voyageurs, Yellowstone, Yosemite (?)

Sea Otter. Sea Otters are the largest members of the weasel family; males reach 58 inches (1,473 millimeters) in length and up to 100 pounds (45 kilograms) in weight. Females are slightly shorter and weigh two-thirds as much as males. The stubby, thick tail is about one-third the total length and is slightly flattened from top to bottom. Besides being larger and found only in salt water, the Sea Otter has a less pointed face than the River Otter and usually has a yellowish-white head. The Sea Otter's huge, webbed hind feet, which may be 8 inches (203 millimeters) long, resemble a seal's flippers. The dark brown hair is composed mainly of underfur that is unbelievably dense; there may be more than 100,000 hairs per square inch (625 square millimeters), and it was the quest for this fine fur that led to the near extermination of the species by about 1910. Unlike many other marine mammals that rely on a layer of fat to insulate them from cold water, Sea Otters depend on air trapped in their fur to protect them from cold. This is so effective that, between air temperatures of $-2°$ to $+70°$ F. ($-19°$ to $+21°$ C.), Sea Otters do not suffer from either heat or cold, but they must periodically be able to dry and groom their fur, often coming out on land to do so. In addition, when it is warm, Sea Otters can lose heat through their large hind flippers, and, in cool but sunny weather, they can absorb heat in the dark skin and webbing.

The original range of the Sea Otter was from northern Japan across the Aleutian Islands and along the Pacific Coast of North America as far south as central Baja California, Mexico. At present, they are thinly distributed from central California northward, in some places as a result of restocking from the main population in the Aleutians, which has grown under protection. There are believed to be about 2,000 Sea Otters in California waters, but less than half that number between there and Alaska. Along the Aleutians, the population may be over 100,000.

Although they are awkward on land, Sea Otters can swim as fast as 5.6 miles (9 kilometers) an hour underwater, undulating their bodies and using their hind feet. On the surface, they may paddle with their front feet and scull with their tail. They live near land, generally in relatively shallow water because they feed on bottom-dwelling animals. They are known to dive as deep as 318 feet (97 meters) and probably can go much deeper. Sea urchins, crabs, and mollusks are their main prey, and they also eat slow-moving, bottom-dwelling fish. They bring their food to the surface in order to eat it, and characteristically float on their back and use their chest as a table. To open crustaceans and clams, Sea Otters carry a rock, held in the armpit when diving, and place it on their chest while floating on their back. They pound the hard-shelled animal against the rock until it cracks. When hunting underwater they use their paws rather than their mouths, to pick up the food.

Sea Otters are gregarious, but the sexes are segregated. The males maintain territories by splashing at other males, keeping an area of about 100 acres (40 hectares) for themselves. However, this may not be true of all Sea Otter populations. Mating takes place in the water, and can be at any time of the year, but mainly in autumn. There is delayed implantation, and the single offspring (twins are rare) is born 6–9 months later, usually in the water. The newborn weighs about 4 pounds (1.9 kilograms), and the mother supports it on her chest; it nurses there as well. The baby continues to nurse until it is almost adult size, although it is offered and eats solid food when it is very young. By two months old, it starts to make its own dives in search of food, but it usually remains with its mother for six to eight months. Generally, a female gives birth every other year. The young females reach sexual maturity when they are four years old; the males, when they are five or six. However, males may not be able to compete for females until they are seven or eight years of age.

When on the water's surface, Sea Otters usually float on their backs and, at night, wrap themselves in strands of kelp to keep from floating away. These animals are a major attraction, but unfortunately, there are no permanent colonies in any of the national parks, although stragglers are sometimes seen at Olympic and Redwood parks. Sea Otters can be readily seen near Monterey, California, and at nearby Point Lobos.

Sea Otter: Channel Islands (E), Glacier Bay (RI), Katmai, Olympic (RI), Redwood

HAIR SEALS

Hair seals differ from sea lions and fur seals in not having any external ear, just a small opening, and in not being able to rotate their hind flippers forward. The flippers of a fur seal or sea lion are not furred, unlike those of a "true" or hair seal. When a hair seal moves on land, it generally raises its hind flippers and pulls itself on its belly with its front flippers, whereas the sea lion and fur seal can walk on their hind flippers.

There are ten species of hair seal in North American waters, but only three are found in any national parks or monuments. These are the Harbor Seal, the Gray Seal, and the Northern Elephant Seal. The Harbor Seal is a small species that occurs in both the Atlantic and Pacific and that is distinguished by its small size—about 5 feet (1.5 meters) in length—and by a gray coat with dark spots. It is common along the coasts. The Gray Seal, an Atlantic species, is much larger, reaching 10 feet (3 meters) in length, and, as its name suggests, it is uniformly gray in color, without spots. The last species is the unmistakable Northern Elephant Seal, a huge Pacific species, which reaches 20 feet (6 meters) in length and weighs as much as 8,000 pounds (3,629 kilograms). It is plain brown in color. The males have an extended snout—a proboscis—that, like an elephant's trunk, hangs over the mouth.

Harbor Seal. Harbor Seals (see plate 8b) live close to shore and often enter rivers and estuaries in search of fish. They are frequently active during the daytime and are thus not a rare sight. Harbor Seals usually travel in small numbers, although they may appear, in places, in aggregations that number in the hundreds. They can dive as deep as 300 feet (91 meters) and generally feed on the species of fish that is most abundant in their area, including some of commercial value. For this reason, and because they damage nets, they are disliked by fishermen. Depending upon the locality, the single pup, weighing about 25 pounds (11 kilograms), is born from March in the south, to June in the north, and on land or on an ice floe. The pup is suckled for about three weeks, but it can swim almost as soon as it is born. The adults mate in midsummer, but implantation of the embryo is delayed, and the actual growth of the fetus takes place mostly in the last seven-and-a-half months before birth. Sexual maturity is probably attained by three or four years of age. In zoos, these seals have lived more than eighteen years.

Harbor Seals can generally be seen in all of the coastal parks except Biscayne Bay and Everglades. In late June, they can readily be seen with pups on ice floes in the upper bays of Glacier Bay National Park.

Gray Seal. On both sides of the Atlantic, but only as far south as Maine, the Gray Seal usually inhabits rocky shores. It has a highly arched "Roman nose" which, in addition to its large size and unspotted coat, distinguishes it from the small, spotted Harbor Seal, which also occurs in these waters. In midwinter, the bulls establish territories on land, and it is on these territories that the cows give birth to their single pup, which is usually 2½ feet (0.8 meter) long and weighs 35 pounds (16 kilograms). At birth, the pup is generally covered with long white hair, which is soon shed. Although they can swim at birth, baby Gray Seals do not do so while in their white coats. The pups gain weight rapidly, as much as 3 pounds (1.4 kilograms) a day; the mother loses twice this much a day while she is nursing. The pups are weaned by three weeks old and are deserted by the mother. As soon as they have molted from their white coats, the pups leave the beach. Gray Seals are a long-lived species, some surviving forty years in zoos and almost as long in the wild.

Gray Seals often fish in groups, diving to 250 feet (76 meters) or more for various species of fish. They can remain underwater for twenty minutes. Although they are not numerous over most of their range, Gray Seals are disliked by fishermen because they damage nets and eat salmon, and also because they are hosts for a parasite that infects codfish. Few Gray Seals venture into U.S. waters, and the only park where they can sometimes be seen is Acadia.

Northern Elephant Seal. Although many kinds of seal were killed for their skins, the Northern Elephant Seal was nearly exterminated for its blubber. These huge animals, which may weigh up to 4 tons (3,629 kilograms), once occurred along much of the Pacific Coast of North America but, by the end of the nineteenth century, had been reduced to about twenty animals on Guadalupe Island, off Baja California, Mexico. Because of stringent protection, the population has grown and has now expanded its range, so that Elephant Seals now breed as far north as San Francisco Bay, and stragglers have been seen in southern Canadian waters. The population now numbers more than 40,000.

Starting in December, the huge bulls establish their territories on gently sloping sand or pebble beaches. They inflate their pendulous snouts, and the resonating bellows they emit can be heard a mile (1.6 kilometers) away. They rear up, jostle one another, and bite savagely at one another's well-padded necks until the holder of the territory is determined. When the females come ashore, they give birth almost immediately. A single, black-haired pup, which weighs about 65 pounds (29 kilograms), is born. The mothers nurse their offspring for three weeks, not feeding themselves, and in the process the young seals gain about 20 pounds (9 kilograms) a day, while the mothers lose some 35 pounds (16 kilograms) a day. By the time the pup is weaned and has molted into a new coat, it is deserted by the mother. Mating takes place about a week after the pup is born, so that the females are pregnant fifty-one weeks a year. By two years old, females can mate, but males do not reach sexual maturity until they are four, and they may be seven or eight before they are large enough to fight for, and hold onto, a territory.

Northern Elephant Seals are mainly nocturnal; during the day they can be seen on beaches, lying close together in the sun. They are relatively docile, but males feel threatened by an upright human figure and sometimes attack; a person crawling on hands and knees can move through a beached herd of elephant seals without much danger. These seals feed mainly on squid and fish. They can dive deeply, to more than 200 feet (61 meters), and remain submerged for a half-hour.

Northern Elephant Seals are now seen occasionally off all the Pacific coastal parks from Olympic to the Channel Islands, and they breed on the latter.

Harbor Seal: Acadia, Channel Islands, Glacier Bay, Katmai, Olympic, Redwood; **Gray Seal:** Acadia; **Northern Elephant Seal:** Channel Islands, Olympic, Redwood

---

CAT FAMILY

Of the seven members of the cat family that are recorded from America north of Mexico, six may have once occurred in some parks or monuments. Three of these, however, were probably never resident and are so rare in the United States that, for all practical purposes, they might be excluded from mention. These are the Jaguar, the Ocelot, and the Jaguarundi. All three species are known mainly from areas of Arizona and Texas that are within a few miles of the Mexican border, and reports of them are few and far between. It is questionable whether there are any breeding populations of these species within the United States. The largest of the three, the Jaguar, may once have wandered more widely in the southwest, perhaps even reaching the southern edge of the Grand Canyon. There are reports that the small, dark-colored, and unspotted cat, the Jaguarundi, lives in the Everglades, but these Jaguarundis are derived from captive animals that were liberated, and they may not have persisted.

The other three cats are the Mountain Lion, Lynx, and Bobcat. They are widespread, sometimes abundant, but never frequently seen. The Mountain Lion may weigh more than 200 pounds (91 kilograms) and be 9 feet (2.7 meters) from tip of nose to tip of tail. It is the only large, unspotted cat in North America. The Lynx and Bobcat are much smaller, short-tailed, and spotted or mottled; they are difficult to differentiate. Their sizes and weights overlap, although generally the Lynx has longer legs, bigger feet, and longer tufts on its ears. The surest way to distinguish them is that the Lynx's tail is tipped with black both above and below, whereas the Bobcat's tail is tipped with black only on top and is white on the underside.

Mountain Lion. Originally, the Mountain Lion (see plate 4b) had the greatest range of any American mammal: from northern Canada to southern South America, and from the Atlantic to the Pacific. It lived high in the mountains, on plains, in swamps, and on coasts, as well as in tropical forests. In North America, it has been eliminated over much of its range and is essentially a western animal now, with some populations in southern

Florida and on the Gulf Coast. It has acquired a great number of common names; the more frequently used ones are Cougar and Puma. It is so secretive that many naturalists spend their careers in areas where Mountain Lions live without ever seeing one.

Mountain Lions are solitary, except during the brief mating period and when cubs are with their mothers. They have large home ranges; a 25-mile (40-kilometer) trip during a night of hunting is not unusual. They feed mainly on deer, stalking their prey cautiously, to get as close as possible before the final leap or dash. The first meal generally consists of parts of intestines, liver, and heart; up to 8 pounds (3.6 kilograms) is consumed. The remainder is cached in order to be eaten later; it is covered with sticks and leaves. It serves as food until it begins to spoil. One Mountain Lion kills an average of one deer a week, but also dines on as many other kinds of mammals as it can catch, including rabbits, hares, marmots, Beavers, and Porcupines. In regions where there is livestock, Mountain Lions kill sheep, calves, and colts, which is why they have been killed off in so many places and are still hated by ranchers in the west.

Although a male and a female may mate at any time of the year, it is usually late winter when they travel together for several weeks and breed. After a gestation period of three months, usually two or three cubs are born. They are well-furred and weigh 1–1½ pound (0.2–0.5 kilograms). Their eyes are closed and do not open for about two weeks. Although the adults are plain colored, the cubs are spotted. The birth site is a den, often in a cave or rock crevice, where some leaves have been carried for bedding. The mother starts to wean the cubs when they are about three months old, and, by six months, the cubs have shed their baby coats and may weigh as much as 45 pounds (20 kilograms). By this time, the maternal den may be abandoned, and the mother and cubs hunt around the home range. The young Mountain Lions remain with their mother for a year, sometimes even two, before moving away to establish their own hunting site, sometimes as much as 100 miles (161 kilometers) from their birth site. Mountain Lions reach sexual maturity at two or three years of age. Because females keep their young with them so long, they may have litters only every two or three years. Males will fight for a female that is in heat.

Mountain Lions are good climbers, and, although they do not usually hunt in trees, they take refuge in them when pursued by hunters' dogs; hunters usually use dogs to help locate them. They are relatively docile cats, and, although they sometimes silently trail a human being for many miles, there are very few instances of unprovoked attacks. Often, it is only when hikers start back along a trail that they discover the tracks in the dust that indicate they had been followed. The Mountain Lion screams loudly, supposedly sounding like a terrified woman, but even this sound is seldom heard.

Although Mountain Lions were supposedly eliminated from most eastern states many years ago, reports of them persist. Many of these are substantiated; a few of them have proved to be of animals that were brought in from western states and released to be hunted; and some have turned out to be of large domestic cats. Nevertheless, in many of the parks, rangers are convinced that Mountain Lions are present, if rare, even in the absence of adequate confirmation. Considering how seldom these animals are sighted, they may persist in some of these parks. There can be no assurance of ever seeing a Mountain Lion, even in places where they are known to live, but perhaps the likeliest locations in which to see one are Everglades and Zion national parks.

Lynx. With its long legs, dense fur, and huge feet, the Lynx gives the impression of an animal larger than it actually is. Rarely weighing much more than 20 pounds (9 kilograms), a Lynx stands 2 feet (0.6 meter) tall at the shoulders and is covered with soft, smoke-gray fur. Its fur is currently much in demand for coats. Its range coincides largely with the great coniferous forest zone of Canada, and, in the United States, it occurs only in Alaska; in a few areas bordering Canada; in the central Rocky Mountains; and in the northwest, as far south in the mountains as central Oregon.

The home range of these cats is tied largely to the abundance of Snowshoe Hares, and, when the hares are numerous, a Lynx may confine itself to not much more than one square mile (2.5 square kilometers) for much of the time. When the hare population is at one of its periodic lows, Lynx may have to hunt over a huge range, sometimes moving more than 100 miles (161 kilometers) in search of anything edible. At these times, starving Lynxes attack and kill snowbound deer, and they even eat old leather boots, as well as birds and any small mammals, such as squirrels, lemmings, and Beavers, that they can capture. When there are few hares, many Lynxes die of starvation.

Solitary and nocturnal most of the year, Lynxes pair briefly in late winter, and then, after mating, the male leaves and the female raises the young alone. The gestation period is about 2 months, and the usual number of young is two. The furred kits are born in a sheltered den, often beneath tree roots. They are blind at birth, but their eyes open at ten days. By two months of age, they start following their mother on her hunts. The young leave their mother to start life on their own in the autumn or, more commonly, remain with her through the winter and disperse the following spring. When a year old, the Lynxes are sexually mature. They are known to have lived as long as fifteen years in zoos, but probably do not survive more than half that span in the wild.

The Lynx's huge, padded feet serve it well in winter, when it can snowshoe across the surface of snow, whereas other animals are encumbered. Although they can climb well, Lynxes hunt mainly on the ground. Except during the years when the Snowshoe Hare population is very low and the Lynxes are starving, Lynxes are seldom seen, and their tracks are generally the main evidence of their presence. The best opportunities to see them are in Alaska, especially at Denali and Katmai.

Bobcat. The Bobcat (see plate 4a), which is approximately the same size as the Lynx, but with shorter legs and smaller paws, has a range that largely complements that of the Lynx. It is found throughout much of the United States, except the central states and parts of the southeast. It is certainly our most common and abundant wild cat, but because it is nocturnal, solitary, and secretive, it is not often seen, even though it may live close to human residences. Its preferred habitat is broken and brushy country in the west, but it also lives in swamps, especially in the southeast. Ordinarily, it wanders over an area not much more than 5 miles (8 kilometers) in diameter but, when food is scarce, may hunt over as much as 50 miles.

Hares and rabbits provide most of the Bobcat's food, but it also hunts squirrels, rats, and mice, and especially ground-nesting birds. Sight and smell are the main senses that it uses in locating prey, generally investigating any thicket or cover that might contain an edible species. Less commonly, deer are killed in winter when they are bogged by snow, and fawns are attacked and killed in the spring. Large amounts of food are covered with sticks and leaves, to be revisited for meals until the food starts to spoil. Because the Bobcat is less dependent on a single source of food than the Lynx, its populations do not undergo such marked fluctuations.

Bobcats are probably territorial; they are usually solitary animals. In late winter, the males commence nightly yowling, much like domestic cats. After a brief courtship and copulation, the pair separates, the male going to seek other females. The kits are born in a den that is usually under a log or rocks, or in a thicket, after a gestation period of 63 days. There are usually two to four in a litter. They are covered with fur and weigh about 12 ounces (340 grams). Although they are born blind, their eyes open in 10 days. They are weaned in two months, having been fed bits of meat by the mother prior to this time. By autumn, the young Bobcats may weigh as much as 10 pounds (4.5 kilograms) and disperse to establish their own ranges. In some areas, the mother may have a second litter early in the fall, and these offspring overwinter with her. Bobcats have lived fifteen years in captivity.

Bobcats occur in many national parks and monuments, but they are not seen in proportion to their abundance. Most observations have been at night, when Bobcats have been seen along roadsides. The Everglades is one of the better places in which to see them.

Jaguar: Grand Canyon (E), Organ Pipe (?), Saguaro (E); **Mountain Lion:** Arches, Badlands, Big Bend, Bryce Canyon, Canyonlands, Capitol Reef, Carlsbad Caverns, Crater Lake, Death Valley, Everglades, Glacier, Grand Canyon, Grand Teton, Great Smoky Mountains (?), Guadalupe Mountains, Joshua Tree (?), Kings Canyon, Lassen Volcanic, Mammoth Cave (?), Mesa Verde, Mount Rainier, North Cascades, Olympic, Organ Pipe Cactus, Redwood, Rocky Mountain, Saguaro, Sequoia, Shenandoah (?), Theodore Roosevelt, Voyageurs (E), Wind Cave, Yellowstone, Yosemite, Zion; **Ocelot:** Big

(continued)

(continued)

Bend; **Jaguarundi:** Everglades (I?), Saguaro (E); **Lynx:** Denali, Glacier Bay, Glacier, Grand Teton Isle Royale, Katmai, Mount Rainier (E), North Cascades, Rocky Mountain (?), Theodore Roosevelt (E?), Voyageurs, Yellowstone; **Bobcat:** Acadia, Arches, Badlands, Big Bend, Bryce Canyon, Valley, Everglades, Glacier, Grand Canyon, Grand Teton, Great Smoky Mountains, Guadalupe, Hot Springs (?), Joshua Tree, Kings Canyon, Lassen Volcanic, Mammoth Cave (E?), Mesa Verde, Mount Rainier, North Cascades, Olympic, Organ Pipe Cactus, Petrified Forest, Redwood, Rocky Mountain, Saguaro, Sequoia, Shenandoah, Theodore Roosevelt, Voyageurs, Wind Cave, Yellowstone, Yosemite, Zion

Manatee

One of the first American mammals that Columbus saw was the "Sea Cow," or Manatee, which he called a mermaid; he said it was not as beautiful as reputed. Manatees are wholly aquatic mammals, which may reach 15 feet (4.5 meters) in length and more than 1,323 pounds (600 kilograms) in weight. They are distantly related to elephants. Their front limbs are modified as flippers, but with toenails; there are no hind legs; the tail is modified to a broad, rounded paddle, which is flattened top-to-bottom. The round head lacks external ears, and the thick upper lip is cleft and adorned with stiff bristles. Manatees were one source of the mermaid legend, partly as a result of the perceptions of sailors who had been at sea for a long time. The breasts of the female Manatees, located on the chest, were thought to resemble those of human beings. These animals inhabit warm water coastlines, mainly in the Caribbean, but also along the Atlantic shore, and live in coastal rivers.

Manatees are social, but the groups are transient; the only strong social bond is between a mother and her calf. They are active day or night, spending most of their time feeding on aquatic vegetation, of which they consume 66–110 pounds (30–50 kilograms) a day. They usually swim slowly from place to place, cruising at 2.5–6 miles (4–10 kilometers) an hour but, for short distances, can reach a speed of 16 miles (25 kilometers) an hour. The tail is the sole means of propulsion, and the main means of steering; the flippers are held at the sides. Manatees live in shallow water and may not be able to dive deeply; the maximum recorded dive was 33 feet (10 meters). Most of their activity takes place 3–10 feet (1–3 meters) below the surface. They have been known to stay submerged for up to sixteen minutes, but four minutes is the average time span between breaths. Manatees are solid-boned and can sink readily, often sleeping on the sea or river bed.

Manatee

These animals are very sensitive to cold and seem unable to tolerate water temperatures much below 46° F. (8° C.), and, when the temperature drops to 56° F. (13° C.), they usually seek warmer waters, sometimes man-made industrial water outlets. After extreme cold spells, dead Manatees have been found, presumably having died of pneumonia or other cold-induced ailments. Cold weather undoubtedly limits their northward distribution. They are virtually hairless but are often covered with algae, diatoms, and barnacles, and they are uniformly gray or brownish in color. Manatees have been killed or injured by motorboats. They are generally scarred with marks where they have been hit by propellers, thus, for their protection, speed limits for boats have been lowered in many places.

Manatees may breed at any time of the year, and mating takes place underwater. The exact gestation period is not known but is believed to be about 13 months. Twins are rare. A baby manatee weighs 25–60 pounds (11–27 kilograms) and is 4½ feet (1.4 meters) long. Birth is presumed to take place in a secluded backwater, and the mother is supposed to raise the baby to the surface to breathe by carrying it on her back. Newborn Manatees can swim immediately but use only their foreflippers at first; they do not gain the use of the tail fluke for several days. The baby nurses underwater and, although it can eat vegetation within a few weeks, may continue to nurse for one to two years. Sexual maturity is probably not reached until the animals are at least three and perhaps five years of age. In captivity, Manatees have lived more than twenty years. In the wild, females do not produce an offspring more often than every other year.

Manatees are docile, harmless, and often playful. They have been killed over much of their range for their meat, oil, or bones and, in U.S. waters, die mainly as a result of motorboat accidents. Manatees are resident in the waters of Everglades and Biscayne national parks, and rangers can usually inform the interested visitor where and how to find the animals.

Manatee: Biscayne, Everglades

Collared Peccary

Piglike in appearance and habits, the Collared Peccary is not directly related to the true pigs, which are Old World in origin and natural distribution and which are placed in a different family. Covered with coarse, bristly, black and white hair, with a pale zone over the shoulders (the "collar"), these peccaries stand about 22 inches (558 millimeters) high at the shoulder and weigh up to 65 pounds (29 kilograms). Among their differences from the "true" pigs, peccaries have short, straight tusks, rarely more than 2 inches (51 millimeters) long, and a musk gland on the surface of their backs. Unlike the Old World pigs, which have four hooves on each foot, Peccaries have four on the front feet, but only three (one of which is a "dew hoof") on the hind. Pigs have a two-chambered stomach, whereas

Collared Peccary

those of Peccaries have three chambers—but neither of these mammals ruminates.

Collared Peccaries are southwestern mammals, found from eastern Texas to central Arizona, but only in the southern parts of these states. They are active at night, as well as early in the morning and evening, traveling in packs of four to thirty animals. When foraging, they grunt softly, the sound probably serving to keep the animals together, and the musk gland also serves for communication by rubbing on brush under which they pass. When the Peccary is excited, the hairs on its back raise, and the gland emits a skunklike odor, which works as an alarm signal to the other animals.

The diet of the Collared Peccary is varied. These omnivores feed on many kinds of berry, fruit, nut, and seed, and they root for tubers. They rob young from the nests of mice, rats, and ground-dwelling birds. They especially like prickly pear fruits and joints; they get moisture from both of these and seem not bothered by the spines. Peccaries also kill lizards and snakes, including rattlesnakes, by jumping on them with their sharp hooves, and then they eat them. Peccaries usually live near water. Areas where peccaries have foraged show signs of rooting, and chunks bitten out of cactus are another indication of their presence. Caves are much desired for shelter, but during midday heat, peccaries bed down in any shady spot, usually a densely vegetated ravine.

Breeding can take place at any time of the year, but little is known of the herd structure or mating behavior. The gestation period is about 4½ months, and usually two well-developed young are born in a sheltered spot. They are reddish brown and have a dark strip down the center of their backs. Within a day or two after the babies are born, they can follow the mother; she leads them back to the herd within three or four days. The young reach sexual maturity in their second year. Peccaries have lived more than twenty years in captivity.

Because they avoid extremes of heat, Collared Peccaries are seen more often during the winter in the parks where they occur, or else at night, but they are usually not too hard to find. Saguaro and Big Bend are two parks where they can usually be seen.

Collared Peccary: Big Bend, Guadalupe Mountains (I), Organ Pipe Cactus, Saguaro

---

Deer Family

All five species of native deer in North America—Elk, Mule and Black-tailed Deer, White-tailed Deer, Moose, and Caribou—are found in one or more national parks and are among the more conspicuous animals in many places. They are characterized by their relatively large size, and by the males (and in Caribou, females as well) having antlers. Antlers, which are not horns, are composed of bone. They are shed each year, in winter, and regrown again during the summer. While they are growing, they are covered with a skin called "velvet," which contains the blood vessels that carry the nutrients for their development. When the antlers are mature, the blood supply is choked off and the "velvet" dries and withers and is then scraped off by the deer. Deer are ruminants, swallowing their food after chewing it only slightly and then, at leisure, regurgitating it and chewing it thoroughly. Their stomachs have four chambers, which function in rumination, and, in the larger segments of the stomach, bacteria and protozoa work to break down the complex chemicals of the leaves, which cannot otherwise be digested.

The two species specifically called deer—the White-tailed and the Mule Deer—are widespread in the United States. The former occurs over all of the eastern two-thirds of the country and as far west as the Pacific Coast in the northwest and to southern Arizona in the southwest. The Mule Deer lives in the west, from Minnesota westward, and ranges north to southern Alaska. The two species can be distinguished by their tails: the White-tailed Deer has a relatively large, wide tail, solid white below, and with no black on the surface, whereas the Mule Deer has a smaller, narrower tail that is always black on the tip, if not the entire top. The antlers of the males differ; the White-tail's antlers have tines that protrude from a single beam, whereas those of the Black-tailed Deer are a series of evenly forked branches.

The largest American deer, the Moose, is unmistakable because it is the only solid-colored, dark brown member of the family in North America. The male's antlers are huge and have a flattened "palm," off which the tines protrude. Moose are mainly coniferous forest animals and occur in the United States (aside from Alaska), only in the northern Rocky Mountains as far south as Colorado, in some of the Canadian border states, and in northern New England. The American Elk, second in size to the Moose, was once widespread in the United States, but has been eliminated from most of its range and now is found only along the Pacific Coast and in the Rocky Mountain area. It is lighter brown in color than the Moose and has a dark neck and a pale rump.

The Caribou, a northern species that lives in Alaska and in the lower states, no longer occurs in Glacier, Isle Royale, and Voyageurs parks. It is the only species of deer whose females also have antlers and the only American species with a dark body and a white neck.

At the beginning of the twentieth century, all the species of the deer family, except the Caribou, were in very low numbers, partly as a result of deforestation and market-hunting (venison was readily sold commercially). Since then, most of the species have increased in number, especially White-tailed Deer in the east, and restocking has returned some kinds of Caribou to areas they formerly inhabited. At present, however, Caribou numbers are considerably lower than they once were, partly as a result of the changed hunting habits of the people of the northland, who depend upon them for food.

In many parks, deer have become quite accustomed to human presence and often remain around hotels and lodges. They can usually be found at dawn along roads, but a few hours later may be much harder to find. Males with antlers can be dangerous. Cow Moose, who are known to be especially protective of their calves, often chase people who venture too close.

ELK

Elk, or Wapiti, as Indians called them, once inhabited much of what is now the United States. They were killed off over most of their original range and now persist mainly in the Rocky Mountains and on the Pacific Coast. Second in size among the deer only to the Moose, a bull Elk may weigh more than 800 pounds (363 kilograms) and stand 5 feet (1.5 meters) tall at the shoulders. The beautiful antlers of the males, whose beams can reach 60 inches (1,524 millimeters), are much sought as trophies. Hunting for meat and for the canine teeth (which were used as fob ornaments for a fraternal order) were the main cause of the elimination of Elk from so many places. With a brown body, a yellowish rump, and a dark mane around the neck, the Elk is readily identifiable wherever it occurs. This is the same species that lives in Eurasia and that is called the Red Deer; the animal that Europeans call "Elk" is the same species that Americans call "Moose."

More than any other member of the deer family in North America, Elk feed on grass and, in common with the other open-country hoofed animals, usually associate in herds that number more than twenty-five. The herd leader is an old cow, not a bull, and, in fact, during the summer the males and females live apart, each in their own herds. Where undisturbed, Elk feed during the day, mainly in the morning and evening, and can often be seen in mountain meadows at these times. At midday, especially when it is hot, Elk seek shade and ruminate, and at night they often seek shelter under trees, although on moonlit nights it is not rare to find them grazing in open meadows.

In mountainous areas, Elk move upward to the higher meadows during the summer to feed on the new grass there. At this time, the males grow their antlers, which are mature in August, at that time, the velvet outer covering is shed. In September, the bulls and cows begin to move to the lower meadows and valleys, where the males attempt to assemble harems, bugling their battle cries and fighting. Their necks are swollen with engorged blood at this time, as a prelude to mating, and bulls that fed peacefully side by side all summer now challenge one another for the cows. They bugle, prance, and then, from a position about 30 feet (9 meters) apart, charge with their heads lowered and crash their antlers together. Usually two or three such encounters settle the issue, with one bull moving off. On rare occasions, antlers interlock, and both bulls starve to death, if they are not killed first by a predator. These matches are generally between bulls of equal size; the smaller animals do not challenge the larger ones. For about a month, the bulls spend all of their time guarding the harem of cows, which may number up to fifty or sixty, and mate with them. Toward

Elk

the end of this breeding season, when the vigilance of the harem bulls diminishes, some of the younger bulls that have remained around the periphery of the harems may manage to mate with cows that have come into heat late. By the end of November, before winter sets in, the bulls try to feed heavily to regain the weight that they lost during the breeding season.

A single calf (twins are rare) is born in spring, after a gestation period of about 8½ months. It is reddish-brown in color and spotted with yellowish white. Within a few hours, the calf is strong enough to move to a hiding place and, for the first week of its life, remains hidden, its neck and head flattened on the ground. The mother returns warily to nurse and, after about a week, the calf is strong enough to follow the mother back to the herd. Although it may nurse for several months, the calf can graze when about a month old and is usually weaned by autumn, at which time it has shed its natal coat and looks like a small adult. At about two-and-a-half years, the females are sexually mature. Bulls are also capable of breeding at this age, but are generally several years older before they are large enough and strong enough to fight for and maintain a harem. Bulls lose their antlers in midwinter, but within a month (usually by March), the new ones have started to grow. The bull Elk with the largest antlers are not necessarily the oldest, because antler development is a result of general nutrition, and old bulls, past their prime, have less-developed antlers than do the bulls in the best condition.

Wintering in valleys and lower slopes, Elk may suffer food shortages and, at this time, often strip bark from trees, especially aspen, and in some places a browse line is evident where the animals have scarred the trees as high as they can reach. In summer in the Rockies, Elk are sometimes hard to find because they are in the high mountain meadows. However, there are usually some readily seen in Yellowstone and Redwood parks, and, in the autumn, Rocky Mountain National Park is a good place to hear them bugling in the evenings.

Elk: Badlands, Bryce Canyon (E), Carlsbad Caverns (RI), Crater Lake (RI), Glacier, Grand Canyon, Grand Teton, Great Smoky Mountains (E), Guadalupe Mountains, Mesa Verde, Mount Rainier (RI?), North Cascades, Olympic, Redwood, Rocky Mountain, Shenandoah (E), Theodore Roosevelt (E), Wind Cave, Yellowstone, Zion

---

MULE DEER

The deer of the western United States have long been considered two species by hunters: Mule Deer (plate 7a) and Black-tailed Deer. Although, in some places, these two kinds of deer seem to behave as species, intermingling but not interbreeding, it is generally established that, because there is a zone of interbreeding in central California, the two are actually subspecies of one species of deer, usually called the Mule Deer. The large ears, which are 5–6 inches (127–152 millimeters) long, give the species its common name. Throughout their western range, Mule Deer occupy a variety of habitats. In the southwest, they are found in the arid desert scrub; in the mountains, in rocky ravines and coniferous forests; and on the coasts, in the humid forests, such as the redwoods. About 3–3½ feet (0.9–1 meter) tall at the shoulder, and weighing as much as 400 pounds (181 kilograms), the largest representatives of the species live in the Rocky Mountains; the coastal ones are smaller. Females are smaller than males and rarely weigh more than 150 pounds (68 kilograms). Compared with the White-tailed Deer, Mule Deer seem stockier and less graceful. They have shorter, narrower tails, which are either tipped with black—as are those of the typical Mule Deer—or which are all black on top in the coastal races that are called Black-tailed Deer. There are notable differences in antler shape between the Mule and White-tailed Deer; the former's antlers fork into equal branches; the latter's antlers consist of a single heavy beam from which the smaller points emerge.

In summer, Mule Deer are rather solitary, except for does and their fawns. Their main periods of activity are early and late in the day, and at

Mule Deer

night. They feed on a variety of plants. In summer, they usually move higher up mountainsides, where they eat grasses and herbs. In winter, they move to lower elevations, where they feed mainly on twigs, buds, and bark from many kinds of tree and shrub. They do not paw through snow for food. Acorns, berries, cactus fruits, and even mushrooms and flowers are a regular part of their diet. They feed actively throughout the day in winter, which is the only time of the year that they may occur together in modest-sized herds.

Autumn is the breeding season, and at this time the bucks fight one another for mating privileges as they attempt to assemble small harems. The animals clash head-on and, after the initial impact of the antlers, shove each other back and forth in a test of strength, usually without injury to either participant. Unlike Elk males, Mule Deer males do not seem to try to keep their small harem, usually three or four does, intact, and the does often wander off. The males lose their antlers in late winter. The spotted fawns are born in late spring, after a gestation period of about 7 months. They are able to stand and walk within minutes after birth and move to a secluded place, where they remain hidden for the first several weeks of their lives. Two fawns are usually born, each of which weighs 5–7 pounds (2.3–3.2 kilograms). However, those does giving birth for the first time usually produce only a single fawn. The young fawns nurse until September, at which time they have shed their spotted coats. They remain with their mother through the winter, until just before she gives birth again. Sometimes, yearlings rejoin the mother and her new fawns the following summer. In captivity, Mule Deer have lived twenty-five years, but their life span in the wild is generally less than half of this.

One characteristic of Mule Deer is a form of bounding called stotting. The animal moves along in a series of stiff-legged jumps, all four feet hitting the ground together in what seems an effortless movement. Although not appearing as graceful or as speedy as the White-tailed Deer, Mule Deer have been known to run 35 miles (56 kilometers) an hour and to make leaps 25 feet (7.6 meters) in length.

Mule Deer are readily seen in most of the parks where they occur, usually early in the morning or late in the day.

Mule Deer: Arches, Badlands, Big Bend, Bryce Canyon, Canyonlands, Capitol Reef, Carlsbad Caverns, Crater Lake, Death Valley, Glacier Bay (I), Glacier, Grand Canyon, Grand Teton, Guadalupe Mountains, Joshua Tree, Kings Canyon, Lassen, Mesa Verde, Mount Rainier, North Cascades, Olympic, Organ Pipe Cactus, Petrified Forest, Redwood, Rocky Mountain, Sequoia, Theodore Roosevelt, Wind Cave, Yellowstone, Yosemite, Zion

---

WHITE-TAILED DEER

Although mainly an eastern species, White-tailed Deer also live in the western states, usually in stream valleys where there are deciduous trees. Over its range, it is highly variable in size. White-tailed Deer from the Florida Keys and the mountains of Arizona are tiny—some adults weigh as little as 50 pounds (23 kilograms), whereas large northern bucks have reached weights of more than 400 pounds (181 kilograms). However, they are fairly uniform in color, with a thin, reddish summer pelage, and a grayish, dense winter coat. The winter hair is such good insulation that snow often rests on a deer's back without melting from its body heat.

In winter, White-tailed Deer may occur together in groups that seek shelter from snow, forming "yards" of packed snow, often in a low-lying place. For the rest of the year, they tend to remain to themselves, although only those in the southwest seem to form small herds. Ordinarily, White-tailed Deer remain in a relatively small area throughout the year, sometimes as little as one square mile (2.6 kilometers), and do not make seasonal migrations, as do the Mule Deer. They prefer forest edge situations, where the trees and brush provide cover; the shrub growth at the edge provides much food. White-tailed Deer feed selectively on a large number of plant species. They eat succulent aquatic plants, grasses, nuts, leaves, and twigs. In winter, they browse on both conifer and hardwood trees. In agricultural areas, they devour cultivated crops of corn, alfalfa and vegetables. They ordinarily start to feed late in the day and may consume 10–12 pounds (4.5–5.4 kilograms) in twenty-four hours. During the night, they may seek a sheltered area and lie down to ruminate and then get up and feed again until daylight.

White-tailed Deer

The mating season varies over the great range of White-tailed Deer. In Florida, there seems to be no restricted season; in the north, the mating season is usually in November; in the southwest, in December or January. Bucks track receptive does and fight with one another for mating privileges. A fight usually consists of a single charge, enmeshing of antlers, and a shoving match, which is usually settled with this one encounter. One buck is shoved backward and eventually breaks the contact and runs away. Does that breed for the first time usually have a single fawn and thereafter usually two, although some have three. The spotted fawns are born in a secluded place, after a gestation period that is a little less than 7 months. They weigh about 4½–5½ pounds (2–2.5 kilograms) and can stand within ten minutes. The fawns are quickly led to a hiding place, where they remain for several weeks; the mother returns to nurse them six or seven times a day. When a month to a month-and-a-half old, the fawns are weaned, but they do not shed their spotted coats until they are about three-and-a-half months. At this time, young males may leave their mother, but female fawns usually remain with her through the winter. Although some young does can breed in their first year, most are one-and-a-half years old before they mate for the first time. In the wild, White-tailed Deer do not usually live more than ten years but, in captivity, have lived twice as long.

Considering that White-tailed Deer were once almost extinct in much of the eastern United States, it is remarkable that today they are among the more common large mammals in so many places, including suburbs within a few miles of large urban centers. Because their natural predators have been eliminated in so many areas, the main control on their populations has been hunting by human beings. However, the killing of bucks has little impact on the production of fawns, because remaining males mate with more females than they would ordinarily. In many places, White-tailed Deer pose a problem. They may be destructive to crops. They may also starve during the winter.

In the national parks, especially those in the east, White-tailed Deer are a common sight. They can usually be observed only at dawn; within an hour or two after sunrise, they have gone into hiding in the woods. Near sundown is another time when they can be seen. They can also be seen around lodges and campgrounds, which they often live near. Shenandoah, Great Smoky Mountains, and Everglades are parks where White-tailed Deer are readily seen.

White-tailed Deer: Acadia, Badlands, Big Bend, Crater Lake, Everglades, Glacier, Grand Teton (E?), Great Smoky Mountains, Guadalupe Mountains (E), Hot Springs, Isle Royale (IE), Mammoth Cave (RI), Mesa Verde (E), Organ Pipe Cactus, Saguaro, Shenandoah (RI), Voyageurs, Wind Cave, Yellowstone (E?)

MOOSE

Up to 7½ feet (2.3 meters) high at the shoulder and 1,800 pounds (816 kilograms) in weight, Moose are the largest members of the deer family. The great palmate antlers of a bull moose may spread 6½ feet (2 meters) from tip to tip and weigh 90 pounds (40 kilograms). Moose are unmistakable; they are dark brown in color, with long, pale legs, and huge, pendulous noses. Except for Alaska, their range in the United States is limited to the Rocky Mountains as far south as Colorado, and to the eastern states bordering Canada. They occur over most of Alaska. Their overall range coincides with that of the coniferous forest, and they are generally found near streams or ponds where there are willows. In summer, they feed on aquatic vegetation, willow leaves, and twigs. In winter, they eat the bark and twigs of many kinds of trees.

Moose are solitary animals, although sometimes several can be seen in the same feeding area along a streamside. The strongest social bond is between the mother and her calf. The mother is fiercely protective and frequently charges people who come too close; she does not hesitate to drive off a wolf. As the mating season approaches in early autumn, bulls may attack anything in sight, including people, cars, and trains. A threat is usually sufficient to cause smaller bulls to leave the area where a large bull is, but fights between equal-sized bulls may end in the death of one. During mating season, the males make a loud bellowing sound, which seems to serve as a challenge to other bulls within earshot; artificial moose-calls are used by hunters to attract males within rifle range.

When a male Moose finds a receptive female, he remains with her for a week or so before leaving to find another. After a gestation period of 8

Moose

months, one or two calves are born, each weighing about 20–25 pounds (9–11 kilograms). Unlike the calves of Elk and deer, these calves are not spotted, but they are not dark brown like their parents, either; they are a light reddish-brown. They nurse at least until winter, and when a year old, a bull may weigh as much as 600 pounds (272 kilograms). The calves are usually driven away by the mother before she gives birth again. Young cows may breed when a year-and-a-half old, but generally do not do so until in their third year. Although bulls are sexually mature earlier, they probably do not breed until they are five or six. Moose may live for twenty years in the wild.

Despite their ungainly appearance, Moose can move swiftly on their long legs; some having been clocked at 35 miles (56 kilometers) an hour. They are excellent swimmers, able to sustain a speed of 6 miles (10 kilometers) an hour for a long time and to outdistance a canoe guided by two people. In winter, their long legs aid them in snow. However, if the snow is deeper than 30 inches (762 millimeters), the animal's belly begins to drag and it must plow with its chest.

In the parks where they live, Moose are often seen in ponds and streams in summer. Glacier, Yellowstone, and Isle Royale are among the better parks in which to find them.

Moose: Acadia, Denali, Glacier Bay, Glacier, Grand Teton, Isle Royale, Katmai, North Cascades, Rocky Mountain (RI), Voyageurs, Yellowstone

CARIBOU

The Caribou (see plate 1a) is unusual in that it is the only member of the deer family whose females, as well as males, have antlers. This northern deer stands about 3½ feet (1 meter) high at the shoulder and weighs up to 700 pounds (318 kilograms). Caribou are the most gregarious of deer, assembling in herds that may number in the tens of thousands at certain times of the year. They are also the most migratory; some populations move 800 miles (1,287 kilometers) from wintering grounds to the places where they spend the summer. Caribou are the American representatives of the species that is called the Reindeer in the Old World and that has been domesticated by people in northern Eurasia. At present, Caribou are found only in the Alaskan parks, having been wiped out in the other parks—Voyageurs, Isle Royale, and Glacier—where they once occurred.

Caribou are characterized by their antlers, which tend to have small "palms" and a tine that projects over their forehead and which, in the males at least, is palmate at the end. The antlers of the females tend to be less ornate, less palmate, smaller, and more spindly than those of the males. The Caribou's body is brown, and the mane, the belly, and underside of the tail are white. Caribou have large hooves, which aid them in moving over boggy tundra in summer and on ice and snow in winter. They are the only deer whose ankles make a loud clicking sound when the animals walk. Just what function this noise serves is unknown. Some have speculated that it helps to keep the herd together, especially in fogs or snowstorms. In other hooved animals whose ankles click, the noise is limited to the largest males and is believed to be associated with social dominance.

In summer, Caribou are on the tundra, where they feed on grasses, sedges, lichens, and the leaves of the low, woody vegetation. Late in summer, they start to move away from the tundra toward their wintering grounds at the edge of the forest but, on approaching them, again move into the tundra for the mating season in late October and early November. At this time, the males assemble small harems of a dozen or so cows and fight to maintain them. As with so many other deer, after the initial charge and clashing of antlers, the fight is mainly a shoving match, which tests the opponents' strength. For a month or more during the breeding season, the bulls do not feed, having put on great amounts of fat during the summer, and are gaunt and depleted when it is over. When the mating period is over, the animals move again to the wintering grounds, which are generally at the edge of the coniferous forest, and the bulls lose their antlers by January. The winter food is mainly lichens, which the animals find by pawing through the snow—and twigs and leaves of willows and other trees, as well as any grasses that can be uncovered.

Generally, female Caribou have a single calf, after a gestation period of about 8 months. It is born on the tundra to which the animals have returned in the spring. The calf weighs 10–15 pounds (4.5–6.8 kilograms) and has a brown, unspotted coat. It can follow the mother within hours after its birth. By August, a calf may weigh 50 pounds (22.7 kilograms) and molts into adult pelage for the winter, but may continue to nurse until spring. Females may breed when a year old, but males probably do not mate until they are two or three, or older. Caribou probably do not live much more than ten years in the wild.

Because of their migratory ways, Caribou are not in the Alaskan parks throughout the year. They are present in Katmai only in the winter, but migrate through Denali Park in the summer and can usually be seen on the tundra there.

Caribou: Denali, Glacier (E), Isle Royale (E), Katmai, Voyageurs (E)

Pronghorn

Speediest of the American hoofed animals, Pronghorns (see plate 7b) are reputed to be able to reach speeds of 60–70 miles (97–113 kilometers) an hour for a few minutes at a time and to run steadily at 20–30 (32–48 kilometers) an hour. Although Pronghorns are generally called antelopes, their relationship to the true antelopes of the Old World is obscure; they seem to have evolved in America. They are usually put in their own family, although recent research has shown that their affinities are with the cattle family. Members of the deer family shed their antlers each year, whereas members of the cattle family do not shed their horns, which consist of a bony core, over which is a sheath of horn (keratin, the same material of which fingernails are composed). Pronghorns are unique in that they are the only mammals with horns of keratin that are shed each year and that also overlie a bony core. Also, while antlers are generally branched, horns are not, except for those of the Pronghorn.

Pronghorns are inhabitants of grasslands and low-brush country, where they rely on their excellent eyesight to espy distant danger. They then utilize their speed to escape. Pronghorns are about 3 feet (0.9 meters) tall at the shoulder and weigh up to 154 pounds (70 kilograms). Both males and females have horns, but usually only those of the males are branched (the "prong"). The males' horns may be as long as 20 inches (508 millimeters), but those of the females are rarely more than 5 inches (127 millimeters) and are often absent altogether. Pronghorns inhabit prairies and plains from the central Dakotas to Mexico, and west to northeastern California and southeastern Oregon and Washington. The Pronghorn is an attractive animal; its general yellow-orange color is marked with brilliant white rump patches, two white bands on the neck, partial black on the snout, and a white belly and side mark. In addition to being slightly larger than the female and having pronged horns, the male can be distinguished by a black mark at the rear edge of the jaw.

The hairs of the rump patch can be raised, to make a brilliant heliograph that serves as an alarm signal. Both males and females also have glands within the rump patch and these, too, serve in communication between animals. Males have additional glands, which are utilized in the marking of territories. The large eyes of the Pronghorns are shaded from the sun by very long lashes, and their ability to see at a distance is excellent.

Pronghorn

From late spring to early fall, adult males establish and hold territories that may be as much as one square mile (2.6 square kilometers) in area. Other adult males, as well as young adults, are kept out of the area by various techniques, the first of which is being stared at by the territory holder and then by various vocalizations. Rarely does territory defense involve an actual fight, but when it does, the combat is fierce, and serious injury or death, usually to the invader, are not uncommon. By the time that females, which have occupied the male territories, come into heat in midsummer or early fall, the male has been displaying to them to keep them within his bounds, and he then mates with them. Bucks that are too young, or otherwise unable, to establish territories of their own remain in bachelor herds and avoid the territorial males. Shortly after the end of the breeding season, the bucks shed their horns, the new ones having already started to grow beneath the sheath of the old ones.

A little over 8 months after mating, female Pronghorns give birth, usually to two fawns. Acutally, more than two eggs may be fertilized, but they are not implanted for about a month afterward, and, during this period, some embryos die. Thus one or two is the normal number of young produced. At birth, the fawns average about 8 pounds (3.6 kilograms) in weight and can stand within a few minutes. By four days of age, they can run swiftly. At birth their coats are more grayish than those of adults. They may remain hidden for several weeks before starting to follow the mother throughout the day, and, as they grow older, they play much among themselves.

Pronghorns feed mainly on forbs and grasses and, in winter, paw through the snow to find food. They may move many miles from their summer range to be able to winter where feeding is better. Sagebrush is an important food during times when snowfall is heavy. Their feeding range is generally not more than a square mile or two.

It is estimated that the original Pronghorn population of the west was 35 million but, by 1924, had been reduced to fewer than 200,000. Since then, because of conservation measures and appropriate management techniques, they now number several million. Pronghorns are readily seen in some of the western parks, especially Wind Cave, Badlands, Petrified Forest, and Yellowstone.

Pronghorn: Arches (E?), Badlands, Big Bend, Bryce Canyon, Canyonlands (E?), Capitol Reef (E?), Carlsbad Caverns (E?), Crater Lake, Grand Canyon, Grand Teton, Guadalupe Mountains (E), Lassen Volcanic, Mesa Verde (E), Organ Pipe Cactus, Petrified Forest, Rocky Mountain (E), Theodore Roosevelt (E), Wind Cave, Yellowstone

---

Bison

From a pre-Columbian population that was estimated to be 60 million, the American Bison (see plate 3b) was killed off until, by 1890, fewer than one thousand were thought to live in the United States. Although the species was saved from extinction, nowhere in the United States does it live as an animal whose existence is not somehow controlled by human beings. The destruction of the great herds that roamed the western plains was, ultimately, as much of a political act as an economic one. To bring the Plains Indians under control of the government, it was deliberately decided to expedite the elimination of Bison, because, by so doing, the Indians would be deprived of their main source of subsistence. Bison bulls may weigh up to 2,000 pounds (907 kilograms) and stand 6 feet (1.8 meters) tall at the shoulder; females are less than half this weight and are one foot (0.3 meter) shorter in shoulder height. From tip of nose to tip of tail, a Bison may be 11.5 feet (3.5 meters), and its horns may span 35 inches (889 millimeters) and each be more than 20 inches (508 millimeters) long and have a circumference at the base of 16 inches (406 millimeters). The shaggy forequarters and head make the animal appear more massive.

Only those who have seen the great herds of hoofed animals in Africa can have an inkling of what the Bison herds of the western plains must have been like in their original state. What we know now of Bison behavior may not be the behavior that governed the structure of the old herds. Now the males remain in separate herds for most of the year, with some bulls being wholly solitary. These are usually old ones that depart voluntarily from the prime groups and, at times, form their own small band of old-timers. When the breeding season starts in July, bulls bellow, paw the ground, and charge head-to-head, taking the impact on their skulls, rather than their horns. They continue battering until one animal gives up and moves off, and the victor then rounds up a small harem of up to twenty cows, with which he mates as they come into heat. In May, about 9 months after mating, the cow leaves the herd for a secluded place, where she gives birth to her single (twins are rare) calf. The newborn is a reddish-yellow color and can stand within a half-hour and walk soon thereafter. The mother keeps it hidden for two or three days before leading it to join the herd. A calf may nurse for nearly a year. By six months of age, the calf's hump, dark coat, and horns have all commenced their development. A cow Bison may breed when she is two years old, more ordinarily when she is three, but bulls are probably more than five before they can obtain a harem. Bison have lived more than twenty-five years.

Bison can walk deceptively fast, about 5 miles (8 kilometers) an hour, and, in a gallop, they can reach 35 miles (56 kilometers) an hour. Under original conditions, the herds migrated north and south 100–200 miles (161–322 kilometers) in the spring and fall each year. When Bison cross rivers, their buoyancy is evident; their heads and forequarters ride out of the water. Many thousands died while crossing rivers such as the Missouri when it was in flood. Many of the highways made across the west, and especially through the mountains, followed Bison trails, and it is still possible to find depressions in the plains that were places where Bison once wallowed in the mud or dust.

Bison are readily seen in Wind Cave, Theodore Roosevelt, Badlands, and Yellowstone national parks.

Bison: Badlands (RI), Capitol Reef (E), Carlsbad Caverns (E), Glacier (E), Grand Tetons, Great Smoky Mountains (E), Guadalupe Mountains (E), Rocky Mountain (E), Shenandoah (E), Theodore Roosevelt (RI), Wind Cave (RI), Yellowstone

---

Mountain Goat

Restricted to the higher, craggy areas of the northern Rocky Mountains and to Cascade and Coast mountains in the west, Mountain Goats (see plate 3a) are stocky, all-white animals. They are not true goats; their closest relatives are other mountain dwellers of Eurasia, including the Chamois. Both sexes have short (to 12 inches; 305 millimeters), smooth, black, pointed horns, which curve slightly backward. Mountain Goats stand 3–3½ feet (0.9–1 meter) tall at the shoulder and weigh 102–300 pounds (46–136 kilograms). Although the Mountain Goat is sometimes confused with the Dall's Sheep, another white animal that occurs within parts of its range, the former has black horns, the latter has brown ones; the goat also has a white beard which may be 5 inches (127 millimeters) long.

The natural range of Mountain Goats—from southwestern Alaska to southern Montana, central Idaho, and north-central Washington—has been extended by introductions of the animals to central Colorado, South Dakota, Oregon, and other areas, including Olympic National Park. Because of the relative inaccessibility of their habitat, it is often said that Mountain Goats are the American game animal whose original distribution has been least affected by the activities of human beings. In their rocky habitat, Mountain Goats are sure-footed, but also cautious and deliberate in their movements. Their hooves are adapted for a more certain grip by being relatively large and almost square and by combining a firm outer edge with a flexible, almost rubbery, inner pad. They are not infallible in their climbing, however, and goats have been known to miss their footing and fall to injury or death.

Mountain Goats are mainly grazers, and most of their diet is composed of grasses, sedges, and herbs. In winter, they feed on grasses, if they can find them, but also eat mosses and lichens and browse on willows and even on pines and firs. They get what water they need by drinking melting snow or, at times, by eating snow. Minerals are also important in their diet, and Mountain Goats are known to frequent salt licks, sometimes traveling relatively long distances to reach them.

Mountain Goats are not highly gregarious. In summer, males are often solitary, and females and their kids are usually in groups that number fewer than ten. At this time of the year, the animals generally remain above timberline but, as winter approaches, move down to lower slopes, where there may be more shelter and less snow. Even when they winter in valleys, Mountain Goats never go far from rocky hillsides, which are their main refuge from predators. Few animals prey on Mountain Goats successfully. The Mountain Lion is one of the few species that can stalk them wherever they roam, although Coyotes, Bobcats, and Black and Grizzly Bears may be able to kill one on occasion. Perhaps the main danger to Mountain Goats is from eagles, which are known to have carried off kids and even to have knocked young animals off cliffs.

When the mating season starts in November, males and females, which have lived apart all summer, come together. Most competition for females involves displays and threats. However, death may easily result from a fight; the pointed horns of the Mountain Goats are dangerous weapons. Mountain Goats do not have the massive skulls and horns of Mountain Sheep and do not, when they fight, clash head-on. Most of the wounds that have been observed are on the rump, where the skin is extra thick. Such wounds result from these animals displaying in a side-to-side, head-to-tail position. Odor plays a major part in their sexual behavior. A special gland located behind the horns is used to mark vegetation and even to mark females. The gestation period is about 6 months, and the kids, usually one or two, are born in May or June, and weigh about 6½ pounds (3 kilograms). The kids are able to stand immediately after birth and, within a week, can follow their mother anywhere. They are usually weaned by September, but remain with the mother through the winter and until the next kid is born. During the summer, females keep the adult males away from themselves and their kids by active aggression toward them, and they seem dominant over males at this time of the year.

Mountain Goats are diurnal and usually feed from daybreak until mid-morning and then again late in the afternoon. However, they may also be active at night. Although Mountain Goats make a variety of sounds, people are not ordinarily close enough to the animals to hear them. In the parks where Mountain Goats occur, they can usually be seen in certain places. Mount Rainier and Glacier are good parks in which to see them.

Mountain Goat: Denali (?), Glacier Bay, Glacier, Mount Rainier, Olympic (I)

Dall's and Mountain Sheep

The two species of wild sheep in North America differ only slightly in size, coloration, and the conformation of the horns. Their ranges do not overlap: Dall's Sheep is a northern species, which lives from northern Alaska to northern British Columbia; the Mountain Sheep (also called Bighorn Sheep) lives from central British Columbia to Mexico, as far east as the front range of the Rocky Mountains, and as far west as California. In Alaska, Dall's Sheep is white, but, in the Canadian part of its range, it may be gray to almost black and is called Fannin's Sheep and Stone's Sheep. The Mountain Sheep are also variable in color; those in the northern part of the range are generally a rather dark brown, whereas those in the southwestern deserts are paler. All Mountain Sheep and Dall's Sheep have a white or cream-colored conspicuous rump patch; white around the muzzle and the eye; and a pale belly. The horns of the two species also differ. Those of the Dall's Sheep are paler in color, less thick at their base, and form a more open spiral, more like a corkscrew, so that they may extend, tip to tip, much wider than those of the Mountain Sheep. Ram Mountain Sheep have thick-based, massive horns, which form a tight curl close to the head; these horns are generally darker brown than those of Dall's Sheep. Tip to tip, a Bighorn Sheep's horns are rarely more than 20 inches (508 millimeters); those of a Dall's Sheep may spread nearly 3 feet (914 millimeters). Bighorns are usually more heavily built than Dall's Sheep, and rams of the former species may reach 300 pounds (136 kilograms), whereas those of the latter reach only 200 pounds (91 kilograms). Ewes of both species are smaller, generally about three-fourths the size of the rams.

Dall's Sheep

American sheep are gregarious; they are usually seen in groups of fewer than a dozen in summer, although, in wintering spots, they may aggregate in numbers of a hundred or more. The adult males are usually away from the females from spring to fall, forming their own bands or roaming alone. The bands of ewes, which consist of adults, newborns, yearlings, and even two-year-olds, are led by an old ewe. These sheep have remarkable eyesight, said to be the equivalent of a human being with eight-power binoculars, and can spot a predator miles away. They are diurnal, feeding mostly in the morning and evening, and resting during the day while chewing their cud. Unlike most other hoofed animals, Bighorns have regular places where they spend the night, often beneath a rock overhang or in the mouth of a cave, and these sites may be used for years.

Sheep are grazers and, in summer, feed on a variety of grasses and sedges in mountain meadows. Vertical movement is characteristic of this species; they move down to lower valleys for the winter, where they may browse on woody plants, if the snow is too deep for them to reach grass. Sheep need water, and even those that live in the arid southwestern states know of, and rely on, every water hole in the vicinity.

Fall is the mating season, and at this time the rams return to the ewe herds and attempt to assemble small harems. Fighting is frequent, always between rams of about equal size, and consists of the rams charging head-first toward one another and colliding horn to horn; their crash can be

Mountain Sheep

heard a mile away. This battering may continue or develop into a shoving match until the weaker ram departs. The gestation period is about 6 months, and usually only one lamb is produced. It weighs about 8 pounds (3.6 kilograms) at birth and can walk within a few hours. By a week old, the lamb starts to nibble grasses and it and its mother have rejoined the flock. The lambs, which may nurse for several months, remain with the ewe herd at least until their second year. At that time, males are evicted and join ram herds. Ewes may breed when two-and-a-half years old, but rams must be at least three-and-a-half before they may have a chance to win some females in fights with larger rams. Rams are usually five or six years of age when they reach their prime. Although sheep have lived up to twenty years in captivity, few survive more than fifteen years in the wild. In the north, wolves are their main predators, and, in the south, Mountain Lions, bears, and other predators, including eagles, kill some sheep, but only on a small scale.

Mountain and Dall's Sheep are dramatic sights and, in some of the parks, have become slightly less wary than in areas where they are customarily hunted. Nevertheless, they are easily frightened, and in places where they are known to visit, the park service often asks tourists to provide space for the animals to pass. Among the parks where there is a reasonable chance of seeing Mountain Sheep are Joshua Tree, Rocky Mountain, Yellowstone, and Glacier. Dall's Sheep are readily seen in Denali in summer.

> **Mountain Sheep:** Arches (E?), Badlands (RI), Big Bend (E), Canyonlands (E?), Capitol Reef (E), Carlsbad Caverns (E?), Death Valley, Glacier, Grand Canyon, Grand Teton, Guadalupe Mountains (E), Joshua Tree, Kings Canyon, Mesa Verde (RI), North Cascades (?), Organ Pipe Cactus, Rocky Mountain, Saguaro (E), Sequoia, Theodore Roosevelt (RI), Wind Cave, Yellowstone, Yosemite, Zion (RI); **Dall's Sheep:** Denali

Exotic Mammals

In some national parks there are mammals that were not part of the original fauna. Among these are nondomestic animals, such as the Mexican Red-bellied Squirrel, Eurasian Wild Boar, Asian Mongooses, and African Barbary Sheep. In addition, some parks have animals that have escaped domestication and are living free; these include Donkeys, Pigs, and Goats. The philosophy of the National Park Service is that exotics should be removed from the parks, but this is not uniformly practiced. Unfortunately, in some places, the presence of exotic mammals is detrimental to the native fauna and flora. Wild Boars, released nearby Great Smoky Mountains National Park for hunting, are extremely destructive in the park. Donkeys in Grand Canyon and Death Valley compete with native Mountain Sheep for water and forage. Goats and Pigs in Hawaii's two national parks have been responsible for major alterations in the environment and for the destruction of native plants on which indigenous birds depend for food. Mongooses in the West Indies and Hawaii have caused the near-extinction of various kinds of ground-nesting bird.

Some parks harbor North American species that were not originally part of their fauna, but that were transplanted from some other area, or contain species that were eradicated by human beings and now have been restocked. These introductions and reintroductions range from the Virginia Opossum on the Pacific Coast, which was not originally a far-western mammal, to the transplantation of Mountain Goats into Olympic Park, where they are not believed to have occurred before. Mountain Sheep have been restocked in a number of places, including Zion National Park, as have Elk, Bison, and White-tailed Deer in some other parks. The reintroduction of species that had been extirpated from an area is part of a general program to "restore" parks to their original (pre-Columbian) state, although some bias on the part of the park service is notable. No attempts have yet been made to restore Grizzly Bears or Gray Wolves to the parks where they were killed off.

The presence of exotic mammals and instances of introduction or reintroduction of species in the parks and monuments has been noted in the account of each park. The life history information about native species appears in the appropriate account. Because the wholly exotic species will eventually be removed from the parks, their life histories are not discussed in this book.

Appendix
Common and Scientific Names

This list follows J. K. Jones, Jr., D. C. Carter, and H. H. Genoways, *Revised Checklist of North American Mammals North of Mexico*, Occasional Papers of The Museum, Texas Tech University, no. 62 (Lubbock: Texas Tech University, 1979), pp. 1–13. With a minor exception, I have followed the common name usage contained here and, for purposes of identification, have also followed the scientific names—except for listing only one species each of Spotted Skunk and Hog-nosed Skunk. Although I do not agree with many of the scientific names that they have used and have published my versions of the names of many of the species before, it is not the purpose of this listing to be a point of digression for classification and nomenclature, but merely to indicate by a standard, published list, a set of scientific names for the common ones that have been used here.

Order Marsupialia—Marsupials
    FAMILY DIDELPHIDAE—NEW WORLD OPOSSUMS
    Virginia Opossum   *Didelphis virginiana*

Order Chiroptera—Bats
    FAMILY MOLOSSIDAE—MOLOSSID BATS
    Brazilian Free-tailed Bat   *Tadarida brasiliensis*

Order Edentata—Edentates
    FAMILY DASYPODIDAE—ARMADILLOS
    Nine-banded Armadillo   *Dasypus novemcinctus*

Order Lagomorpha—Lagomorphs
    FAMILY OCHOTONIDAE—PIKAS
    Collared Pika   *Ochotona collaris*
    Pika   *Ochotona princeps*
    FAMILY LEPORIDAE—HARES AND RABBITS
    Brush Rabbit   *Sylvilagus bachmani*
    Marsh Rabbit   *Sylvilagus palustris*
    Eastern Cottontail   *Sylvilagus floridanus*
    New England Cottontail   *Sylvilagus transitionalis*
    Nuttall's Cottontail   *Sylvilagus nuttalli*
    Desert Cottontail   *Sylvilagus audubonii*
    Swamp Rabbit   *Sylvilagus aquaticus*
    Snowshoe Hare   *Lepus americanus*
    Northern Hare   *Lepus timidus*
    White-tailed Jack Rabbit   *Lepus townsendii*
    Black-tailed Jack Rabbit   *Lepus californicus*
    Antelope Jack Rabbit   *Lepus alleni*

Order Rodentia—Rodents

FAMILY APLODONTIDAE—MOUNTAIN BEAVER

Mountain Beaver	*Aplodontia rufa*

FAMILY SCIURIDAE—SQUIRREL

Eastern Chipmunk	*Tamias striatus*
Alpine Chipmunk	*Eutamias alpinus*
Least Chipmunk	*Eutamias minimus*
Yellow-pine Chipmunk	*Eutamias amoenus*
Townsend's Chipmunk	*Eutamias townsendii*
Merriam's Chipmunk	*Eutamias merriami*
Cliff Chipmunk	*Eutamias dorsalis*
Colorado Chipmunk	*Eutamias quadrivittatus*
Red-tailed Chipmunk	*Eutamias ruficaudus*
Gray-footed Chipmunk	*Eutamias canipes*
Long-eared Chipmunk	*Eutamias quadrimaculatus*
Lodgepole Chipmunk	*Eutamias speciosus*
Panamint Chipmunk	*Eutamias panamintinus*
Uinta Chipmunk	*Eutamias umbrinus*
Woodchuck	*Marmota monax*
Yellow-bellied Marmot	*Marmota flaviventris*
Hoary Marmot	*Marmota caligata*
Olympic Marmot	*Marmota olympus*
Harris' Antelope Squirrel	*Ammospermophilus harrisii*
White-tailed Antelope Squirrel	*Ammospermophilus leucurus*
Texas Antelope Squirrel	*Ammospermophilus interpres*
Townsend's Ground Squirrel	*Spermophilus townsendii*
Richardson's Ground Squirrel	*Spermophilus richardsonii*
Uinta Ground Squirrel	*Spermophilus armatus*
Belding's Ground Squirrel	*Spermophilus beldingi*
Columbian Ground Squirrel	*Spermophilus columbianus*
Arctic Ground Squirrel	*Spermophilus parryii*
Thirteen-lined Ground Squirrel	*Spermophilus tridecemlineatus*
Mexican Ground Squirrel	*Spermophilus mexicanus*
Spotted Ground Squirrel	*Spermophilus spilosoma*
Franklin's Ground Squirrel	*Spermophilus franklinii*
Rock Squirrel	*Spermophilus variegatus*
California Ground Squirrel	*Spermophilus beecheyi*
Round-tailed Ground Squirrel	*Spermophilus tereticaudus*
Golden-mantled Ground Squirrel	*Spermophilus lateralis*
Cascade Golden-mantled Ground Squirrel	*Spermophilus saturatus*
Black-tailed Prairie Dog	*Cynomys ludovicianus*
White-tailed Prairie Dog	*Cynomys leucurus*
Utah Prairie Dog	*Cynomys parvidens*
Gunnison's Prairie Dog	*Cynomys gunnisoni*

FAMILY SCIURIDAE—SQUIRREL *(continued)*

Gray Squirrel	*Sciurus carolinensis*
Western Gray Squirrel	*Sciurus griseus*
Abert's Squirrel	*Sciurus aberti*
Fox Squirrel	*Sciurus niger*
Arizona Gray Squirrel	*Sciurus arizonensis*
Red Squirrel	*Tamiasciurus hudsonicus*
Douglas' Squirrel	*Tamiasciurus douglasii*

FAMILY CASTORIDAE—BEAVERS

Beaver	*Castor canadensis*

FAMILY CRICETIDAE—NEW WORLD RATS AND MICE

Eastern Woodrat	*Neotoma floridana*
Southern Plains Woodrat	*Neotoma micropus*
White-throated Woodrat	*Neotoma albigula*
Desert Woodrat	*Neotoma lepida*
Arizona Woodrat	*Neotoma devia*
Stephen's Woodrat	*Neotoma stephensi*
Mexican Woodrat	*Neotoma mexicana*
Dusky-footed Woodrat	*Neotoma fuscipes*
Bushy-tailed Woodrat	*Neotoma cinerea*
Round-tailed Muskrat	*Neofiber alleni*
Muskrat	*Ondatra zibethicus*

FAMILY ERETHIZONTIDAE—NEW WORLD PORCUPINES

Porcupine	*Erethizon dorsatum*

Order Odontoceti—Toothed Whales

FAMILY ZIPHIIDAE—BEAKED WHALES

Goose-beaked Whale	*Ziphius cavirostris*

FAMILY PHYSETERIDAE—SPERM WHALE

Sperm Whale	*Physeter macrocephalus*

FAMILY DELPHINIDAE—DELPHINIDS

Bottle-nosed Dolphin	*Tursiops truncatus*
Northern Right-whale Dolphin	*Lissodelphis borealis*
Pacific White-sided Dolphin	*Lagenorhynchus obliquidens*
Killer Whale	*Orcinus orca*
Grampus	*Grampus griseus*
False Killer Whale	*Pseudorca crassidens*
Common Pilot Whale	*Globicephala melaena*
Short-finned Pilot Whale	*Globicephala macrorhynchus*
Harbor Porpoise	*Phocoena phocoena*
Dall's Porpoise	*Phocoenoides dalli*

Order Mysticeti—Baleen Whales

FAMILY ESCHRICHTIIDAE—GRAY WHALE

Gray Whale	*Eschrichtius robustus*

FAMILY BALAENOPTERIDAE—RORQUALS

Fin Whale	*Balaenoptera physalus*
Minke Whale	*Balaenoptera acutorostrata*

FAMILY BALAENOPTERIDAE—RORQUALS (continued)

Blue Whale	*Balaenoptera musculus*
Hump-backed Whale	*Megaptera novaeangliae*

Order Carnivora—Carnivores

FAMILY CANIDAE—CANIDS

Coyote	*Canis latrans*
Red Wolf	*Canis rufus*
Gray Wolf	*Canis lupus*
Red Fox	*Vulpes vulpes*
Kit Fox	*Vulpes macrotis*
Swift Fox	*Vulpes velox*
Gray Fox	*Urocyon cinereoargenteus*
Insular Gray Fox	*Urocyon littoralis*

FAMILY URSIDAE—BEARS

Black Bear	*Ursus americanus*
Brown and Grizzly Bear	*Ursus arctos*

FAMILY OTARIIDAE—EARED SEALS

Northern Fur Seal	*Callorhinus ursinus*
Guadalupe Fur Seal	*Arctocephalus townsendii*
Northern Sea Lion	*Eumetopias jubatus*
California Sea Lion	*Zalophus californianus*

FAMILY PROCYONIDAE—PROCYONIDS

Ringtail	*Bassariscus astutus*
Raccoon	*Procyon lotor*
Coati	*Nasua nasua*

FAMILY MUSTELIDAE—MUSTELIDS

Marten	*Martes americana*
Fisher	*Martes pennanti*
Ermine	*Mustela erminea*
Least Weasel	*Mustela nivalis*
Long-tailed Weasel	*Mustela frenata*
Black-footed Ferret	*Mustela nigripes*
Mink	*Mustela vison*
Wolverine	*Gulo gulo*
Badger	*Taxidea taxus*
Spotted Skunk	*Spilogale putorius*
Striped Skunk	*Mephitis mephitis*
Hooded Skunk	*Mephitis macroura*
Hog-nosed Skunk	*Conepatus mesoleucus*
River Otter	*Lutra canadenis*
Sea Otter	*Enhydra lutris*

FAMILY PHOCIDAE—HAIR SEALS

Harbor Seal	*Phoca vitulina*
Gray Seal	*Halichoerus grypus*
Northern Elephant Seal	*Mirounga angustirostris*

FAMILY FELIDAE—CATS
Jaguar   *Felis onca*
Mountain Lion   *Felis concolor*
Ocelot   *Felis pardalis*
Jaguarundi   *Felis yagouaroundi*
Lynx   *Felis lynx*
Bobcat   *Felis rufus*

Order Sirenia—Sea Cows
FAMILY TRICHECHIDAE—MANATEES
Manatee   *Trichechus manatus*

Order Artiodactyla—Even-toed Ungulates
FAMILY TAYASSUIDAE—PECCARIES
Collared Peccary   *Dicotyles tajacu*
FAMILY CERVIDAE—CERVIDS
Elk   *Cervus elaphus*
Mule and Black-tailed Deer   *Odocoileus hemionus*
White-tailed Deer   *Odocoileus virginianus*
Moose   *Alces alces*
Caribou   *Rangifer tarandus*
FAMILY ANTILOCAPRIDAE—PRONGHORN
Pronghorn   *Antilocapra americana*
FAMILY BOVIDAE—BOVIDS
Bison   *Bison bison*
Mountain Goat   *Oreamnos americanus*
Mountain Sheep   *Ovis canadensis*
Dall's Sheep   *Ovis dalli*

References

MAMMAL IDENTIFICATION

Burt, William H., and Grossenheider, Richard P. *A Field Guide to the Mammals.* 3d ed. Boston: Houghton Mifflin, 1976.

This excellent book contains fine color paintings of mammals, clear distribution maps, photographs of skulls, and concise life-history information. The scientific names are not current.

Palmer, Ralph S. *The Mammal Guide.* Garden City, N.Y.: Doubleday, 1954.

This guide is illustrated with color paintings of mammals, but not as clearly as in Burt and Grossenheider. The distribution maps are clearer than those in Whitaker's book, (see below), but not as up-to-date as those in Burt and Grossenheider's. However, considering that the text is more than twenty-five years old, it is superior to either of these two guides.

Whitaker, John O., Jr. *The Audubon Society Field Guide to North American Mammals.* New York: Alfred A. Knopf, 1980.

This field guide is illustrated with color photographs, which can aid in identifying the larger mammals. The distribution maps are too small, but considerable life-history information is presented.

MAMMAL LIFE-HISTORY

American Society of Mammalogists. *Mammalian Species.*

This is a series of individual loose-leaf articles, which summarize, in condensed and scientific fashion, the information known about a given species. As of January 1980, 144 separate articles by various experts had been published; further articles are expected to be produced at a rate of about twenty per year. At present, thirty of the articles deal with the national park species discussed in this book.

Cahalane, Victor H. *Mammals of North America.* New York: Macmillan, 1947.

This is certainly the most readable account of the lives of North American mammals. No other recent book is comparable, although Palmer's book contains much of the same information in condensed fashion.

MAMMAL SIGN

Murie, Olaus J. *A Field Guide to Animal Tracks.* 3d ed. Boston: Houghton Mifflin, 1974.

This guide, well-illustrated with black-and-white drawings, serves to identify the sign that mammals leave of their presence—tracks, manure, cuttings, and nests. It is a detailed, accurate, and most useful account.

NATIONAL PARKS

Matthews, William H., III. *A Guide to the National Parks.* Garden City, N.Y.: Doubleday/Natural History Press, 1973.

This is the best single paperback volume on the national parks, dealing mainly with their geology, but also with some basic information on the flora and fauna, trails, facilities, and tips for the traveler.

The National Park Service.

Each of the national parks will send, on request, information about the park, including its special features, accessibility, accommodations, and climate. A supplementary list of available literature is often sent, and, on request, many parks will provide a free list of the mammals and birds that are present. The visitor center at each park can usually provide an excellent introduction to the park. In the visitor center, one can find maps, books, pamphlets, and other information—in the form of slide shows, museum exhibits, and special exhibits. Thus one can obtain an excellent introduction to the park. The address of each park is given at the beginning of each park account in this book.

Index of Mammal Descriptions

Page numbers refer to major discussions of mammals, which are listed under their commonly used names. Boldface numbers indicate illustrations.

THE JOHNS HOPKINS UNIVERSITY PRESS

Mammals of the National Parks

This book was composed in Trade Gothic text and display type by
FotoTypesetters, Inc., from a design by Lisa S. Mirski. It was printed on
60-lb. Glatfelter Offset paper, and the hardcover edition was bound in
Holliston's Kingston natural finish cloth by Universal Lithographers, Inc.
The manuscript was edited by Jane Warth.